An Intermediate
Course in Algebra
An Interactive Approach

An Intermediate Course in Algebra
An Interactive Approach

Alison Warr

Catherine Curtis

Penny Slingerland

Mt. Hood Community College

THOMSON

BROOKS/COLE

Australia • Canada • Mexico • Singapore • Spain • United Kingdom • United States

Publisher: Emily Barrosse
Executive Editor: Angus McDonald
Marketing Strategist: Julia Downs-Conover
Developmental Editor: James LaPointe
Project Management: TSI Graphics
Production Manager: Alicia Jackson
Cover Credit: Mt. Hood, Oregon (© Stuart Westmoreland/Stone Images)

An Intermediate Course In Algebra: An Interactive Approach 1/e, Alison Warr, Catherine Curtis, and Penny Slingerland

ISBN 0-534-43673-0
Library of Congress Catalog Card Number: 131479

For information about our products, contact us:
Thomson Learning Academic Resource Center
1-800-423-0563
http://www.wadsworth.com

For permission to use material from this text, contact us by
Web: http://www.thomsonrights.com
Fax: 1-800-730-2215
Phone: 1-800-730-2214

This material is based upon work funded in part by the National Science Foundation under Grant No. DUE-9454627.

Any opinions expressed are those of the authors and not necessarily those of the National Science Foundation.

Some material in this work previously appeared in *Interactive Mathematics III* 1996 Preliminary Edition, copyright © 1996 by Alison Warr, Catherine Curtis, and Penny Slingerland. All rights reserved.

Printed in Canada
10 9 8 7 6 5

Contents

Appendices

Preface

Overview

This overview will acquaint you with the features of this book and concepts covered as well as our philosophy in teaching mathematics. Following that, we will give you a view of the classroom that works with this philosophy and this text. Included are discussions of how the class is organized, what goes on in the classroom, the instructor's role, and the students' role. We will also address some of the major changes between this textbook and preliminary versions.

Keep in mind that what we are describing is the structure that has worked for us. Some of the elements are crucial in making this text work with students. Each of us, however, adapts the situation to our own comfort level and experience.

Features of the Text

Activity Sets.   Nearly every section of the text begins with an Activity Set. These activities are designed to guide students to discover, at an introductory level, the major concepts of the section. While some of the activities can be done by students at home, they are most effective when done in class in teams. The team often ensures success; what one student cannot figure out, a team most often can.

Discussions.   In this portion of the text, the concepts of the section are discussed, when possible, in the context of applications. It is crucial that the students read the discussion part of the textbook. Looking briefly at the examples will not be sufficient for understanding and success. This textbook should be read by students with paper, pencil, and calculator in hand.

Definitions.   As the concepts are presented, new vocabulary is introduced and important ideas are highlighted.

Examples.   As mentioned above, the concepts of the section are discussed, when possible, in the context of applications. This means that some examples in the textbook are longer and more involved than in a traditional textbook. As students read the text, it is very important that they work through each and every example.

Strategy Boxes.   Throughout the text, strategy boxes are included in appropriate places. These boxes are intended to provide problem-solving strategies for skill-oriented tasks. There is an index of these strategies in the back of the book.

Problem Sets.   The problem sets include skill problems, conceptual questions, applications, and open-ended questions. In these problem sets, we are asking students to communicate in writing their thought processes, conjectures, comparisons, and mathematical arguments.

Cumulative Reviews.   There are three cumulative reviews in the book. These sections allow students to test their understanding of the concepts learned in the previous chapters. Students must realize that although a concept may have been tested when a chapter was completed, the skills learned are needed throughout the course.

Topics

The topics in this textbook will look somewhat different from a more traditional intermediate algebra text. They are in line with the National Council of Teachers of Mathematics Standards and the American Mathematics Association of Two Year Colleges Crossroads document.

Your choices of topics will be driven by your own curriculum. We hope that you will consider including some of the less familiar topics. As topics are added to the curriculum, other topics need to be omitted or de-emphasized. We are constantly asking ourselves, "What do students need to know and when do they need to know it?" Some traditional topics have been moved to other places in our curriculum based on this question. One example is the moving of completing the square to precalculus when students are studying transformations. Difficult, but rewarding, discussions are needed to make these hard choices. We are in continual dialogue with our colleagues, in and out of the mathematics division, as to the use of topics down the road and how our decisions may affect students in their future classes. We feel we are making responsible choices but welcome continued dialogue on this issue.

We encourage you to cover the sections in the order that they appear in the book. Many sections rely on concepts and skills learned in previous sections. However, there are a few exceptions. Section 9.1 may be covered anytime after Section 1.2. Chapter 6 may be done anytime after Chapter 5. Many people may choose to do this chapter just before Chapter 10, and it is not necessary to do Chapter 6 before Chapter 7. Chapter 8 can be done anywhere in the course. Sections 3.2, 4.4, 4.5, and 7.6 are optional, and no other sections of the textbook rely on the content in these sections.

Course Content.   At Mt. Hood Community College, we use Chapters 0 through 8, omitting sections 3.2, 4.4, 4.5, 7.6 and Chapter 6 in a 5-credit, ten-week course that meets six hours per week. This course is the prerequisite for precalculus. It serves both transfer- and technical-professional students.

Pacing of the Course.   One section of this textbook does not correspond to one class period. It is organized around concepts rather than time.

Changes from 1996 Preliminary.   In its preliminary versions, this textbook was titled Interactive Mathematics III. Following are the major changes from the 1996 Preliminary Edition.

- Many of the activity sets have been rewritten based on instructor and student input.
- The discussion portion of the textbook is more complete and includes many more straightforward examples.
- The problem sets include more basic problems related to the section, and the number of skill problems has increase.
- Most sections include problems from previous sections to integrate review throughout the book.
- The numbers in the abstract problems are more reasonable.

- The total number of problems in the problem sets has increased and we have added three cumulative review problem sets.
- There are more answers in the back of the book.
- Each chapter ends with a chapter summary of the key concepts learned in the chapter.
- Some of the topics in the book have been reordered to create some spiraling of topics. Previously Chapter 2 contained everything on linear functions. Now those concepts are in Chapters 2 and 4. Additionally the chapter on exponents has been split so that integer exponents occur in Section 3.1 and the remaining exponents are in Chapter 5. The emphasis on exponents has shifted to using exponents to solve power equations.
- Two new appendices have been added: Unit Conversion and Solving Linear Equations.

Desired Outcomes

We have tried to focus on a set of desired outcomes in writing this textbook and in planning what goes on in our classroom. To achieve these results, pedagogical changes have taken place in the classroom. The desired results and pedagogical changes are listed next. Some of these items are discussed in more detail later in the Preface.

Desired Student Results

- **Patterning**
  Mathematics is the study of patterns. Students' patterning skills are developed, giving them an additional tool to investigate and solve problems.

- **Multiple Models**
  Students are introduced to algebra with verbal descriptions, numerical models, graphical models, and equations. This approach emphasizes the connections between the models from the beginning.

- **Verification**
  Students are required to verify simplification results and solutions to equations and formulas. Students must always check the reasonableness of their results and accuracy of their models. These processes empower students in their learning of mathematics.

- **Communication**
  Students are expected to communicate their ideas, processes, and understanding orally and in writing.

- **Problem Solving**
  Applications are fully integrated. Students are expected to clearly communicate their mathematical processes, results, and verifications.

- **Interpretation**
  Mathematics is an abstract system used to represent situations. Whether the information is presented graphically, numerically, or algebraically, it is only useful if it can be correctly interpreted in the particular scenario. Therefore, students are continually asked to interpret information represented mathematically.

Pedagogical Changes

- **Visual Learning**
  Visual learning is incorporated throughout the curriculum to address the needs of visual learners.

- **Collaborative Learning**
  Students benefit from communicating mathematics, sharing different approaches, and being more actively involved. They attain a deeper understanding by explaining concepts to others. Teamwork creates a more positive learning environment in class; students enjoy attending.

- **Integrated Review**
  Once a concept has been introduced, related problems are sprinkled throughout the course.

Technology

A graphing caluculator is required for this course. Our textbook integrates appropriate calculator technology without trying to teach specific keystrokes for specific calculators. Our goal is to teach mathematics; the technology is taught as it applies to the mathematics. This means that we do not spend time up front on the calculator, but rather, we teach various features as the need arises. In this textbook, graphs are presented as we expect students to present them on written work, even when the calculator is used to determine details of the graph. This means that none of the graphs will look like they do on a calculator screen.

Student Involvement

This course requires students to be much more actively involved in their learning than they may have been before. One of the problems that sometimes surfaces is the complaint that they do not have enough time to give the two hours per credit per week that is needed for outside of class work. We all are encountering more students who are trying to work, raise a family, and go to school. Students need to be realistic about how much they can take on at one time. One exercise that we have found helpful is to do the following reality check with the class early in the term. It serves to bring the discussion out in the open and is a real eye-opener for many students. We try to get through to them that taking the course once with enough time devoted is much less time consuming in the long run, less frustrating, and cheaper than taking the course three times!

Reality Check

1. Number of hours of work per week   _____
2. Number of hours of sleep per night times 7 days per week   _____
3. Number of hours spent eating and bathing per day times
   7 days per week   _____
4. Number of credit hours this term times three   _____
5. Number of hours commuting each week   _____
6. **Total of 1–5**   _____
7. Total number of hours in a week   _____
8. Number of hours remaining in a week for family, household
   chores, entertainment, exercise, etc.   _____

A Typical Day in the Classroom

Student teams are essential to the success of the program. To see our philosophy of teams and how they are used, read Chapter 0. We try to set up the class teams as soon in the term as possible.

Beginning of Class.   We feel it is essential at the beginning of each class period to have time set aside for students, in their teams, to compare their homework. We circulate during that time to keep teams focused and to find which questions need to be addressed for the entire class. Many minor questions will be handled by students in their teams. One of the reasons that this is a crucial part of class time is that very few answers are included in the back of the textbook. This is a chance for students to compare and see where they disagree. We want students to become confident problem solvers, able to verify in many ways that they are correct. We give them many methods for verifying results throughout the term.

While this process sometimes takes a lot of time, it is well worth it. Student understanding is improved, and the instructor can quickly assess the progress of the class. Also, time used this way cuts down on the time spent answering homework questions in front of the whole class on problems that only a handful of students may have.

Active Student Involvement.   What happens next will depend on where the class ended the previous class period.

If a new section is being introduced that includes an activity set, students would now begin that in their teams. These activity sets are designed to guide students to discover, at an introductory level, major concepts of the section. While some of the activities can be done by students at home, most are more effective when done in class in teams.

During an activity, students are expected to do several things.

- They are to be actively involved with their teammates following the directions in the activities.
- They are then expected to make their best effort in discussing their observations of patterns and in making conjectures. This takes some training for many of them, who are used to being told to "work alone and keep your eyes on your own paper."
- They are expected to rearrange their chairs to easily work with one another. Unless you are fortunate enough to have tables for students to work around, chairs should be formed into a circle to allow everyone to hear and to participate. Some will resist this—but you should insist.
- They are expected to take notes on their efforts that allow them to contribute to the class follow-up of the activity. They will not always have the precise mathematical language. Let them find ways to say it in their own words and supply the correct new vocabulary during the follow-up discussion sessions.

While students are working on the activities, the instructor is busy circulating through the room watching and listening to what the teams are doing. Different actions by the instructor may be appropriate.

- The instructor can watch for students who are off topic and give them a gentle nudge to get back on task.
- If a team is going way off the intended course, the instructor can ask questions to get them focused and back on track.
- If a team has a major misconception, it may be advisable to correct them.
- The instructor needs to practice reading the students to see how they are feeling and what their individual level of tolerance for exploration is. When some students direct a question to the instructor, it is appropriate to put the question back in the lap of the team for them to grapple further with. With other students, that action may cause them to put up a wall and tune out. An answer or partial answer with a focused question may be more appropriate. There is no easy way to know the correct intervention. Time and practice will make it easier!
- If it becomes clear that several teams are all having the same question or point of confusion, it may be appropriate for the instructor to pull the class back together for a couple of minutes to clarify the point or the directions.

When the activities are taken seriously, they give the student a sense of ownership of the concepts. They will have experienced the mathematics, not just memorized someone else's findings. Even when students do not reach all of the conclusions intended in the activities, attempting the activities gives students a feel for the questions that the section hopes to answer. Knowing what questions to anticipate gives the student some insight and confidence as they begin reading the section.

The activities are not designed to build particular skills. They are designed to get students involved in doing mathematics and creating their own understanding. Students should not be held responsible for particular results from the activities but for their genuine effort in doing what the activities asks them to do. A great deal of learning occurs even when the final result is not what the instructor might have had in mind. The effort by the team to describe their findings and to make generalizations, forces the students to use the language of mathematics in communicating with each other and in writing down their conclusions. The activities lay the groundwork for the discussions that follow.

Follow-up Discussion.   When most teams have had time to finish the activity, the instructor needs to pull the class back together for an all-class follow-up of the activity. This follow-up discussion after an activity is the key to the success of the activity.

During the follow-up, the instructor directs the discussion making sure to include several elements.

- The instructor solicits input from the students based on their observations. This is a good time to point out the variety of responses from student teams. If a team is reluctant to share something you know they observed, they can sometimes be coaxed or you can share their observations with the class.

  At times two teams will come up with seemingly different but equivalent observations. This is an ideal time for the instructor to show mathematically that they are equivalent. It also reinforces the fact that more than one right answer is possible. There are times when a student or a team will be adamant that one of their incorrect observations is correct. This is a wonderful time for a well-selected counter-example. If you are unable to come up with one on the spot do not hesitate to promise to bring one in the next day—and then don't forget. The instructor is put on the spot much more often in this type of classroom than with a lecture followed by a predictable discussion. There is a definite loss of control of just where a discussion may lead.

- Next the instructor should elaborate on the student input to make sure all the points of the objectives are covered.

- This is the time that the instructor should offer the mathematical justification for the observations and conjectures. Students must realize that two to four observations of a pattern do not form enough evidence to formulate a rule in mathematics.

- During the discussion, the instructor should provide students with correct mathematical language.

The instructor should be prepared to conclude the discussion with a problem or example for the students to do that reinforces the concepts covered in the activity.

During class time, there is a constant movement back and forth between students working on problems or activities and instructor-led discussions or explanations. This is true whether or not there is an activity.

Whatever the class has worked on during the period, the instructor should plan time to pull things together before students leave for the day. The class should leave at a common point of understanding.

Homework Assignments.   At the end of the class period, students are given an assignment to be done for the next class meeting. This will usually include both a reading assignment and appropriate problems from the problem set. What can be assigned may change based on how far the class got during a certain day. Instructor flexibility is the key!

Reading the Text.   Students should be expected to carefully read the discussion and examples in the textbook. In this portion of the textbook, the concepts of the section are discussed, when possible, in the context of applications. This means that some examples in the textbook tend to be longer and more involved than in a traditional textbook. It is crucial that the students read the discussion part of the textbook. Looking briefly at the examples will not be sufficient for understanding and success! The textbook was written to be read by students with paper, pencil, and calculator in hand. They should read in detail, answering any questions that are asked, working through the examples as they read them, and making notes in the margins. And most important, they should write down any questions that they have and be sure that they get their questions answered. As the concepts are presented, new vocabulary is introduced and important ideas are highlighted. Each presentation concludes with a summary of the key points of the section.

Problem Sets.   The problem sets generally have a few straightforward skill problems. Because we want the problem sets to give students a chance to apply the problem solving tools that they are developing in the course, the problems often do not just mimic the examples in the discussion. The problem sets include conceptual questions, applications, and also open-ended questions. In these problem sets, we are asking students to communicate in writing their thought processes, conjectures, comparisons, and mathematical arguments. We want our students to see mathematics differently. In the presentation of the material and in the problem sets, problems are presented numerically, graphically, and algebraically with an emphasis on the connection between the three models.

Selected Answers.   Many of the answers included in the back of the textbook are those that cannot be checked or verified by some other method. This textbook models checking and verifying results using one or more of the three models (numeric, graphic, or algebraic).

Most answers for problems that can be checked or verified are not included in the back of the book for three main reasons.

1. Many students gain a false sense of confidence by working backwards from the answer. Because the exams they will be taking and life in general do not start with the answers, the skills that they are developing of solving a problem by working backwards will not transfer to real life situations.

2. Checking and verifying problems are important skills to develop and use. If answers are included for problems that the students can verify themselves, students will rely on the answers in the back rather than learning and practicing these important skills.

3. When students are able to check or verify their results themselves they gain confidence in their mathematical ability.

Supplements

Instructors who adopt this text may receive, free of charge, the following items.

An Instructors' Resource Manual is available to all instructors. This manual provides complete solutions to all the exercises in the problem sets. Additionally, the manual contains additional material from the authors on how to effectively teach from this text.

The Printed Test Bank contains prepared tests for each chapter of the text. Final exams are also included. For each section of the text, sample problems are provided that can be used to give students additional practice.

The Computerized Test Bank includes all the test bank questions and allows instructors to prepare quizzes and examinations quickly and easily. Instructors may also add questions or modify existing ones. A gradebook feature is available for recording and tracking student's grades. Instructors have the opportunity to post and administer a test over a network or on the Web. A user-friendly printing capability accommodates all printing platforms.

Students using *An Intermediate Course in Algebra: An Interactive Approach* may purchase the Students Solution Manual. This manual contains complete solutions to all problem numbers or letters in blue. There are often multiple ways to solve problems; however, the Student Solution Manual does not include all methods. The problems that are blue in the textbook also have some portion of the question answered in the back of the book.

A Web site (URL to come later) has specifically been created for the first edition of *A First Course in Algebra: An Interactive Approach*. This Web site offers additional resources to instructors and students in conjunction with the adoption of the text.

Acknowledgements

We would like to express our appreciation to the Mt. Hood Community College Mathematics Department who agreed to use these materials before the project was complete. Their support and feedback has been invaluable.

Thanks to Steve Bernard, Wini Benvenuti, Bill Covell, Gary Grimes, Harold Hauser, Teresa Kuntz, Pamela Matthews, Mike McAfee, Kory Merkel, Maria Miles, Paul Porch, Gina Shankland, Frank Weeks, Sara Williams, and Steve Yramategui.

We are also grateful to all of our part-time colleagues who have used preliminary editions of this text and offered comments. Additionally, we wish to thank the faculty from other departments who shared their expertise in helping us to write realistic applications from other disciplines. Others who have offered moral support and advice are Betty Brace, Brenda Button, Lynn Darroch, Andres Durstenfeld, Gretchen Schuette, Mike Shaughnessy, Bert Waits, and Bob Wesley.

Thanks to the hundreds of students who have used these materials in their beginning stages. Their constructive input has led to many improvements.

There are many mathematics instructors across the United States who have reviewed previous versions of these materials. We thank them for their comments and suggestions.

Nancy Alexander, Grand Valley State University
Gian Mario Besena, Eastern Michigan University
Connie Carruthers, Scottsdale Community College
Linda Cave, Western Washington University
Oiyin Pauline Chow, Harrisburg Area Community College
Mary Clarke, Cerritos College
Alice Grandgeorge, Manchester Technical and Community College
Sandy Lynn, Foothill College
Mike Mallen, Santa Barbara Community College
Camile McKayle, University of the Virgin Islands
Susan Nelson, Casper College
Kamilia Nemri, Spokane Community College
Margaret Perrie, University of Wisconsin—Oshkosh
Bernadette Sandruck, Howard Community College
Ann Steen, Santa Fe Community College
Patrick Wagener, Los Medanos College

In addition, there were numerous mathematics instructors who class tested previous versions of these materials. Many of the revisions have come about in response to their recommendations, making the textbook much more useful and student friendly. We thank them for their comments and suggestions.

Asheville-Buncombe Technical Community College: Betsy Hester, Sharon Killian

Casper College: Susan Nelson

Chattanooga State Technical and Community College: Derek Lance

Chemeketa Community College: Ken Anderson, Dorothy Beaufait, Kathy Kinman, Marveen McCready, Susan Poston, Doug Rasmussen

Cincinnati State Technical and Community College: Jan Hoeweler, Joan Jackson

College of the Desert: Karen Tabor

Columbia College: Virginia Gray, Larel Grindy, Maryl Landess

DeAnza College: Miska Chudolowsky, Karl Schaffer, Don Rossi, Carol Olmstead

Eastern Connecticut State University: Richard Broomfield

Foothill College: Sandy Lynn, Helen Magneson, Kathy Perino

Grand Rapids Community College: Jim Chesla

Hartnell College: April Allen

Highline Community College: Diana Bender, Karen Frank

Howard County Community College: Bernadette Sandruck, Paula Milkowicz

J. Sargeant Reynolds Community College: Donna Jawcenovitch, Bernadette Kocyba

Los Medanos College: C.R. Messer, Myra Snell, Patrick Wagener

Manchester Technical and Community College: Alice Grandgeorge

Montgomery Community College: Donna Beverly, Virginia Morgan

Mountain Empire Community College: Chris Allgyer, Robert Rhea

Napa Valley College: Chris Burditt

Northeast State Technical College: Maggie Flint, Jacki Jonas

Palomar College: Annette Parker, Dan Clegg

San Jose City College: Spencer Shaw

Solano Community College: Robert Scott

Spokane Community College: Kamilia Nemri

State University of New York at Cobleskill: Diane Geerken, Charlotte Grossbeck

University of Texas at El Paso: David Harvey

Valencia Community College—East Campus: Debbie Garrison, Judy Jones

Our appreciation and thanks go to all those at Santa Fe Community College for their support of this project. They have been instrumental in the development and success of the project.

We would like to thank the National Science Foundation for their support that made significant work on this project possible.

We also wish to acknowledge our editors and their assistants at Saunders College Publishing. In particular, we want to thank our current editors, James LaPointe and Angus McDonald, for their support, humor, and understanding through the final stages.

A special thank-you to everyone at TSI Graphics, and especially to Donna Cullen-Dolce for her hard work, concern, and understanding.

We are grateful to Lauri Semarne who examined (and corrected where necessary) the activity sets, examples, and problem sets.

We know we may have inadvertently missed others who have had an impact. Please know that we appreciate all the help we have received.

Finally, we wish to express our heartfelt appreciation to Jack, Mike, Erik, and Beth for supporting, enduring, and encouraging us through the development of this project.

Alison Warr
Cathy Curtis
Penny Slingerland

April 2000

Chapter Zero

An Introduction to Learning Mathematics, Problem Solving, and Teamwork

The goals of this text may be different from those of others you have used in the past. Don't be surprised if you find the presentation of the material and the overall content unique as well. The purpose of this chapter is to communicate the goals of the text and to give you some insight into our thoughts about learning mathematics and the reason we approached the writing of this textbook as we did.

Learning Mathematics

One of the major goals of this textbook is to help you become a better problem solver. This does not mean only solving equations and typical mathematics problems but also being able to look at unfamiliar situations, choosing strategies, and applying mathematics to solve these problems.

A second goal of this textbook is to present mathematical ideas in meaningful ways. Mathematics is a useful tool in solving problems in many different fields, and we want you to see this value in mathematics.

Many of you may have learned mathematics in a lecture format. From our experience we have found several drawbacks to a mathematics course taught primarily from a lecture. Many students become bored by mathematics taught in this format and consider the subject matter as a collection of abstract, disjointed facts that they are unable to use in real-life situations. Therefore, they feel that mathematics is an unimportant hoop for them to jump through to receive a degree.

A more important issue is the quality of learning and understanding that students achieve through this approach. Many students who take mathematics in a lecture format comment that "It looks so easy when you do it, but I can't." In a lecture, an instructor typically shows the students many examples and works each of them out correctly. Of course it looks simple! The instructor knows exactly how to do the mathematics. It is much like watching carpenters work. They know exactly how to perform the skills and have hours of experience. It is an entirely different situation when an inexperienced person tries to perform the job. You do not know where you are going to have difficulty until you try. Mathematics is similar. Until you dig into the material, you may not know what you can do or where you have questions.

For this reason, this textbook expects you to actively participate in learning mathematics. In doing so you will not only learn the skills necessary to perform specific tasks, but also obtain an understanding of the concepts behind the mathematics and know when and where to apply the concepts and skills.

Learning anything new takes effort. Learning mathematics is no exception. In using this textbook you will be asked to work. You will be involved in doing mathematics both in and outside the classroom. You will be asked to discuss mathematics with your classmates and your instructor. To be successful in this course, you cannot be a passive recipient of mathematical facts. We are asking you to do more than put in time doing assignments. We are asking you to *Think!* You will be asked to think about the mathematics that you are doing, as well as what is going on in your mind as you do mathematics. You will need to decide when to struggle with a mathematics problem and when to ask for help. In essence, we are asking you to take responsibility for your own learning.

In return, we promise that this textbook will provide you with the opportunity to learn mathematics in a meaningful way. If you use it properly and your course is consistent with the principles on which the textbook is written, then you will begin to claim these mathematical ideas as your own. Mathematics will no longer be a mystery or a mere collection of facts. It will be a collection of principles and concepts that you can understand. You will find yourself less frequently staring at a collection of symbols not knowing how to proceed. More often you will see many possibilities in choosing where to start a problem. Finally, we believe that you will begin to see the use, value, power, and beauty of mathematics.

Discovery Learning

In this textbook, we adopted a discovery approach to learning. Most sections begin with an Activity Set. These activities were written with the assumption that you will be working on them with other members of your class. Each activity has been written to provide you with the opportunity to "dig into" the mathematics. Although we do not expect that you will learn all the mathematics through these activities, we do expect you to be actively involved with others in your class and to make your best effort in discussing your observations of patterns and in making conjectures. Many activities include several parts. Some parts ask you to perform certain tasks, such as make a table, draw a graph, or simplify an expression. Other parts ask you to summarize your observations or make a conjecture. Making observations or conjectures can be difficult. However, this is the most important aspect of the activity. Doing this will greatly enhance your understanding of the mathematics. Be sure to give it your best effort.

Following a set of activities, your instructor will typically lead a discussion of the observations and conjectures made by the class. This discussion may include vocabulary, generalizations to formulas, or applications of the ideas learned in the activities. Even when you do not reach all of the conclusions intended in the activities, attempting the activities will make this discussion more meaningful and useful.

How to Use This Text

We have tried to produce a textbook that

- demands greater understanding and less routine manipulation
- covers less material in greater depth
- presents concepts numerically, graphically, and algebraically
- develops concepts through commonsense investigations rather than abstract definitions
- incorporates the use of graphing technology and expects this technology to be available at all times
- is written to be read
- is written for students to discover concepts, not as a reference for those who know the concepts

We wrote this textbook with you, the student, in mind. We believe that the combination of doing discovery activities, participating in class discussions, and actively reading the textbook will lead to your success in learning mathematics. Most sections of this textbook begin with activities, are followed by a text discussion, and end with a problem set. Some of the discussions read like a standard textbook, but there are some important differences. Many of the examples emphasize mathematics in context. We de-emphasize rote learning of mathematics. Questions are raised and left to you, the reader, to answer. In addition, appendices in the back of the textbook include review material.

To successfully use this book, you must read it with a pencil and calculator in hand. The activities at the beginning of each section *should* be done first; however, you are not penalized for looking ahead. Once you have done the activities, you should participate in any class discussion that may follow, read the textbook in detail, answer any questions that are asked, work through the examples as you read them, and make notes in the margins. Most importantly, write down any questions that you have, and be sure that you get your questions answered.

Mathematics was created to solve real-world problems like balancing a checkbook or determining the load that a roof can bear. Because mathematics has many different applications, it is not possible to learn specific methods or templates to solve every type of application. Instead we need to learn problem-solving skills that will allow us to tackle all sorts of problems. For this reason, you will not find worked examples for every type of problem you will encounter in the problem sets. In addition, for a given problem, many different solution processes are often correct. We encourage you to discuss your solutions with your classmates. As you see different ways of approaching and solving problems, you will learn new strategies that can be used on future applications.

What Is a Good Problem Solver?

One of the goals of this course is to make you a better problem solver, but what is a good problem solver? The following is a list of suggestions for becoming a better problem solver written by students who successfully completed a course in problem solving.

How to Become a Better Problem Solver

- Accept the challenge of solving a problem.
- Take time to explore, reflect, think, . . .
- Look at the problem in a variety of ways.
- If appropriate, try the problem using simple numbers or break the problem into smaller steps.
- Develop good problem-solving helper skills. Don't give solutions; instead provide meaningful hints.
- Write up your solutions neatly and clearly enough so that you could understand your solution if you reread it in 10 years.
- Help others by giving hints. You will find that you develop new insights.
- Don't hesitate to take a break. Many problems require an incubation period. But remember to return to try again!
- Be persistent. Don't give up!
- Don't just sit there, do something!

When solving a problem in mathematics (or elsewhere), you may begin by asking "How do I get started?" There is not one answer to this question but several possible ways to get started. The following is a list of strategies for trying to solve a problem. These are just the beginning. As you solve more and more problems, you will be adding to this list.

Write Down Everything You Know
Write down what you know about the problem and the question you are trying to answer. This can help organize the information.

Draw a Picture
Can you draw a picture of the problem situation? Often you may find that in the process of drawing and labeling a picture you gain some insight into relationships that can lead to the solution of the problem, or at least to the next step in the process.

Make a Table of Values
A table of values can give you some concrete information from which to discover patterns. In the textbook, you will see how to use this powerful tool.

Guess

If you don't know where to begin, go ahead and guess. Many problems have been solved by someone saying "What if . . . ?" Once you make a guess, your next step is to determine whether or not your guess is correct. This can be done by substituting your guess into the problem situation. If you guessed correctly, great! If you did not, you may find the insight that you need to solve the problem, or you may be able to eliminate several possibilities.

Make a Physical Model

For many of us, we need to physically see the situation to understand what is happening. This may mean we need to represent the problem using concrete objects. For example, in Problem 7, "Moving Dots," in the Problem Set at the end of this chapter you may want to represent the solid dots with pennies and the striped dots with dimes. Then you can physically try solving the problem using the coins. Other times you may want to build a model. For example, if the problem is about a swimming pool, it may be helpful to construct a model of the swimming pool out of cardboard.

Throughout this course, you will encounter many mathematical problems. Some you will be able to solve with what you already know or what you are learning. Others you will need to develop more ideas and tools to solve. However, *the time to conclude that you are unable to solve a problem is after you have given it a royal try—not before.*

As you struggle with various problems throughout the course, the most important thing for you to remember is not to think of these problems in terms of success and failure. When you are exploring new ideas and the results that you get do not completely solve the problem, think of this as an opportunity to learn and improve your understanding. Often a strategy that does not lead to the solution will lead to new strategies.

Why Work in Teams?

In this class, you may be expected to work in a team. Students often ask why they are being asked to do this. There are many reasons for working in a team in this class. Working in a team gives you a safe environment to try out new ideas and problem-solving strategies. By working with your team, you will learn to communicate more clearly your ideas and your understanding of concepts. Your facility with the language of mathematics will improve as you get more practice using it. You will also learn to listen carefully to other people's ideas. You will be challenged to explain concepts so others can understand them. The give and take of ideas will improve your own reasoning and critical thinking abilities.

In addition, many recent school-to-work studies call on schools to teach students more of the skills that business and industry find their employees need. The general ability to work with others and communicate clearly, orally and in writing, are always on the lists of desired skills. Other expectations are that workers know how to develop new skills, can manage themselves, and are responsible for their actions. Working in teams in class gives you one more place to develop and refine these skills. There is more to school than learning specific academic skills.

Production Team Versus Learning Team

Although many of the skills learned by working in teams are appropriate for both school and work settings, many important differences exist. In a workplace team, the team may have been formed by pulling together people with different skills and expertise to work together to produce a product. In this situation, the team may do some initial planning together. They may then go off to produce their individual part of the project before the team comes back together to integrate the separate pieces into the finished product. Each member of the team relies heavily on the differing expertise of the team members. Although the team members count on each other for support and troubleshooting, they may not be working closely together every day. This type of team could be called a **production team.**

In the classroom, everyone on the team is expected to learn the same concepts. The product they are producing is mutual understanding of the course material—not a consumer product for others. It is important that all members of the team gain a full understanding of all concepts. The main point of the team is to help in the learning process. We refer to this type of team as a **learning team.** We need to keep in mind that in this class we are forming learning teams and not production teams.

Team Selection

The manner in which teams are selected will vary from class to class and instructor to instructor. Many instructors look for a balance of gender, previous experience, age, and motivation. Other considerations that may be taken into account are times you are available to study outside of class or what kind of graphing calculator you have. The object is to get a mix that allows all of you to have a good learning experience.

What Makes a Team Work?

For a team to be effective, it is important that everyone knows what is expected of him or her. Often ground rules for team behavior are established to help with this process. A sample set of ground rules is included in the following box.

Ground Rules for Teams

- Attend class regularly, and come prepared.
- Stay focused on the team task.
- Work cooperatively with other team members.
- Reach a team decision for each problem.
- Make sure each person on the team understands the solution before the team moves on.
- Listen carefully to others, and try to build on their ideas.
- Share the leadership of the team.
- Make sure that everyone participates and no one dominates.
- Take turns recording team results.

It is important that team members agree on the ground rules for the class. It is the responsibility of team members to monitor each other in following the agreed-on rules. Although the instructor can help in this process, it is not possible for the instructor to be in all teams at all times.

Team Building

Being an effective team member takes some practice. The problems at the end of this section are designed for you to practice team-building skills in a nonthreatening environment. As you work on these activities, keep in mind the ground rules. Remember that the goal is to learn to work together and communicate clearly so that you can use these skills to help each other effectively learn mathematics.

Many of these activities may not seem mathematical to you. They are all in some way related to thinking and visualizing skills that are important in mathematics. Sometimes you will want to work together on a problem from the start—brainstorming approaches and strategies. At other times, you may want to spend some time working individually before you share your ideas. Be aware of the needs of your team members as well as your own preferences. Some of the activities that follow will be easier to perform with some concrete props—coins or pieces of paper, for instance. Don't hesitate to use any tools that make the problem solving or explaining easier.

Last Words of Wisdom

Don't get discouraged if you don't always feel successful. Frustration is a natural part of any learning experience. Jump in, get involved, and enjoy yourself! Solving problems is a positive experience.

Problem Set 0

1. Tennis Tournament.   One hundred people are playing in a single elimination, singles tennis tournament. How many matches need to be played to determine the winner?

2. Politician Puzzle.   Imagine a political convention attended by 200 politicians. Each politician is either honest or crooked. We are given the following facts:

   • At least one of the politicians is honest.
   • Given any two of the politicians, at least one of them is crooked.

   Determine how many of the politicians are honest and how many are crooked. Convince another team that your conclusion is correct.

3. Count the Triangles.   Count all of the triangles in the figure.

4. Constructing Pentominoes.   A pentomino is a figure made of five congruent squares. Adjacent squares must have a full side in common.

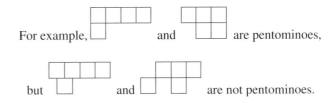

For example, [figure] and [figure] are pentominoes, but [figure] and [figure] are not pentominoes.

Pentominoes are considered identical if they are just rotations or reflections of one another. Sketch all possible pentominoes. Discuss how you are sure you have found them all.

5. Toothpick Problems.   Use the following two rules to solve the problems.

    i. No toothpicks are to be broken.

    ii. All of the toothpicks are part of the figure described.

a. Use 15 toothpicks to make the figure shown here. Move two toothpicks to form a pattern of five identical squares.

b. Use nine toothpicks to make the pattern shown here. Move four of the toothpicks to form a pattern of five triangles.

c. Move two toothpicks in the following pattern to form seven identical squares.

d. Using the figure from part c, remove six toothpicks to create a figure consisting of two congruent squares.

e. Use 15 toothpicks to make the following pattern. Transform it into two squares in two ways: first create two squares by moving four toothpicks; then create two squares by moving five toothpicks.

f. Using 15 toothpicks, create a figure consisting of two congruent squares.

g. Using six toothpicks, create a figure consisting of four congruent squares.

h. Place three toothpicks side by side, ║║║. Add two more toothpicks to make eight. Magic is not necessary but open-minded thinking is!

6. Cube Fold.

a. Which of the following patterns can be folded into a closed cube?

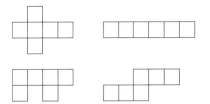

b. Find two more patterns that can be folded into a closed cube.

c. Determine all of the patterns made of six squares that can be folded into a closed cube.

7. Moving Dots Problem.

Start with the arrangement of dots shown here:

Try to produce the following arrangement

given these three restrictions:

   i. The solid dots can move only to the right, and the striped dots can move only to the left.

   ii. A dot can move into an adjacent space.

   iii. A dot can jump over *one* other dot—either solid or striped—into an empty space.

Once you can do the problem, work together on a way to clearly communicate your solution in writing. Assume that the solution is being written for someone who is familiar with the rules of the problem but has not necessarily solved it.

8. Following Directions.　Apply the directions given in each step to your result from the previous step.

a. Print the words GEOMETRY IS FUN without a space between the words.

b. Change every E to an S.

c. Move the second vowel from the right to the first position from the left.

d. Exchange the second and seventh letters from the left.

e. Replace the 14th letter in the alphabet with the 12th letter in the alphabet.

f. Replace the letter M with the letter preceding it in the alphabet.

g. Insert the letter I between the first and second consonants from the left.

h. Reverse the order of the third and fourth consonants from the right. (Consider Y to be a consonant.)

i. Replace the letter R with the first vowel following it in the alphabet.

j. Exchange the letters O and G.

k. Replace the letter that is to the left of the letter F with the letter that is to the left of F in the alphabet.

l. Replace the letter G with the first vowel in the alphabet.

m. Insert spaces after the second letter, the fourth letter, and the eighth letter.

What is your conclusion?

9. Crossing the Enchanted Forest.   Four apprentice magicians must pass through an enchanted forest on their way to a Magicians' Conference in Minneapolis. They stop at the edge of the forest and discover that they were careless in their packing and have only one magic wand among the four of them. The wand is only strong enough to protect two magicians at a time and, while in the forest, the wand must at all times be in the hand of one of the magicians. One of the magicians can cross the forest in 5 minutes, the second takes 10 minutes, the third needs 20 minutes, and the fourth magician needs 25 minutes. (When going together, they must travel at the rate of the slower of the pair.) From the time they begin, they have only 60 minutes to pass through the forest or they will be turned into frogs! Describe how the four magicians can successfully pass through the forest. (No tricks are necessary!)

Chapter One

Functions and Modeling

Mathematics can be used to solve many problems in society, in business, and in our everyday lives. This chapter will introduce you to a specific type of mathematical relation, a function. We will explore the different mathematical models that can be used to represent functions: verbal, numerical, graphical, and algebraic. The concepts from this chapter are integrated throughout the text.

1.1 Tables, Graphs, and Equations

Activity

Set

1.1

When we are working with relationships in two variables, it is helpful to decide which variable depends on the other. For example, in the formula for the area of a circle, $A = \pi r^2$, the area clearly *depends* on the radius. Therefore, we call the area A the dependent variable and the radius r the independent variable.

> Definition _____
>
> A **dependent variable** is a variable whose value depends on the other variable's value. The other variable is known as the **independent variable.**

In a table, the independent variable is represented first. This means that the independent variable is either in the leftmost column or in the top row, depending on the orientation of the table.

Independent Variable	Dependent Variable

or

Independent Variable	
Dependent Variable	

When graphing, the independent variable is graphed on the horizontal axis and the dependent variable on the vertical axis.

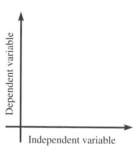

1. In each of the following situations, decide which variable is the dependent variable and which is the independent variable.

   a. You plan to purchase hardwood flooring. The variable C represents the cost of the flooring, and f represents the number of square feet of flooring.

   b. Janet needs to worm her dog. W represents the weight of her dog, and M represents the number of milligrams of medication.

   c. Kramer is a senior in college. S represents the number of hours spent studying for classes, and C represents the number of credits he is taking.

2. The following graph depicts the monthly payment for a loan in terms of the amount of the loan, assuming the loan is paid off in 5 years at an 8.9% annual interest rate. *Use the graph* to answer the following questions.

   a. Under these conditions, how much would your monthly payment be if you want to borrow $3000? $6000? $7200? $22,500?

   b. If you know that you can afford a $150 monthly payment, approximately how much money can you borrow?

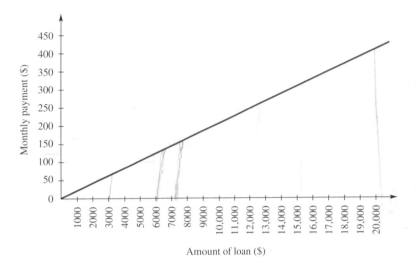

Amount of loan ($)

3. The table below relates the monthly payment to the original amount of a loan for 5 years at an 8.9% annual interest rate. *Use the table* to answer the following questions.

   a. Under these conditions, how much would your monthly payment be if you want to borrow $3000? $6000? $7200? $22,500?

   b. If you know that you can afford a $150 monthly payment, approximately how much money can you borrow?

Amount of Loan ($)	Monthly Payment ($)	Amount of Loan ($)	Monthly Payment ($)
2,000	41.42	12,000	248.52
3,000	62.13	13,000	269.23
4,000	82.84	14,000	289.94
5,000	103.55	15,000	310.65
6,000	124.26	16,000	331.36
7,000	144.97	17,000	352.07
8,000	165.68	18,000	372.78
9,000	186.39	19,000	393.49
10,000	207.10	20,000	414.20
11,000	227.81		

4. The equation $R = 0.02071A$ gives the monthly payment for a five-year loan at 8.9% annual interest in terms of the amount of the loan, where R is the amount of the monthly payment and A is the amount of the loan.

   Use the equation to answer the following questions.

   a. Under these conditions, how much would your monthly payment be if you want to borrow $3000? $6000? $7200? $22,500?

   b. If you know you can afford a $150 monthly payment, approximately how much money can you borrow?

5. In Activities 2, 3, and 4 you answered the same questions using three different models: a numerical model (the table), a graphical model (the graph), and an algebraic model (the equation). Describe in words the advantages and disadvantages of each model in this situation.

6. **Weiser, Idaho.** A long-distance telephone company in Oregon charges $0.48 for the first minute and $0.28 for each additional minute for a phone call from Portland, OR, to Weiser, ID.

   a. Identify the independent and dependent variables.

   b. Complete the following table.

Length of Call (Minutes)	Cost of Call ($)
1	
2	
3	
4	
5	
10	
20	
30	
m	

   c. Draw a graph of the cost of a call in terms of the length of the phone call in minutes.

   d. Write an equation for the cost of the phone call in terms of the number of minutes. Verify that your equation is correct by substituting $m = 1$, $m = 2$, and $m = 3$ into your equation, and compare with your table.

   e. Use your equation to determine the cost of a phone call from Portland, OR, to Weiser, ID, that is 7 minutes long. How long can you talk for $5?

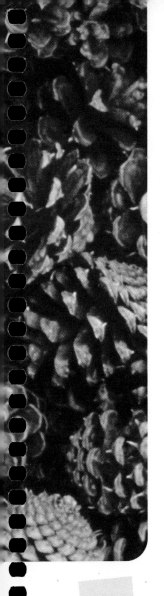

7. Interest.   Zoe has $1500 to put in a savings account. The account earns annual interest of about 4%.

NOTE: An increase of 4% assumes we are starting with 100% and adding 4%. Therefore, an increase of 4% is equivalent to 104% of what we start with. As an example, if we start with $100 and want to increase this by 4% we can multiply by 104%, that is,

$100 * 1.04   (*Reminder:* 104% = 1.04)

= $104

a. Complete the following table.

Number of Years the Money Is in the Account	Amount of Money in Savings Account ($)
0	
1	
2	
3	
4	
t	

b. Graph the amount of money in terms of the number of years.

c. Write an equation for the amount of money in terms of the number of years.

d. How much money will be in the account after 7 years?

e. How long will it take for the money to grow to $2000?

Discussion 1.1

In this section we will be looking at mathematical modeling. That is, we will take a problem situation and model that problem mathematically. We will consider three basic models in this section: a numerical table of values, a graph, and an algebraic equation. In the next three examples we will look at the same problem situation but use a different mathematical model each time.

Example 1

Car Rental: Using a Table.   A car rental company in Salem, Oregon, charges $30 per day for a midsize car with unlimited mileage. Most of their midsize cars average 20 miles per gallon, and the price of unleaded fuel is $1.21 per gallon. Assume that the car is rented for one day, the gas tank is full when the car is rented, and the gas tank is full when the car is returned.

a. What do you define as the dependent variable? What would you define as the independent variable?

b. Make a table of values for cost to rent a car in terms of the number of miles driven.

c. Use your table to determine how much it will cost to drive 200 miles.

d. Use your table to determine how far you can drive for $50.

Solution   a. The cost *depends* on the number of miles driven. Therefore, the cost is the dependent variable, and number of miles driven is the independent variable.

b. In this example, we want to create a table with the number of miles listed first. The cost to rent the car is determined by adding the daily cost of renting the car to the cost of the gasoline. The cost of the gasoline is $1.21 per gallon, and a midsize car needs one gallon of gas for every 20 miles. So for every 20 miles, it costs an additional $1.21.

Number of Miles	Cost ($)
20	$30 + 1.21 = 31.21$
40	$30 + 2 * 1.21 = 32.42$
60	$30 + 3 * 1.21 = 33.63$
80	$30 + 4 * 1.21 = 34.84$
100	$30 + 5 * 1.21 = 36.05$

Do we have enough values in our table? We need to be able to answer two questions from the table, the cost to drive 200 miles and how far we can drive for $50. Neither of these can be answered yet. We need to add more values to the table. Perhaps we should use a larger increment for the number of miles driven. In the table we created so far, the number of miles increased by 20. Therefore, we multiplied 1.21 by a number 1 higher than the previous entry. If we now create a table with larger increments we need to decide what number to multiply 1.21 by. Because the car averages 20 miles per gallon, we need to determine the number of 20-mile units we are driving. For example, suppose the number of miles is 100. There are five 20's in 100, so we multiply 1.21 by 5. To mathematically determine the number of 20's in a number, we divide that number by 20. Following is the extended table.

Number of Miles	Cost ($)
20	$30 + 1.21 = 31.21$
40	$30 + 2 * 1.21 = 32.42$
60	$30 + 3 * 1.21 = 33.63$
80	$30 + 4 * 1.21 = 34.84$
100	$30 + 5 * 1.21 = 36.05$
200	$30 + 10 * 1.21 = 42.10$
300	$30 + 15 * 1.21 = 48.15$
400	$30 + 20 * 1.21 = 54.20$

c. From our table we see that it costs $42.10 to drive the car 200 miles.

d. From our table we see that we can drive between 300 and 400 miles if we have $50. If we want to find out more precisely how many miles we can drive for $50 we can try several values between 300 and 400 miles and compute the cost.

In problems like that in Example 1, it is clear which variable depends on the other. In some situations it may not be obvious. If knowing either of the variables allows you to determine the other one, you can just assign one of the variables as the dependent variable and the other as the independent, and then proceed.

Example 2   Car Rental: Using a Graph.   A car rental company in Salem, Oregon, charges $30 per day for a midsize car with unlimited mileage. Most of their midsize cars average 20 miles per gallon, and the price of unleaded fuel is $1.21 per gallon. Assume that the car is rented for one day, the gas tank is full when the car is rented, and the gas tank is full when the car is returned.

    a. Make a graph of the cost to rent a midsize car in terms of the number of miles driven in a day.

    b. Use the graph to determine how much it will cost to drive 200 miles.

    c. Use the graph to estimate how far you can drive for $50.

Solution   a. Because the independent variable is graphed on the horizontal axis, the number of miles driven is the label of the horizontal axis. The dependent variable is graphed on the vertical axis, so the cost is graphed on the vertical axis. We can graph the data from the table we made in Example 1.

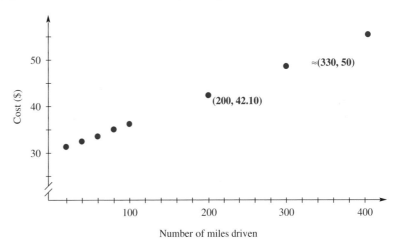

From the graph, we can see that the cost increases as the number of miles driven increases. It appears that all of the points lie in a line. If we connect these points with a line we can use the line to answer the questions in parts b and c.

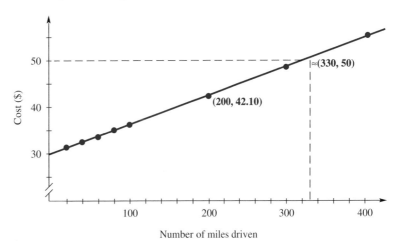

    b. Because we actually plotted the point (200, 42.10), we know that it costs $42.10 to drive the car for 200 miles.

    c. To determine how far we can drive for $50 we draw a horizontal line across at $50 and estimate the corresponding mileage. In the preceding figure we can see that we can drive about 330 miles for $50.

In the previous example, it is clear which variable depends on the other. In other situations it may not be obvious. When asked to graph one variable *A* in terms of another *B*, it is a convention to consider *A* the dependent variable because its value depends or is given "in terms of" the other variable. Then *B* is the independent variable. Therefore, *A* is graphed on the vertical axis and *B* on the horizontal axis.

> A **graph of *A* in terms of *B*** is a graph of the relationship between *A* and *B*, where the vertical axis represents *A*, the dependent variable, and the horizontal axis represents *B*, the independent variable.

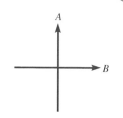

To write an equation of one variable *A* in terms of another *B*, we again consider *A* the dependent variable because its value depends or is stated "in terms of" the other variable.

> To write an **equation for *A* in terms of *B*** means to write an equation where *A* is equal to an expression in terms of *B*. That is,
>
> *A* = *expression in terms of B*

Example 3

Car Rental: Using An Equation.   A car rental company in Salem, Oregon, charges $30 per day for a midsize car with unlimited mileage. Most of their midsize cars average 20 miles per gallon, and the price of unleaded fuel is $1.21 per gallon. Assume that the car is rented for one day, the gas tank is full when the car is rented, and the gas tank is full when the car is returned.

   a. Write an equation for the cost of driving a midsize car in terms of the number of miles driven.

   b. Use your equation to determine how much it will cost to drive 200 miles.

   c. Use your equation to determine how far you can drive for $50.

Solution   a. Because we wrote numerical expressions in our table in Example 1, we can use the table to help us write an equation. We start by writing an expression for the right-hand column. Then, we are able to turn the expression into an equation.

Look at each of the expressions. What is constant in each? What changes? How does the input compare with the values that change?

Number of Miles	Cost ($)
20	30 + 1.21 = 31.21
40	30 + 2 * 1.21 = 32.42
60	30 + 3 * 1.21 = 33.63
80	30 + 4 * 1.21 = 34.84
100	30 + 5 * 1.21 = 36.05

In each expression, 30 and 1.21 stay constant. Therefore, the algebraic expression we are trying to write has 30 as the first term and 1.21 as a factor of the second term. In each expression, the second term contains a number that changes. We need a variable to represent this amount. Our input variable is the number of miles driven. From the table we see that the input in the left column does not equal the changing value in the right column. Each time we had to divide the number of miles by 20 before we multiplied by 1.21. So, if we let m represent the number of miles driven, we can write an expression for the right-hand column.

Number of Miles	Cost ($)
20	$30 + 1.21 = 31.21$
40	$30 + 2 * 1.21 = 32.42$
60	$30 + 3 * 1.21 = 33.63$
80	$30 + 4 * 1.21 = 34.84$
100	$30 + 5 * 1.21 = 36.05$
. . .	. . .
m	$30 + \left(\dfrac{m}{20}\right) * 1.21$

To make this expression into an equation, we let the variable C represent the cost to rent a midsize car for a day. Then our equation is

$$C = 30 + \left(\frac{m}{20}\right) * 1.21$$

where C = cost to rent a midsize car for one day

m = number of miles driven

Whenever we write an equation for a problem situation, we need to check the equation before we use it to answer more questions. To check our equation we substitute a couple of values for m from our table. Let's substitute $m = 40$ and $m = 60$ into our equation.

$$30 + \left(\frac{40}{20}\right) * 1.21 = 32.42 \qquad ✔$$

$$30 + \left(\frac{60}{20}\right) * 1.21 = 33.63 \qquad ✔ \qquad \text{Both of these values match the output values in our table. This means that the equation checks.}$$

Throughout this text, the check mark (✔) is used to indicate when an equation, expression, or inequality is correct. If something does not check, we indicate it with the symbol ✘.

b. To determine the cost to drive the car 200 miles, we substitute $m = 200$ into the equation and find C.

$$C = 30 + \left(\frac{200}{20}\right) * 1.21$$

$$= 42.10$$

The cost to drive 200 miles is $42.10. This agrees with our previous value.

c. To determine how far we can drive for \$50, we substitute $C = 50$ into our equation and find m.

$$50 = 30 + \left(\frac{m}{20}\right) * 1.21$$

$$50 = 30 + 0.0605m \qquad \text{Simplify the right side of the equation.}$$

$$20 = 0.0605m \qquad \text{Subtract 30 from both sides of the equation.}$$

$$\frac{20}{0.0605} = \frac{0.0605m}{0.0605}$$

$$330 \approx m \qquad \text{Round down to stay } under \text{ \$50.}$$

To verify our result, substitute 330 into the original equation for m.

$$\textbf{VERIFY} \qquad C = 30 + \left(\frac{330}{20}\right) * 1.21.$$

$$= 49.965 \qquad ✔$$

We can drive the midsize rental about 330 miles in one day for \$50.

In Example 3, we saw how to use the table to write the equation for the problem situation. Which of the following two tables makes it easier to write an equation that models the table?

x	y
0	33
1	43
2	57
3	75

x	y
0	$2 * 2^2 + 25 = 33$
1	$2 * 3^2 + 25 = 43$
2	$2 * 4^2 + 25 = 57$
3	$2 * 5^2 + 25 = 75$

For most people it is easier to write an equation if the expression is given, not just the result. In the table on the right, it is easier to see what values are constant and what values change. Can you write an equation now? It is still challenging. Once you guess, substitute $x = 2$ and $x = 3$ into your equation. Do you get 57 and 75? If not, can you fix your equation?

This table is challenging because each expression is 2 times the square of a changing number plus 25, but the value that changes is not equal to the input value. Can you see that it was always two more than the input value? Therefore the equation is

$$y = 2 * (x + 2)^2 + 25$$

To verify this we can substitute in $x = 2$ and $x = 3$.

$$2 * (2 + 2)^2 + 25 = 57 \qquad ✔$$
$$2 * (3 + 2)^2 + 25 = 75 \qquad ✔$$

These results match our table values.

Example 4

The Fixed Area Rectangular Pen. A rectangular pen must be built to enclose an area of 300 square feet. How can such a pen be built? How can such a pen be built so as to minimize the amount of material required? Make a table for several possible widths, then sketch a graph of the length in terms of the width and a graph of the perimeter in terms of the width. Write an equation for the perimeter in terms of the width.

Solution   We can construct several possible rectangles. First we sketch some possible rectangles, all with an area of 300 square feet. Suppose that the width of the rectangle is 5 ft. Then the length of the rectangle must be 60 ft. If the width is 10 ft, then the length must be 30 ft. If the width is 15 ft, then the length must be 20 ft, and so on.

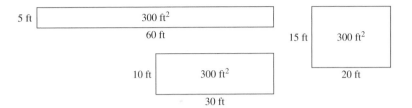

The next step is to organize our information. We use a table starting with the width at 5 feet and increasing the values of the width in 5-foot increments. Our table needs to include width, length, area, and perimeter. The formula for the area of a rectangle is $A = l * w$. Because the area of this pen is constant, we can find the length by dividing the area by the width.

Width (ft)	Length (ft)	Area (sq. ft)	Perimeter (ft)
5	$\dfrac{300}{5} = 60$	300	$2 * 5 + 2 * 60 = 130$
10	$\dfrac{300}{10} = 30$	300	$2 * 10 + 2 * 30 = 80$
15	$\dfrac{300}{15} = 20$	300	$2 * 15 + 2 * 20 = 70$
20	$\dfrac{300}{20} = 15$	300	$2 * 20 + 2 * 15 = 70$
25	$\dfrac{300}{25} = 12$	300	$2 * 25 + 2 * 12 = 74$
30	$\dfrac{300}{30} = 10$	300	$2 * 30 + 2 * 10 = 80$

From the table we can make some observations. The length decreases as the width increases. The perimeter starts out large, decreases to 70 feet, and begins to increase. It appears that the pen that requires the least amount of material has a width between 15 and 20 feet. So we can try some widths between 15 and 20 feet.

Width (ft)	Length (ft)	Area (sq. ft)	Perimeter (ft)
16	$\dfrac{300}{16} = 18.75$	300	$2 * 16 + 2 * 18.75 = 69.50$
17	$\dfrac{300}{17} \approx 17.65$	300	$2 * 17 + 2 * 17.65 \approx 69.30$
18	$\dfrac{300}{18} \approx 16.67$	300	$2 * 18 + 2 * 16.67 \approx 69.34$
19	$\dfrac{300}{19} \approx 15.79$	300	$2 * 19 + 2 * 15.79 \approx 69.58$

If we are only considering whole-number values for the width, we choose the width to be 17 feet because this gives us the minimum perimeter. From the table we can see that if the width is 17 feet, then the length is about 17.65 feet. If we want the length to be a whole number, we round the length to 18 feet. This gives us a pen whose area is slightly larger than the desired 300 square feet. If we want to be more precise than this, we can continue this numerical process until we reach our desired precision.

Next, we graph the length in terms of the width. This indicates that the length is the dependent variable and the width is the independent variable. This agrees with how we created the preceding table. However, you may have noticed that in this problem it would have made sense to select either the width or the length as the independent variable. The following graph is the graph of the length in terms of the width for a rectangular pen with area 300 square feet.

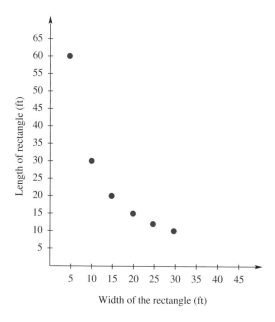

We can see that the length decreases as the width increases, but we also see that the length decreases rapidly for small values of the width and then decreases more slowly as the width increases.

Next, we will graph the perimeter in terms of the width.

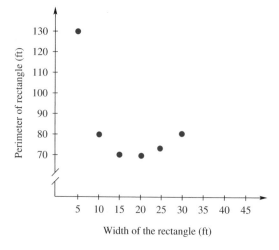

From the graph we can see that the perimeter decreases as the width increases to a certain point and then the perimeter begins to increase again. If we connect the points with a smooth curve we can *estimate* the perimeter for any rectangle. For example, the perimeter is about 90 feet when the width is 8 feet. We can also *estimate* when the perimeter will be at a minimum. It appears that the perimeter is about 69 feet. This occurs when the width is about 17 feet.

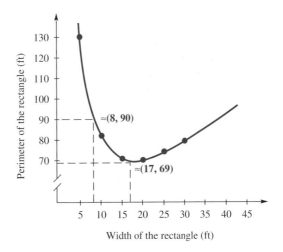

Width of the rectangle (ft)

Finally, we need to write an equation for the perimeter in terms of the width. In the following table, we can see that the perimeter is equal to the sum of twice the width and twice the length. Therefore, we can write the equation $P = 2W + 2L$, where P is the perimeter in feet, W is the width in feet, and L is the length in feet. But this equation is in terms of the width *and* the length.

Next, we can use the table to write an equation for the length in terms of the width and then use this to write the equation for the perimeter in terms of the width only.

Width (ft)	Length (ft)	Area (sq. ft)	Perimeter (ft)
5	$\dfrac{300}{5} = 60$	300	$2 * 5 + 2 * 60 = 130$
10	$\dfrac{300}{10} = 30$	300	$2 * 10 + 2 * 30 = 80$
15	$\dfrac{300}{15} = 20$	300	$2 * 15 + 2 * 20 = 70$
20	$\dfrac{300}{20} = 15$	300	$2 * 20 + 2 * 15 = 70$
25	$\dfrac{300}{25} = 12$	300	$2 * 25 + 2 * 12 = 74$
30	$\dfrac{300}{30} = 10$	300	$2 * 30 + 2 * 10 = 80$

We see that $L = \dfrac{300}{W}$. We can replace L in the equation $P = 2W + 2L$. Therefore, $P = 2W + 2\left(\dfrac{300}{W}\right)$. This can be rewritten as $P = 2W + \dfrac{600}{W}$. We need to verify that this equation is correct by substituting $W = 25$ and $W = 30$ into the equation.

VERIFY

W	$2W + \dfrac{600}{W}$	
25	$2 * 25 + \dfrac{600}{25} = 74$	✔
30	$2 * 30 + \dfrac{600}{30} = 80$	✔

Because these values match our previous table, our equation is verified.

Many of the numerical answers in the examples were approximate. When we read an answer from a graph, the result is always an estimate, unless the point is a labeled point. Using tables, we can continue to guess and check to get a more precise answer, but the question is how much precision is necessary? Even when solving a problem algebraically, the result on the calculator may display many digits. Are they all needed? The answer is, usually not. In this course, as you solve applications, try to think about what is a *reasonable* answer. Sometimes the context of the problem helps you decide. For example, money is usually rounded to the nearest cent. If you are working with measurements, all computed lengths should match the precision of the original measurements. That is, if two sides of a triangle are measured to a tenth of a centimeter, the calculation for the third side should be rounded to a tenth of a centimeter. This guideline applies when the measurements are all linear. If you are computing area, volume, or other formulas, and you are uncertain where to round a result, you can round to the number of significant digits in the given measurements. This usually produces a reasonable result.

We have now seen three ways that we can model a problem situation. We can use a **numerical model,** which often consists of a table of values. A table provides concrete values from which to analyze the behavior of the situation. We can use a **graphical model,** which consists of one or more graphs. A graph provides a visual model from which to analyze the behavior. We can use an **algebraic model,** which consists of an equation or an expression. Algebra provides us with a mathematical language to describe patterns. We can also see that these three models are intimately related. The graph is a visualization of the table. The equation is a generalization of the table. The algebra is a mathematical description of the graph. What advantages and disadvantages do you see in each of these models? Throughout the remainder of the text we will be investigating these three models, the relationship between them, and how they can be used to solve problems and discover important patterns.

In the problem set in this section, we will sometimes be asking you to model each problem using all three techniques so that you can become familiar with the advantages and disadvantages of each. However, throughout the book, you will often have to make the decision on which model or models to use. More than one correct decision will usually be possible. For suggestions on problem solving strategies, refer to Chapter 0.

Problem
Set
1.1

1. In each of the following situations, decide which variable is the dependent variable and which is the independent variable.

   a. Your garden is overgrown with weeds, and you hire a neighbor to weed it. Let A represent the square footage of your garden and H represent the number of hours it will take to weed the garden.

   b. You plan to pay the neighbor by the hour for weeding the garden. Let P represent the payment for weeding the garden and H represent the number of hours it will take.

   c. Let S represent the average speed you drive between your home and the college. Let T represent the time it takes for you to drive to college from home.

   d. The American Association of Pediatricians has set guidelines for daily calorie intake for infants and children. Let A be the age of a male child and C be the recommended number of daily calories.

   e. When a cup of coffee is placed on a table, it cools. Let T be the temperature of the coffee and M be the number of minutes it has been sitting on the table.

2. In each of the following tables, observe the pattern and use the pattern to complete the table.

   a.

w	A
1	$10 + 1$
2	$10 + 2$
3	$10 + 3$
4	
w	

   b.

x	y
0	4
1	$4 - \dfrac{1}{20}$
2	$4 - \dfrac{2}{20}$
3	
x	

   c.

t	S
1	3
2	$3 * 3$
3	$3 * 3 * 3$
4	
t	

   d.

n	M
2	$2^2 + 5 * 2$
4	$4^2 + 5 * 4$
6	$6^2 + 5 * 6$
8	
n	

3. In each of the following tables, observe the pattern and use the pattern to complete the table.

   a.

x	y
1	$5 * 3$
2	$5 * 4$
3	$5 * 5$
4	
x	

   b.

t	S
1	$8 - 2 * 10$
2	$8 - 4 * 10$
3	$8 - 6 * 10$
4	
t	

   c.

p	q
5	3
10	$3 * 3$
15	$3 * 3 * 3$
20	
p	

   d.

r	P
1	$0.40 * 5^2 + 2.5$
2	$0.40 * 6^2 + 2.5$
3	$0.40 * 7^2 + 2.5$
4	
r	

4. a. Both of the following tables represent the same relationship. Which table is more useful for writing the expression for *y* in terms of *x?* Explain why.

   b. Write an expression for *y* in terms of *x*.

x	y
0	5
1	$5 * 3 = 15$
2	$15 * 3 = 45$
3	$45 * 3 = 135$
x	

x	y
0	5
1	$5 * 3 = 15$
2	$5 * 3 * 3 = 45$
3	$5 * 3 * 3 * 3 = 135$
x	

5. a. Both of the following tables represent the same relationship. Which table is more useful for writing the expression for *A* in terms of *t?* Explain why.

   b. Write an expression for *A* in terms of *t*.

t	A
0	200
1	$200 * 1.1 = 220$
2	$200 * 1.1 * 1.1 = 242$
3	$200 * 1.1 * 1.1 * 1.1 = 266.2$
t	

t	A
0	200
1	$200 + 0.1 * 200 = 220$
2	$220 + 0.1 * 220 = 242$
3	$242 + 0.1 * 242 = 266.2$
t	

6. Checking Account.   To keep a checking account, a local bank charges $3 per month plus 12¢ per check.

   a. Make a table for several possible monthly charges (What are some reasonable choices for the number of checks you might write?).

   b. Graph the data in your table.

   c. Write an equation for the monthly charge in terms of the number of checks written per month.

   d. What is the monthly charge to write 10 checks? 27 checks? 0 checks?

   e. If you receive a monthly charge of $4.92, how many checks did you write? Could your monthly charge ever be $4.90? Why or why not?

7. Wildspring Resorts rents in-line skates. They charge $10.00 for the first 4 hours and $1.25 for each additional half hour.

   a. Make a table of values that models this relationship.

   b. Draw a graph.

   c. Write an equation for the cost to rent in-line skates.

   d. If you rent the skates for 8 hours, how much are you charged?

   e. How long can you rent the skates for $15?

8. Paper Tear.   Start with a single piece of paper. Tear the sheet in half and stack the pieces. Then tear the stack in half and stack the halves. Repeat this process.

   a. Make a table of values for the number of sheets in a stack after several tears. What is the dependent variable? What is the independent variable?

   b. Draw a graph for the relationship between the number of sheets and the number of tears.

   c. Write an equation for the number of sheets in terms of the number of tears.

   d. If it were physically possible to continue this tearing process indefinitely, how many tears would you have to make before the stack were at least as tall as you? Assume that a stack of 500 sheets is 2 inches high.

   e. Is the stack of sheets of paper from part d exactly your height or is it taller than you? If it is taller than you, determine how much taller it is (in inches).

9. The Largest Rectangular Pen.   Chris has bought 50 feet of fencing to build a rectangular pen.

   a. Make a table of several possible rectangular pens using the 50 feet of fencing. Include in your table the length, width, perimeter, and area of each.

   b. Draw a graph representing the length in terms of the width and a graph representing the area in terms of the width.

   c. Write an equation that represents the length in terms of the width and an equation that represents the area in terms of the width.

   d. Determine how the pen should be built to maximize the enclosed area. Can you use more than one of your models to answer this question? Explain.

10. The Garden.   You are constructing a rectangular garden in a field bordered by a river. You have purchased 124 feet of fencing to put around three sides of the garden. The fourth side of the garden borders the river as seen in the following sketch. Determine the length and width that will provide the largest planting area. (Be sure to thoroughly explain your process in solving this problem.)

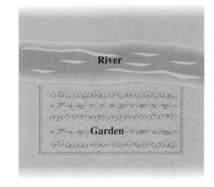

11. The College Education Account.   Chris and Terry have decided to invest $10,000 in a savings account for their daughter's college education. The savings account pays an annual interest of 4.5% compounded annually. We will refer to this as the College Education Account.

    a. Make a table that gives the amount of money in the account for the first few years.

    b. Graph the data in your table.

    c. Write an equation for the amount of money in the College Education Account in terms of the number of years.

    d. Chris and Terry's daughter is now 5 years old and she will begin college when she is 18. How much money will she have for her education?

    e. Do you think this will be enough money for her to attend college for four years?

12. Many Moles. Carl has an infestation of about 500 moles in his cattle field. By trapping the moles, Carl estimates that he can reduce the mole population by about half each year.

   a. Write an equation for the population of moles in terms of the number of years.

   b. How many moles are left after five years?

13. Windless Officers. The Windless Wind Surfing club needs two officers (a chairperson and a secretary). If the club consists of two people, how many ways can the chairperson and the secretary be chosen? If the club consists of three people, how many ways can the chairperson and the secretary be chosen?

   a. Make a table for the number of ways to choose the chairperson and secretary for several different club sizes.

   b. Graph the data in your table.

   c. Write an equation for the number of ways to choose the chairperson and secretary in terms of the number of people in the club.

   d. How many ways can a chairperson and secretary be chosen if 20 people are in the club?

   e. How many ways are there to choose a chairperson and secretary if 49 people are in the club?

14. Comparing Checking Accounts. Your bank is offering three different checking account plans. Plan A charges 25¢ per check. Plan B charges $12\frac{1}{2}$¢ per check plus a monthly fee of $4. Plan C requires you keep a minimum of $500 in the account. You decide that Plan C is not an option for you.

   a. If you plan to write only ten checks a month, what plan should you choose? If you plan to write only 50 checks a month, what plan should you choose?

   b. Write an equation for the monthly cost for checking using Plan A in terms of the number of checks written.

   c. Write an equation for the monthly cost for checking using Plan B in terms of the number of checks written.

   d. How many checks would you need to write for Plan B to be the best deal?

15. What is the volume of figure 100 in the sequence if each cube has a side length of 1 centimeter?

Figure 1   Figure 2   Figure 3

16. a. In the following figures, each small square is a tile. For example, Figure 1 has two tiles. How many tiles (both shaded and unshaded) are in Figure 100 of the following sequence?

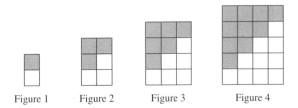

Figure 1   Figure 2   Figure 3   Figure 4

b. How many tiles are in Figure 100 of this sequence?

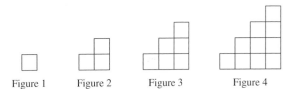

Figure 1   Figure 2   Figure 3   Figure 4

c. Use your results from parts a and b to determine the sum $1 + 2 + 3 + \cdots + 100$.

d. Use your results from parts a and b to write a formula for the sum $1 + 2 + 3 + \cdots + n$.

17. a. Find the distance between the following pairs of points. Round any approximate distances to the nearest tenth. (*Hint:* Plot the points on coordinate axes, and sketch a right triangle using your points.)

   i. $(1, 1)$ and $(4, 5)$ **iii.** $(2, 4)$ and $(7, 0)$

   ii. $(0, 0)$ and $(^-3, 7)$ **iv.** $(^-3, 1)$ and $(4, ^-5)$

b. Describe in words a process for finding the distance between any two points.

1.2 Introduction to Functions

Discussion 1.2

In this section, we will continue to look at mathematical models that represent relationships between two quantities. We say a relationship exists between two variables if knowing the value of one variable allows us to determine the value or values of the other. We saw in Section 1.1 that a relationship between two quantities can be given with a formula containing two variables, as data in a table, or as points on a graph.

In this section we will introduce the idea of a function. A function is a special type of relationship between the independent and dependent variables.

> **Definition**
>
> A **function** is a rule that assigns a *single output value* for every input value. The input is the independent variable, and the output is the dependent variable.

In the following examples, we will explore how to determine if a relationship is a function. We will look at relationships modeled numerically, graphically, and algebraically.

Example 1   Which of the following tables represent functions?

a.

Input	Output
10	15
20	25
30	35
40	45

b.

Input	Output
1	12
2	8
3	5
2	10

c.

Input	Output
3	20
4	30
5	20
6	10

Solution

a. In the table in part a, we see that each of the four inputs is different and each has a single output associated with it. Therefore, this table represents a function.

b. The table in part b gives us two different outputs for an input of 2. Because a function must have a single output for each input, this relationship is not a function.

c. The table in part c has four different inputs. Each input has a single output. The inputs of 3 and 5 both give us an output of 20. The definition for a function does not say that all of the outputs need to be different. Because each input gives us a single output, this is a function.

Example 2

Which of the following graphs represent functions?

a.

c.

b.

d.
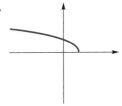

Solution a. On a graph, the input values are along the horizontal axis. For the relationship to be a function, each of those input values must have a single output value. This means only one point on the graph can correspond to any given input value. On the first graph, we pick an arbitrary input value and look at how to identify the output.

On the graph we picked a point on the horizontal axis, marked by the small square. Then, we drew a vertical line from that point. The vertical line intersects our graph in just one point. This point is marked by a dot. The point of intersection represents an ordered pair of the input and the corresponding output value.

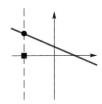

This point is the only point on the graph that has the input we picked. We can see that this is true no matter what input value we select. Therefore, each input has only one output and we conclude that this graph represents a function.

The process we used in part a is very useful in deciding if a relationship presented graphically represents a function. We call this process the vertical line test.

Vertical Line Test

The vertical line test can be used to determine if a graph represents a function.

- If any vertical line intersects the graph of the relationship in at most one point, the graph represents a function.

Equivalently, this can be stated as follows.

- If a vertical line can be drawn that intersects the graph of a relationship in more than one point, the relationship is *not* a function.

---

b. On the second graph we drew two vertical lines. The line on the left does not intersect the graph. This indicates that this input is not part of the relationship. The other vertical line intersects the graph in two points. According to the vertical line test, this graph represents a relationship that is not a function.

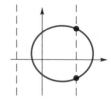

c. On the third graph we drew three vertical lines. No matter which input value we select for the vertical line to go through, it intersects the graph in only one point. Applying the vertical line test, we conclude that this graph represents a function.

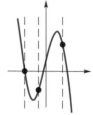

d. On the fourth graph we drew two vertical lines. The line on the right does not intersect the graph. The line on the left intersects the graph in just one point. If we imagine moving the vertical line along the horizontal axis, it would always intersect in one point or not at all. Applying the vertical line test, we see that this graph represents a function.

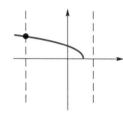

Example 3

Determine whether each equation is a function.

  a. $y = 2x + 1$ b. $y^2 = x$

Solution   a. The equation $y = 2x + 1$ says that for each input x we double the input and add 1 to get the output. Clearly, this gives us just one output value for each input.

We can conclude that $y = 2x + 1$ represents a function.

b. For the equation $y^2 = x$, let's pick an input of $x = 9$. To find the output, we must solve the equation $y^2 = 9$. This equation has two solutions, $y = 3$ and $y = ^-3$.

Because two outputs correspond to an input value of 9, we can conclude that $y^2 = x$ does not represent a function.

We have seen ways to decide if a given relationship is a function. We looked at examples in which the relationships were given numerically, graphically, and algebraically.

It is sometimes helpful to think of a function as a machine. For each input value into the function machine, the function determines the output value. Consider the function defined by the equation $y = 2x - 3$. The process of the function machine looks like this:

Input	Function	Output
Any number	*Double the number and then subtract 3 from the result*	*A number*

  15   →   (2 * Input − 3)   →   27

This input–output process can be repeated for as many input values as you need.

Example 4

Pick a number. Double it, then add 9. Add your original number to your result. Divide by 3. Add 7 to your quotient. Subtract your original number from the sum.

Pick a different number and repeat the process. Is your result the same? Does this process work with any number?

Solution   Let's look at this first as a function machine. We try both 6 and $^-5$ as inputs.

Input	Function		Output
A number, x	*Step-by-step process*		*A number,* y

	Double it.	$2 * 6 = 12$
	Then add 9.	$12 + 9 = 21$
$6 \rightarrow$	Add the original input.	$21 + 6 = 27$
	Divide by 3.	$\frac{27}{3} = 9$
	Add 7.	$9 + 7 = 16$
	Subtract the input.	$16 - 6 = 10$

$\rightarrow \quad 10$

	Double it.	$2 * {}^-5 = {}^-10$
	Then add 9.	$^-10 + 9 = {}^-1$
$^-5 \rightarrow$	Add the original input.	$^-1 + {}^-5 = {}^-6$
	Divide by 3.	$\frac{^-6}{3} = {}^-2$
	Add 7.	$^-2 + 7 = 5$
	Subtract the input.	$5 - {}^-5 = 10$

$\rightarrow \quad 10$

We have an output of 10 for both of the inputs we tried. We can't conclude that we will always get 10 from only two examples. Let's try to look at the problem in a more general way by modeling the situation algebraically. We write an algebraic equation for the output *y* in terms of the input *x*.

	Double it.	$2x$
	Then add 9.	$2x + 9$
	Add the original input.	$(2x + 9) + x$
$x \rightarrow$	Divide by 3.	$\dfrac{(2x + 9) + x}{3}$
	Add 7.	$\dfrac{(2x + 9) + x}{3} + 7$
	Subtract the input.	$\left[\dfrac{(2x + 9) + x}{3} + 7\right] - x$

$\rightarrow \quad \left[\dfrac{(2x + 9) + x}{3} + 7\right] - x$

We see that when we input *x*, the output is the expression

$$\left[\frac{(2x + 9) + x}{3} + 7\right] - x$$

This can be written as an equation.

$$y = \left[\frac{(2x + 9) + x}{3} + 7\right] - x$$

Next, we can simplify the expression on the right-hand side of the equation by performing the indicated operations.

$$y = \left[\frac{(2x + 9) + x}{3} + 7\right] - x$$

$$y = \left(\frac{3x + 9}{3} + 7\right) - x$$

$$y = (x + 3 + 7) - x$$

$$y = 10$$

The output y is *always* 10! This is a special type of function called a constant function.

The definition of a function states that for each input a single output occurs. The definition does not state that two different inputs cannot have the same output. As Example 4 demonstrates, many input values can be paired with the same output value.

Definition _____

A **constant function** outputs the same value for all input values.

What does the graph of the constant function in Example 4 look like? Does it satisfy the vertical line test?

We saw that a **function** is a rule such that for each **input value** a single **output value** is determined. The input is the **independent variable,** and the output is the **dependent variable.** It is possible to decide if a relation is a function by determining the number of output values each input value has. This can be done by observing a complete table of values for the function, using the **vertical line test** on a graph, or by analyzing the equation.

Problem
Set
1.2

1. Solve the following equations. Check your solutions.
    a. $3(5 + y) - 4(5 - y) = 20$
    b. $25(C - 31) = 0.2C$
    c. $3a + 4 = 8 - a$
    d. $2R + 3(5R - 6) = R + 4$
    e. $3v + \frac{2}{3} = \frac{v}{6}$
    f. $4(6 + 3.2w) + 7 = 60.2$

2. Write an equation for each of the following. Use y as the dependent variable and x as the independent variable. Then evaluate each function for the input values 16, 9.4, 0, and $^-10$.
    a. The output is the opposite of the input.
    b. The output is the product of 5 and the sum of the input and 6.
    c. The output is the sum of 5 and the product of the input and 6.
    d. The output is always 100.

3. Each of the following parts describes the step-by-step process for a function machine. Write an algebraic function for each one. Use y as the dependent variable and x as the independent variable. Evaluate the function for the listed inputs.

   a. Pick a number. Multiply by 3. Square the result.

      Evaluate for inputs of 4 and $^-10$.

   b. Square the input, then multiply the result by 3.

      Evalute for inputs of 4 and $^-10$.

   c. Find the average of the input, 6, and 9.

      Evaluate for inputs of 0 and $^-3$.

   d. Pick a number. Output 25.

      Evaluate for 0 and 5.

   e. Pick a number. Double it. Subtract 5. Divide by 4. Add the original number.

      Evaluate for 6 and $^-8$.

4. For each of the following functions, describe the operations and the order in which they are applied, that is, write the step-by-step process for a function machine.

   a. $y = 3x + 6$ c. $y = (x + 3)^2$

   b. $y = \dfrac{x + 35 + 34 + 40}{4}$ d. $y = 45$

5. Determine which of the following tables represents functions.

a.

Input	Output
12	15
15	0
$^-5$	10
12	45

b.

Input	Output
10	22
20	38
30	45
40	50

c.

Input	Output
8	8
9	6
10	5
11	6

6. Determine which of the following graphs represents functions.

a.
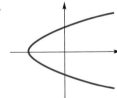

b.

7. Growth of the World Population.   The following graph shows the world population estimates for several different years. Is the world population a function of the year? Explain how you made your decision.

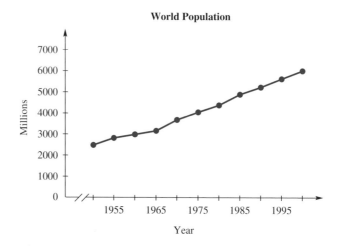

8. Effective Heart Rate.   The following graph can be used to determine the effective heart rate a person should maintain during aerobic exercise, such as running, swimming, and biking. Is pulse rate a function of age? Explain how you made your decision.

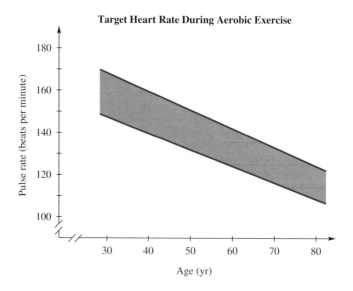

9. Annual Rainfall.   The following table displays the yearly rainfall in the Olympic National Forest, as measured at Quinalt Ranger District. Is the total annual rainfall a function of the year? Explain how you made your decision.

ANNUAL RAINFALL				
Year	Total Annual Rainfall (in.)		Year	Total Annual Rainfall (in.)
1973	116.78		1983	176.50
1974	141.64		1984	155.25
1975	170.21		1985	96.90
1976	113.99		1986	147.50
1977	139.34		1987	112.50
1978	100.40		1988	133.60
1979	146.40		1989	142.25
1980	134.50		1990	173.30
1981	154.80		1991	141.25
1982	159.80			

10. Jazz Run.   The following table displays the ages and finish times for the 8-kilometer Jazz Run. Is finish time a function of age? Explain how you made your decision.

Age	Finish Time (min:sec)		Age	Finish Time (min:sec)
47	37:39		28	24:44
31	37:48		20	26:15
31	38:35		47	27:27
15	39:18		18	26:16
31	39:59		39	29:34
36	40:27		17	31:17
9	40:46		16	31:57
31	41:06		17	32:14
24	41:44		34	32:32
34	42:06		17	33:38
12	42:32		31	35:58
65	43:43		36	34:38
47	50:28		42	35:12
30	54:09		46	36:09
39	1:06:40		11	36:42
32	36:37		32	36:37

11. Create a function for which the input values of the function are January through December. Look up the information for the average rainfall in your home town. Make a table of values for this function, and graph the function.

12. Bring in an example of a function from a current magazine or newspaper. Identify the independent and dependent variables in your example.

13. Healthful Weights.   The following table displays the heights and the corresponding ranges of weights considered to be healthy for adults. Is "healthy weight" a function of height? Explain how you made your decision.

Height (in.)	Healthy Weight (lbs)
60	95–125
61	98–129
62	101–133
63	105–138
64	108–142
65	111–147
66	115–151
67	118–156
68	122–161
69	126–165
70	129–170
71	133–175
72	137–180

14. Soccer Team Statistics.   The graph displays the number of goals scored in the 1999 Women's World Cup Soccer Tournament versus the number of shots on goal, that is, the number of goals attempted. Is the number of shots on goal a function of the number of goals scored? Explain how you made your decision.

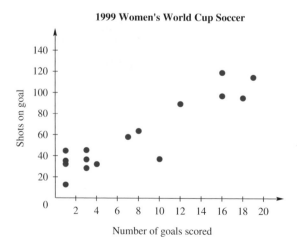

1999 Women's World Cup Soccer

15. Perimeter.   The perimeter of a rectangular yard is 105 feet. Generate a table of values for width versus length. Write an equation for the length in terms of the width. Is this a function? Explain how you decided.

16. Drinking and Driving.   The following table relates a person's weight to the number of drinks he or she consumes in a 2-hour period. We define a relationship for which a person's *weight* is the independent variable and the *number of drinks* he or she can consume in a 2-hour period so as to fall in the "Be Careful" category is the dependent variable. Is this relationship a function? Why or why not?

DRINKS BEFORE DRIVING										
Body Weight (lbs)	**Number of Drinks in a 2-Hour Period**									
100	1	2	3	4	5	6	7	8	9	10
120	1	2	3	4	5	6	7	8	9	10
140	1	2	3	4	5	6	7	8	9	10
160	1	2	3	4	5	6	7	8	9	10
180	1	2	3	4	5	6	7	8	9	10
200	1	2	3	4	5	6	7	8	9	10
220	1	2	3	4	5	6	7	8	9	10
240	1	2	3	4	5	6	7	8	9	10
	Be Careful 0–0.05%			Driving Impaired 0.05–0.10%				Do Not Drive over 0.10%		

17. a. We can define a relationship that assigns to each person's name his or her age. In this case, the name is the independent variable and the age is the dependent variable. Is this relationship a function?

   b. Describe three real-world relations that are functions. Identify the independent and dependent variables.

   c. Describe three real-world relations that are not functions. Identify the independent and dependent variables.

18. U.S. Postal Rates.   Use the following table to decide if the relationship between the weight of first class mail and the U.S. postage rates is a function. Explain how you decide.

U.S. POSTAGE RATES FIRST CLASS MAIL			
Weight (oz)	**Rate ($)**	**Weight (oz)**	**Rate ($)**
1	0.33	7	1.65
2	0.55	8	1.87
3	0.77	9	2.09
4	0.99	10	2.31
5	1.21	11	2.53
6	1.43	12	2.75

19. Each of the following tables shows the outputs for five different inputs. Try to decide how the output is being determined. Describe the function rule in words. Use the rule to give the outputs for the last two inputs.

a.

Input	Output
Monkey	Y
Kangaroo	O
Crocodile	E
Bear	R
Watermelon	N
Eagle	
Rabbit	

b.

Input	Output
Boat	A
Gadfly	F
Racquet	Q
Square	R
Term	S
Pool	
Root	

c.

Input	Output
Difference	3
Addition	2
Multiply	2
Product	1
Division	3
Quotient	
Estimate	

d.

Input	Output
Cue	R
Jay	K
Sea	D
Bee	C
Tea	U
Are	
Why	

e.

Input	Output
Cucumber	3
Onion	15
Garlic	7
Tomato	20
Radish	18
Corn	
Broccoli	

f.

Input	Output
October	3
May	1
August	2
February	4
April	2
January	
December	

20. a. Complete the following table, plot the ordered pairs, and connect the points.

x	$y = 0.02x^2 + 1.9x - 5$
$^-2$	
0	
2	
4	
6	

b. Does the graph of the function you drew in part a appear to be a straight line? Explain how you arrived at your decision.

c. Using the same function, $y = 0.02x^2 + 1.9x - 5$, complete the following table, plot the ordered pairs, and connect the points.

x	y
$^-125$	
$^-100$	
$^-75$	
$^-50$	
$^-25$	
0	
25	

d. Does the graph of the function you drew in part c appear to be a straight line? Explain how you arrived at your decision.

e. What do you think is the point of this problem?

Chapter One Summary

Mathematics provides us with a useful tool for representing problem situations. In this chapter we focused on problem situations that can be modeled with two variables. We started by identifying the independent and dependent variables. The **dependent variable** is the variable whose value depends on the other variable's value. The other variable is the **independent variable.**

Problem situations can be modeled numerically, graphically, and algebraically. Numerical information about a problem can be organized into a table. By plotting points from a table, we can see a visual pattern of the relationship between the two variables. By writing an equation, we generalize the pattern found in the table or from the graph.

In **creating a table of values,** the independent variable is represented first. This means that the independent variable occurs either in the leftmost column or in the top row, depending on the orientation of the table. At this point we do not know how many numbers to include in a table to be able to see a complete picture of the relationship between the two variables. We want to include several values in a table, but even then we cannot be sure that we have a complete picture. In this course we will study several different types of mathematical relations. Learning about these different relationships enables us to *know* if we have a complete picture. In future courses you will add to this list of relations.

In **creating a graph,** the independent variable is graphed on the horizontal axis and the dependent variable is graphed on the vertical axis. The horizontal axis must be scaled in equal increments; likewise, the vertical axis must be scaled in equal increments. However, the scale used on the horizontal axis may be different from the scale used on the vertical axis. The scale should be clearly indicated on each axis. Each axis should include a label indicating what it represents. If a break is to be used, it may be used only at the beginning of the axis and must be clearly indicated.

To **graph an equation,** we make a table of values, plot the points from the table, and connect the points with a smooth curve. At this point, we do not know what the graph of a given equation will look like, so we like to choose several values for a table. Even when we are finished we cannot be sure that we have a "complete" graph. We will begin to study the graphs of specific equations later in this course.

In this chapter we looked at relationships in tables, on graphs, and through equations. Each view of a relationship has advantages and disadvantages. Perhaps one of the more challenging tasks is to create an equation from a problem situation. In Section 1.1, we saw that it is often helpful to use a table to create an equation. We begin by using the problem situation to make a table. In the table, we *write the expressions* used to determine the values, not just the results. When writing numerical expressions in a table, remember not to do any of the computations from the previous step. These expressions can then be generalized into an equation. This may seem difficult at first, but should become easier as you practice.

In this chapter we saw many examples of relationships between two variables that were modeled in different ways. We learned that some of these are a special type of relation called a function.

Definition _____

A **function** is a rule that assigns a *single output value* for every input value. The input is the independent variable and the output is the dependent variable.

We learned how to determine if a relation was a function from each of the three mathematical models.

To determine whether a relation is a **function from a table of values,** we look to see if each input value produces a single output. If an input value is paired with two or more different output values, then the relation is not a function. Remember that it is okay for more than one input value to have the same output value.

To determine whether a relation is a **function from a graph,** we draw, or imagine drawing, vertical lines through the graph. Any vertical line should intersect the graph in at most one point. If you can draw a vertical line that intersects the graph in more than one point, the relation is not a function. This is referred to as the **vertical line test.**

To determine whether a relation is a **function from an equation,** we substitute a value for the independent variable, then solve for the dependent variable. If more than one solution arises for the dependent variable, the relation is not a function. If an equation is not a function, we only need to find one case for which more than one solution occurs for the dependent variable.

Throughout this chapter we saw that a problem situation can be modeled in three ways. We can model a situation with a table of values, with a graph, or with an equation. All three of the models are related. How you enter a problem depends on how the information is given to you and what representation you are most comfortable with.

Chapter Two

Linear Functions

In this chapter, we will look at linear functions and lines in general. We will learn how to recognize them and study their special characteristics from numerical, graphical, and algebraic models. Recognizing these important functions from any of the three models and learning to switch between models will facilitate problem solving. Linear functions are especially important because a wide range of applications can be modeled linearly. The meaning, solution, and application of systems of linear equations will be an additional focus of the chapter.

2.1 Slope

Activity Set 2.1

1. **i.** Complete the following tables. Show your substitution in the first row of each table.

   ii. For each equation, draw a graph by plotting points from your table.

a.

| Input | Output | Change |
x	$y = 2x - 6$	in Output
$^-2$	$2(^-2) - 6 = ^-10$	
		$+2$
$^-1$	$^-8$	
0		
1		
2		

b.

| Input | Output | Change |
x	$y = ^-3x + 6$	in Output
$^-2$		
$^-1$		
0		
2		
4		

c.

| Input | Output | Change |
x	$y = (x - 1)(x + 1)$	in Output
$^-2$		
$^-1$		
0		
1		
2		

d.

| Input | Output | Change |
x	$y = \dfrac{2x - 1}{3}$	in Output
$^-4$		
$^-1$		
2		
5		
8		

e.

| Input | Output | Change |
x	$y = 0.5 * 2^x$	in Output
1		
2		
3		
4		

f.

| Input | Output | Change |
x	$y = \dfrac{1}{x}$	in Output
1		
2		
3		
4		
5		

2. **a.** Which equations from Activity 1 graph as straight lines?

   b. Based on your observations, describe how you can determine from a table of values whether the data will lie in a straight line.

   c. Based on your observations, describe how you can determine from an equation whether the equation will graph as a straight line.

3. Recall that the slope of a line is defined to be the ratio of the vertical change to the horizontal change.

$$\text{slope} = \frac{\text{vertical change}}{\text{horizontal change}}$$

Looking at your graphs from Activity 1, determine the slopes for those graphs that are linear.

Discussion 2.1

In doing the activities, you may have noticed that the graphs of equations in the form $y = mx + b$, where m and b are constants, are lines. In fact, any equation in the form $y = mx + b$ will graph as a straight line and is called a **linear equation.** The equation in Activity 1d also graphed as a straight line, but is not in the form $y = mx + b$. However, we can rewrite the expression on the left by dividing each term in the numerator by 3. Then the equation is in this form.

$$y = \frac{2x - 1}{3}$$

$$y = \frac{2x}{3} - \frac{1}{3}$$

$$y = \frac{2}{3}x - \frac{1}{3}$$

This is now in $y = mx + b$ form.

You may also have noticed from the activities that the tables that graphed as lines had a constant change in output when the change in input was constant. We could restate this by saying that the *ratio* of the change in output to the change in input is constant. This constant ratio is a unique feature of all linear graphs.

> *Linear Relationship from a Table*
>
> For a **linear relationship,** the ratio $\dfrac{\text{change in output}}{\text{change in input}}$ is constant.

To determine if a relationship is linear when we are given numerical information, we can look at the ratio $\frac{\text{change in output}}{\text{change in input}}$. If it is constant, the relationship is linear. This ratio allows us to decide without graphing the data.

Example 1

Without graphing the points, determine if the following tables graph as straight lines. Then use a graph to verify your findings.

a.

x	y
−3	−3
−2	−2
−1	0
0	3

b.

x	y
−2	7
0	4
2	1
4	−2

c.

x	y
0	1
1	4
2	7
4	13

Solution

a. In table a, the y values increase by a different amount when the x values increase by 1. Therefore, the ratio of the change in output to the change in input is not constant, and the points from this table do not graph in a straight line.

x	y	Change in y
−3	−3	
		+1
−2	−2	
		+2
−1	0	
		+3
0	3	

Let's plot these points to verify our results visually.

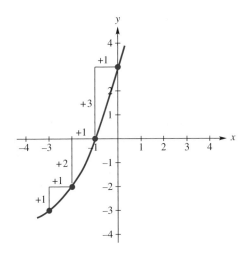

Notice that the change in y values is represented as a vertical change and the change in x values is represented as a horizontal change.

b. In table b, the y values decrease by 3 each time the x values increase by 2. Because the ratio of the change in output to the change in input is constant, the relationship described by the table graphs as a straight line.

x	y	Change in y
−2	7	
		−3
0	4	
		−3
2	1	
		−3
4	−2	

We plot these points to verify our results visually.

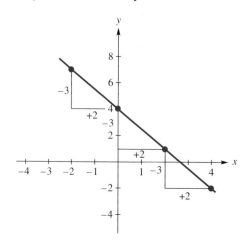

As we noted in part a, the change in *y* values is represented as a vertical change and the change in *x* values is represented as a horizontal change. The constant ratio of the change in *y* values to the change in *x* values for table b is $\frac{-3}{2}$.

c. In the first three rows of table c, the output increases by 3 when the input increases by 1. In the third row, the output increases by 6 when the input increases by 2. Looking at the ratio of the change in output to the change in input, we see that the ratios of 3 to 1 and 6 to 2 are equal.

$$\frac{\text{change in output}}{\text{change in input}} = \frac{3}{1} = \frac{6}{2} = 3$$

Therefore, this table of values will graph as a straight line.

Change in x	x	y	Change in y
	0	1	
$1 - 0 = {}^+1$			$^+3$
	1	4	
$^+1$			$^+3$
	2	7	
$^+2$			$^+6$
	4	13	

Again, we use a graph to verify our result.

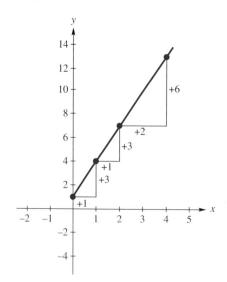

In Example 1, we saw that the change in output was represented on the graph by a vertical change and the change in input was represented by a horizontal change. So, for any line, the constant ratio of the change in output to the change in input is the same as the ratio of the vertical change to the horizontal change on a graph, that is

$$\frac{\text{change in output}}{\text{change in input}} = \frac{\text{vertical change}}{\text{horizontal change}}$$

We call this ratio the slope.

Definition

The **slope** of a line is defined as the *ratio of the vertical change to the horizontal change*. That is,

$$\text{slope} = \frac{\text{vertical change}}{\text{horizontal change}}$$

The vertical change is often referred to as the **rise,** and the horizontal change is referred to as the **run.** Using this terminology, we have

$$\text{slope} = \frac{\text{rise}}{\text{run}}$$

Example 2   Find the slope of the following lines.

a.

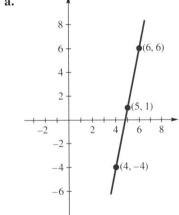

b.

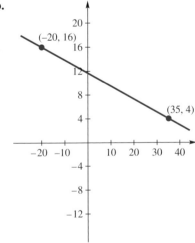

Solution   a. Three points are given on graph a. We select the points (5, 1) and (6, 6) to find the slope of this line. Moving from the point (5, 1) to the point (6, 6), the vertical change is $+5$ units (up 5 units) and the horizontal change is $+1$ unit (right 1 unit). Therefore,

$$\text{slope} = \frac{\text{vertical change}}{\text{horizontal change}} = \frac{5}{1} = 5$$

The slope of this line is 5.

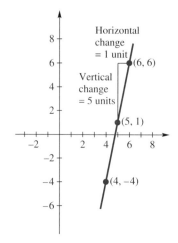

b. In graph b, two points are given, the point $(^-20, 16)$ and the point $(35, 4)$. Moving from the point $(^-20, 16)$ to the point $(35, 4)$, the vertical change is $^-12$ units (down 12 units), and the horizontal change is $+55$ units (right 55 units). In other words, the rise is $^-12$ units, and the run is $+55$ units. Therefore,

$$\text{slope} = \frac{\text{rise}}{\text{run}} = \frac{^-12}{55}$$

The slope of this line is $\frac{^-12}{55}$.

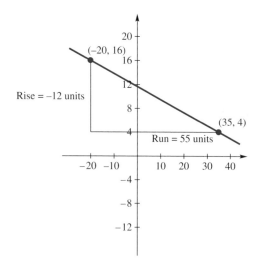

In part a of the previous example we chose the points (5, 1) and (6, 6) to determine the slope of a line. What would have happened if we had chosen a different set of points? Let's see what happens if we had selected the points $(4, ^-4)$ and $(6, 6)$ as seen in the following graph. Moving from the point $(4, ^-4)$ to $(6, 6)$ the vertical change is $+10$ units (up 10 units) and the horizontal change is $+2$ units (right 2 units). Therefore,

$$\text{slope} = \frac{\text{vertical change}}{\text{horizontal change}} = \frac{10}{2} = 5$$

The slope is the same as our result in Example 2a.

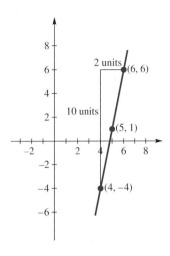

Example 3

a. Sketch the graph of a line that passes through the point $(^-2, ^-5)$ and has a slope of $\frac{3}{4}$. Label two points on the line.

b. Sketch the graph of a line that passes through the point $(0, 1)$ and has a slope of $^-4$. Label two points on the line.

Solution

a. First we plot and label the given point, $(^-2, ^-5)$. From this point we use the slope to find the next point. The slope is $\frac{3}{4}$. This means that we have a rise of 3 units up and a run of 4 units to the right. Our next point is $(2, ^-2)$. We can repeat this process to find a third point. Two points determine a line so only two points are truly needed. It never hurts to find one extra point to help you sketch the line. Then draw the line through these three points.

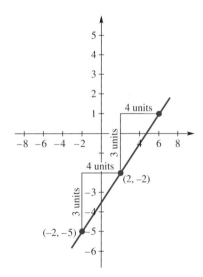

b. First we plot and label the given point $(0, 1)$. From this point we use the slope to find the next point. The slope is $^-4$. As a ratio this can be written as $\frac{-4}{1}$. This means we have a vertical change of $^-4$ units (down 4 units) and a horizontal change of 1 unit to the right. Our next point is $(1, ^-3)$. We repeat this process and then draw the line connecting these three points.

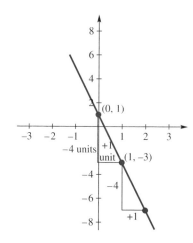

Definition

A **horizontal intercept** is a point where the graph crosses the horizontal axis. A **vertical intercept** is a point where the graph crosses the vertical axis.

When asked to find the intercepts, we need to find both the horizontal and vertical intercepts.

Example 4

What are the intercepts of the following graphs?

a.

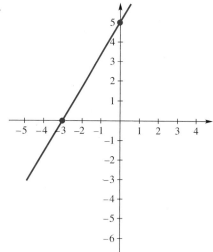

b.

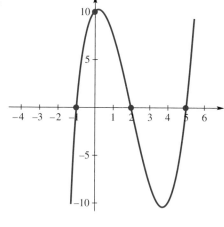

Solution   a. Because this line crosses the horizontal axis once at the point $(^-3, 0)$, the horizontal intercept is the point $(^-3, 0)$. This line crosses the vertical axis once at the point $(0, 5)$, so the vertical intercept is $(0, 5)$.

b. This graph crosses the vertical axis once. It has a vertical intercept of $(0, 10)$. It crosses the horizontal axis three times. Therefore, it has three horizontal intercepts, $(^-1, 0)$, $(2, 0)$, and $(5, 0)$.

Example 5   A garbage facility in your community has already collected 6500 tons of garbage. The following graph shows the predicted amount of garbage that the dump will collect over the next five years. Use the graph to answer the following questions.

   a. How many tons of garbage will be in the dump one year from now? Two years? Five years?

   b. Determine the slope of this line. What are the units associated with the slope? What does the slope tell you about the problem situation?

   c. What is the relationship between the vertical intercept of the line and the problem situation?

   d. Assuming the predictions remain true, and the capacity of the dump is 8200 tons, how many years will the community be able to use this dump?

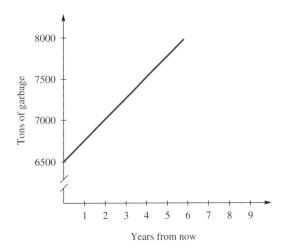

Solution   a. Drawing in vertical and horizontal lines can help us read values from a graph more accurately.

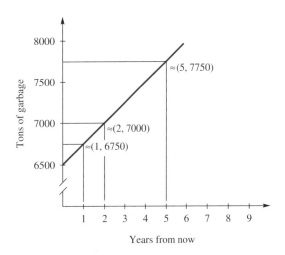

   After one year, there will be about 6750 tons of garbage in the dump. After two years, there will be about 7000 tons of garbage, and after five years, about 7750 tons.

   b. Because we do not know the exact coordinates of any points, we cannot determine the slope exactly. However, we can approximate the value of the slope by using any two of the points we identified in part a.

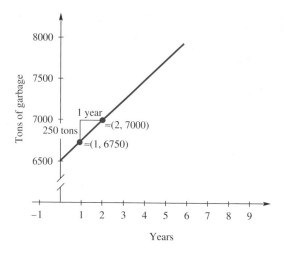

$$\text{slope} = \frac{\text{vertical change}}{\text{horizontal change}} \approx \frac{250}{1}$$

Because slope is rise over run, we look at the units associated with each axis to determine the units for the slope. The units on the vertical axis are tons, and the units on the horizontal axis are years. Therefore, the units associated with the slope are $\left(\frac{\text{tons}}{\text{year}}\right)$, or tons per year. If we put this information together with the numerical value of the slope, we see that the slope is about 250 tons per year. This means that the amount of garbage in the dump is increasing at the rate of approximately 250 tons per year.

c. The vertical intercept is the point where the line intersects the vertical axis. This is the point (0 years, 6500 tons). This point means that after zero years, or right now, 6500 tons of garbage are in the dump.

d. We need to extend the graph to answer this question. We can see that the point ≈ (7 years, 8200 tons) lies on this line. This means that after about 7 years the dump will reach 8200 tons of garbage, which is its capacity.

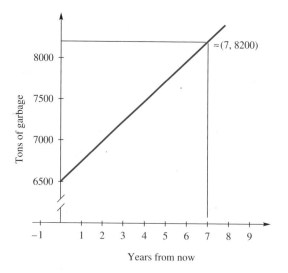

In this section, we reviewed the definition of the slope of a line. We saw that the slope of any line is constant. We can use that fact to decide if a table of values graphs as a straight line. We saw that using the units associated with the vertical and horizontal axes is helpful when we are interpreting the meaning of the slope and the intercepts of graphs.

Problem
Set
2.1

1. Solve the following equations. Check your solutions.

   a. $\dfrac{4w - 19}{2} + \dfrac{w + 1}{10} = {}^-1$

   c. $\dfrac{3}{w} + 5 = \dfrac{1}{2w}$

   b. ${}^-4.2A + 6.5 = 2.3A - 53.3$

   d. $\dfrac{10}{10 + z} = 35$

2. Solve each literal equation for the indicated variable. Verify your solutions.

   a. $4y - 25 = 7 - x$ for y

   c. $3R(4 - K) = 2 - R$ for R

   b. $\dfrac{24p - 3m}{2} = 52$ for m

   d. $\dfrac{W - T}{4T} = \dfrac{1}{8}$ for T

3. For each graph, determine the slope of the line.

   a.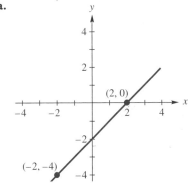

   b.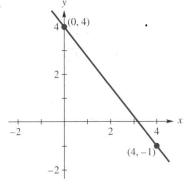

   c.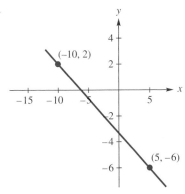

   d.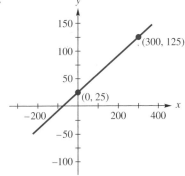

4. Without plotting points, determine if the tables graph as straight lines. For each table that graphs as a straight line, determine the slope of the corresponding line.

   a.

x	y
0	$^-1$
1	5
2	11
3	17

   b.

x	y
1	1
2	4
3	9
4	16
5	25

   c.

x	y
$^-2$	0.25
0	1.00
2	4.00
4	16.00

   d.

x	y
0	125
5	100
10	75
20	25
30	$^-25$

5. **Smokers' and Nonsmokers' Insurance.** The following table shows the costs of term life insurance for males who smoke and for those who do not smoke, at various ages.

   a. Graph the cost of the insurance in terms of age for both smokers and nonsmokers on the same set of axes. In other words, draw one coordinate system, and graph both relationships (costs for smokers and costs for nonsmokers) on that single coordinate system.

   b. Is either graph linear? If yes, what is the slope of the linear graph(s)? What does the slope tell us?

TYPICAL INSURANCE PREMIUMS ($)		
Age	Rates for Nonsmokers	Rates for Smokers
30	198	400
35	336	560
40	474	720
45	612	960
50	750	1290

6. For each part, draw the graph of the line that satisfies the given conditions. On each graph, label two points on the line.

   a. Passing through the point $(^-2, 1)$, with a slope of $\frac{1}{4}$

   b. Passing through the point $(0, 0)$, with a slope of $\frac{^-2}{5}$

   c. Passing through the point $(0, 7)$, with a slope of $^-3$

   d. Passing through the point $(0, ^-250)$, with a slope of 25

   e. Passing through the point $\left(\frac{^-1}{2}, \frac{3}{4}\right)$, with a slope of $\frac{1}{2}$

7. **a.** On the same axes, draw the graphs of the following:

   The line passing through the point $(0, ^-5)$, with a slope of 4

   The line passing through the point $(^-2, 3)$, with a slope of 4

   The line passing through the point $(4, 0)$, with a slope of 4

   b. What is the relationship between the three lines you drew in part a?

8. **a.** On the same axes, draw the graphs of the following:

   The line passing through the point $(0, 0)$, with a slope of $\frac{1}{3}$

   The line passing through the point $(0, 0)$, with a slope of 1

   The line passing through the point $(0, 0)$, with a slope of 3

   b. How does a line with a slope of 5 compare with the graphs you drew in part a? How does a line with a slope of $\frac{1}{2}$ compare with the graphs you drew in part a?

9. The lines pictured here have slopes of $\frac{3}{4}$, 2, and 4. Without measuring, identify which line has each slope.

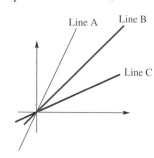

10. Determine the slope of each of the following lines.

a.

b.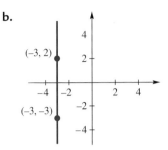

c. Draw any horizontal line. Label two points on your line.

  Determine the slope of your line.

  What is the slope of any horizontal line?

d. Draw any vertical line. Label two points on your line.

  Determine the slope of your line.

  What is the slope of any vertical line?

11. Which coordinate of the ordered pair representing the horizontal intercept is equal to zero? In other words, if the ordered pair (a, b) represents the horizontal intercept of a graph, must $a = 0$ or $b = 0$?

12. Spring Break.   Yoav is traveling home for spring break. He has plotted his trip on the following graph. Use the graph to answer the following questions.

a. How far is it to Yoav's destination after he has been driving for 1 hour? 2 hours?

b. Determine the slope of this line. What does the slope tell you about the problem situation?

c. What is the relationship between the vertical intercept of the line and the problem situation?

d. What is the relationship between the horizontal intercept of the line and the problem situation?

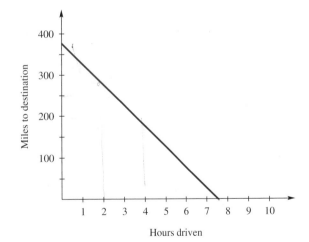

13. Energy Requirements.   The following graph shows the approximate daily energy requirements for males as recommended by many experts to maintain health. Use the graph to answer the following questions.

  a. What is the daily energy requirement for a male child who is 5 years old? 16 years?

  b. Determine the slope of this line. What does the slope tell you about the problem situation?

  c. Extend the line, and use it to determine the daily energy requirement for a male who is 19 years old. The actual recommendation from experts is that a 19-year-old male requires approximately 3000 calories daily. What does this mean about the following graph?

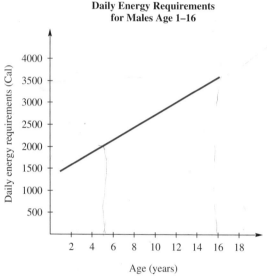

**Daily Energy Requirements
for Males Age 1–16**

14. Spending Money.   Cory budgeted $420 for spending money for next semester. She plans to withdraw $27.50 per week from this amount as her weekly spending allowance. Complete a table similar to the following one to show how much is left each week after she withdraws her allowance. Use the table to answer the questions.

  a. Is this table linear? If yes, determine the slope. What are the units of the slope? What does the slope mean in this problem situation?

  b. What is the vertical intercept? What are the units of the vertical intercept? What does the vertical intercept mean in this problem situation?

  c. What is the horizontal intercept? What are the units of the horizontal intercept? What does the horizontal intercept mean in this problem situation?

  d. A semester at Cory's school is 15 weeks plus a finals week. What do you recommend Cory do regarding her plan?

Week	Amount in Account ($)
0	420.00
1	
2	
3	
4	
5	
6	
7	
8	
9	
10	
11	
12	
13	
14	
15	
16	

2.2 Graphs of Linear Functions

1. a. For each graph, make a table of values that are represented by dots on the graph. Assume that all dots are points with integer coordinates.

  b. For each table of values from part a, write an equation that models the pattern in the table. Check your equation.

  c. Use the graph or table to determine the slope and vertical intercept of each line. Fill in the blanks in the sentences following the graphs.

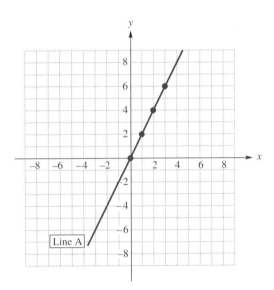

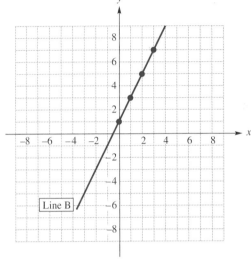

x	y

x	y

The equation of line A is _____.

The slope of line A is _____.

The vertical intercept
of line A is _____.

The equation of line B is _____.

The slope of line B is _____.

The vertical intercept
of line B is _____.

2. Based on your responses to Activity 1, make a conjecture about the relationship between the equation for a line $y = mx + b$ and the slope and vertical intercept of that line.

3. a. Determine the slope and vertical intercept of line C. Assume that all dots are points with integer coordinates.

   b. Use your conjecture from Activity 2 to write the equation for this line.

   c. Check your equation by substituting points from the graph. You need to substitute at least two points from the graph. Did your conjecture hold true?

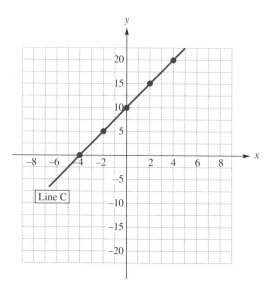

Line C

4. For each line graphed, write the equation for the line and check your equation.

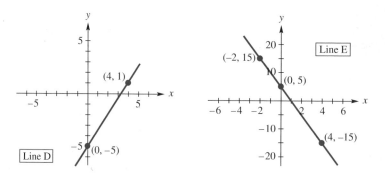

5. On the following graph, which line has a larger slope?

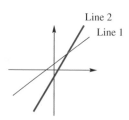

Line 2
Line 1

6. If two lines are parallel, what can you say about their equations?

7. **a.** Answer the following questions *without graphing* the equation.

What is the slope of the line $y = 3x + 50$?

What is the vertical intercept of the line $y = 3x + 50$?

b. Graph the equation $y = 3x + 50$ using your calculator. Check your answers from part a using your graph. Explain how you did this.

c. What is the slope of the line $y = mx + b$?

What is the vertical intercept of the line $y = mx + b$?

8. **a.** Answer the following questions without graphing the equation.

What is the slope of the line $y = 15 - 2x$?

What is the vertical intercept of the line $y = 15 - 2x$?

b. Graph the equation $y = 15 - 2x$ using your calculator. Check your answers from part a using your graph.

9. **a.** Write the equations of two different lines that pass through the origin.

b. Write the equations of two different lines that have a slope of 5.

10. Write the equation of each line. Check your equations.

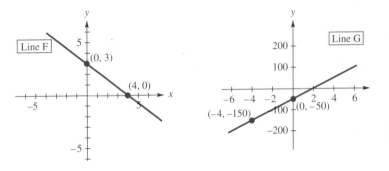

Discussion 2.2

In the activity set, we found that each linear graph could be modeled with an equation of the form $y = mx + b$. On further investigation, we found that the coefficient of x represents the slope of the line and the constant term represents the vertical coordinate of the vertical intercept. That is, the line $y = mx + b$ has slope m and vertical intercept $(0, b)$. For this reason, the equation $y = mx + b$ is called the **slope–intercept equation** of a line. We can also write this as $y = $ slope $* x + $ vertical intercept.

The Slope–Intercept Equation of a Line

$y = $ slope $* x + $ vertical intercept

or

$y = mx + b$ where $m = $ slope, and

$(0, b) = $ vertical intercept

Linear functions can be modeled verbally, numerically, graphically, and algebraically. A problem is usually presented with one of these four models. We will see that it is important to be able to determine each model from any given model. For example, if a linear function is presented with a table of values, can we determine its equation and graph? If an equation of a linear function is given, can we draw its graph?

Knowing the relationship that we discovered in the activity set between a linear function and its equation allows us to translate linear information among the four different models. The details of these translations will be discussed in this and remaining sections of this chapter.

In the first example, we will look at writing an equation of a line from its graph.

Example 1

For the given graph, do the following.

    a. Determine the slope and identify the intercepts.

    b. Write an equation of the line and check your equation.

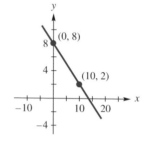

Solution

a. Two points are given, $(0, 8)$ and $(10, 2)$. Between these two points, the vertical change is $^-6$ and the horizontal change is 10. Therefore,

$$\text{slope} = \frac{\text{vertical change}}{\text{horizontal change}} = \frac{^-6}{10} \quad \text{or} \quad \frac{^-3}{5}$$

The vertical intercept is $(0, 8)$. The horizontal intercept appears to be $\approx (13, 0)$; the exact coordinates are uncertain from the graph. To summarize:

$$\text{slope} = -\frac{3}{5}$$

$$\text{vertical intercept} = (0, 8)$$

$$\text{horizontal intercept} \approx (13, 0)$$

b. In the activity set, we saw that an equation of a line is $y = \text{slope} * x + \text{vertical intercept}$. Therefore, an equation of this line is

$$y = -\frac{3}{5}x + 8.$$

To check our equation, we need to be sure that the points $(0, 8)$ and $(10, 2)$ lie on the line with equation $y = -\frac{3}{5}x + 8$. For $(0, 8)$ to be on the line means that the ordered pair is a solution to the equation. To check, let $x = 0$ and $y = 8$ in the equation.

    CHECK

$$8 = -\frac{3}{5} * 0 + 8 \qquad \text{Substitute } x = 0 \text{ and } y = 8 \text{ into our equation.}$$

$$8 = 8 \quad ✔ \qquad \text{Evaluate each side of the equation. The point } (0, 8) \text{ satisfies the equation.}$$

Therefore, $(0, 8)$ lies on the line with equation $y = -\frac{3}{5}x + 8$.

Similarly, we check to see if $(10, 2)$ is a solution to our equation.

CHECK

$$2 = -\frac{3}{5} * 10 + 8 \qquad \text{Substitute } x = 10 \text{ and } y = 2 \text{ into our equation.}$$

$$2 = 2 \quad \checkmark \qquad \text{Evaluate each side of the equation. The point } (10, 2) \text{ satisfies the equation.}$$

Therefore, $(10, 2)$ lies on the line with equation $y = -\frac{3}{5}x + 8$. Because both points lie on the line that is represented by our equation, we can be sure that our equation is correct.

In Example 1, we were not able to determine the horizontal intercept exactly from the graph. Now that we have written an equation, we can determine the exact coordinates. We know that the second coordinate (or y-coordinate) of any horizontal intercept must be 0. Therefore, to find the horizontal intercept, we set $y = 0$ in our equation and solve for x.

$$0 = -\frac{3}{5}x + 8 \qquad \text{To find a horizontal intercept, set } y = 0.$$

$$5 * 0 = \left(-\frac{3}{5}x + 8\right) * 5 \qquad \text{Multiply both sides of the equation by 5.}$$

$$0 = -3x + 40 \qquad \text{Add } 3x \text{ to both sides of the equation.}$$

$$3x = 40$$

$$x = \frac{40}{3} \qquad \text{or} \qquad 13\frac{1}{3}$$

Because $x = 13\frac{1}{3}$ when $y = 0$, the horizontal intercept is the point $(13\frac{1}{3}, 0)$. Notice, this agrees with our earlier estimate.

Example 2   Draw the graph of the line whose equation is $y = 100 - 5x$.

Solution   There are three main ways to draw the graph of the line $y = 100 - 5x$. Because two points determine a line, we only need to find two points that lie on this line.

Option 1   One way to find these two points is to make a table.

x	$y = 100 - 5x$
0	$100 - 5 * 0 = 100$
5	$100 - 5 * 5 = 75$
10	$100 - 5 * 10 = 50$

We can draw a graph by plotting these points. You may have noticed that the table includes three points when only two are necessary. It never hurts to have a spare.

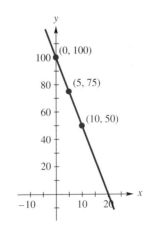

Option 2   Alternatively, we can find two points on the line by using the relationship between the equation and its graph. To do this, we must be able to identify the slope and vertical intercept of the line from its equation, $y = 100 - 5x$. It is helpful to rewrite $y = 100 - 5x$ in slope intercept form as $y = {}^-5x + 100$. In this form we can see that the slope is $^-5$ and the vertical intercept is $(0, 100)$. Therefore, we can plot two points by starting at the vertical intercept and using the slope to find a second point.

Because the slope is $^-5$, this is equivalent to the fraction $\frac{-5}{1}$. So, from the vertical intercept at $(0, 100)$ we need to move down 5 units and to the right 1 unit to find a second point on the line. This takes us to the point $(1, 95)$. We can stop here because two points determine a line, or we can continue this process. The next point is $(2, 90)$ as shown on the following graph.

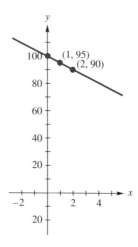

Notice that this graph looks very different from the previous one. This is because the horizontal axes are scaled differently in the two graphs. A slope of $^-5$ is quite difficult to plot accurately on the first scale. However, the first scale presents a more complete representation of the graph because both intercepts can be seen.

To use the slope to find a second point when the horizontal axis is scaled in 5-unit increments, we need to rewrite the slope as a different equivalent fraction. Many fractions are equivalent to $^-5$, for example, $\frac{-10}{2}, \frac{-15}{3}$, and so on. We need to choose one that is easy to plot on our axes. Because 10 is an easy number to work with, we can rewrite our slope as an equivalent fraction whose denominator is 10.

$$\text{slope} = {}^-5 = \frac{-5}{1} = \frac{-50}{10}$$

From $(0, 100)$ we need to move down 50 units and right 10 units to determine a second point. This takes us to the point $(10, 50)$. Similarly, we can proceed to the point $(20, 0)$.

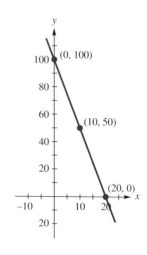

Option 3   A third option is to determine the two intercepts. From the equation $y = 100 - 5x$, which can be rewritten as $y = {}^-5x + 100$, we know that the vertical intercept is $(0, 100)$. To find the horizontal intercept we set $y = 0$ and solve for x.

$$0 = {}^-5x + 100 \qquad \text{To find a horizontal intercept, set } y = 0.$$

$$5x = 100 \qquad \text{Add } 5x \text{ to both sides.}$$

$$x = 20$$

The horizontal intercept is the point $(20, 0)$. To draw the graph, we plot the points $(0, 100)$ and $(20, 0)$. Then, draw a line through them.

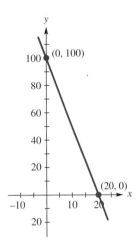

Example 3

Graph the equation $4x - 3y = 16$.

Solution   First we need to decide if the equation graphs as a line. If it does, then we only need to find two points, because two points determine a line. If the equation does not graph as a line, then we need to plot several points to get an idea of the shape of the graph.

So far, all of the equations we have seen that graph as lines have been of the form $y = mx + b$. If we can we rewrite $4x - 3y = 16$ in slope–intercept form then we will know that it represents a linear function.

We can see that $4x - 3y = 16$ can be rewritten in slope–intercept form, and therefore graphs as a line. If you are not sure how, solve the formula for y. In fact, any equation of the form $Ax + By = C$, where A and B are nonzero, can be rewritten in slope–intercept form and, therefore, graphs as a line. Later in this section, we will consider cases where A or B are zero.

Now that we know $4x - 3y = 16$ graphs as a line, let's consider the three options demonstrated in Example 3. We will discuss the process for each option and then pick the one that seems most efficient.

The first option involved making a table of values. To do this, it is beneficial to solve the equation for one of the variables. Typically, we solve for y. The resulting equation includes fractions.

The second option involved using the slope and vertical intercept to determine two points on the line. Again, to use this method, we need to solve the equation for y.

The third option involved using the equation to determine the vertical and horizontal intercepts. Because the first coordinate of any vertical intercept is equal to 0, we can find the vertical intercept by letting $x = 0$ and finding the corresponding value for y. Similarly, we can find the horizontal intercept by setting $y = 0$ and finding the corresponding value of x. Notice that we do not need to rewrite the original equation in slope–intercept form to use this option.

Let's graph this equation using the third option. To find the vertical intercept, set $x = 0$ and find y.

$$4 * 0 - 3y = 16$$

$$-3y = 16$$

$$y = -\frac{16}{3} \quad \text{or} \quad -5\frac{1}{3}$$

The vertical intercept is the point $\left(0, -5\frac{1}{3}\right)$.

To find the horizontal intercept, set $y = 0$ and find x.

$$4x - 3 * 0 = 16$$

$$4x = 16$$

$$x = 4$$

The horizontal intercept is the point $(4, 0)$.

Now that we have the intercepts, we can graph the line by plotting the intercepts and drawing a line through these two points.

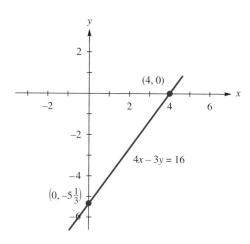

Example 4

Electric Bill.   A family living in Sandy, Oregon, receives their electricity from PGE (Portland General Electric). PGE charges a monthly service charge of $5.50. Recently this customer received a bill of $68.41 for 1080 kWh (kilowatt-hours) of electrical power.

  a. Assuming that the relationship is linear, write an equation for the cost of electricity in terms of the amount of electricity used in kWh. (Even though the bill that each customer pays is rounded to the nearest cent, the rate that PGE charges is to a thousandth of a cent. Therefore, keep all significant digits until you need to determine a monthly bill.)

  b. Use your equation to determine the monthly bill for 1500 kWh.

  c. What are the units associated with the slope of the line? What does the slope tell you about the problem situation?

  d. What is the relationship between the vertical intercept of the line and the problem situation?

Solution   a. From the problem statement, we know that 0 kWh costs \$5.50 and 1080 kWh costs \$68.41. Because the cost depends on the amount of electricity used, the amount of electricity is the independent variable, and the cost is the dependent variable. To start, let's plot these two points. Then, because we are assuming that this relationship is linear, we draw a line through our two points.

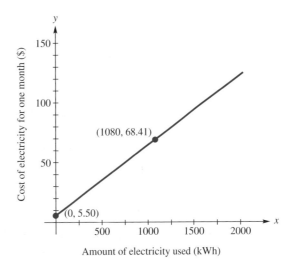

To write the equation of a line, we need to know the slope and the vertical intercept. The vertical intercept we can see from the graph. It is the point $(0, 5.50)$. We can also find the slope from the graph by determining the ratio $\frac{\text{rise}}{\text{run}}$.

Going from point $(0, 5.50)$ to point $(1080, 68.41)$, the graph rises 62.91 and runs 1080. Therefore, the slope is

$$\text{slope} = \frac{62.91}{1080}$$
$$= 0.05825$$

Now that we know the slope and the vertical intercept, we can write the equation.

   Let h = the amount of electricity used in kWh and

   C = the cost, in dollars, of electricity for one month.

Then the equation for this line is

   $C = 0.05825h + 5.50$

   b. To determine the cost of 1500 kWh of electricity, we need to substitute $h = 1500$ into our equation.

   $C = 0.05825 * 1500 + 5.50$
   $C = 92.875$

Because the cost is in dollars, we should round our results to the nearest cent. Therefore, it costs \$92.88 for 1500 kWh of electricity.

We can also see this answer agrees with our graph.

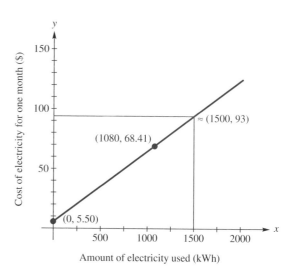

c. In part a, we found the slope of the line to be 0.05825. To determine the units associated with the slope, we must look at the units on each axis. Slope is rise over run. The units on the vertical axis (rise) are dollars, and the units on the horizontal axis (run) are kWh. Therefore, the units associated with the slope are dollars per kilowatt-hour $\left(\frac{\text{dollars}}{\text{kWh}}\right)$. If we put the units together with the numerical value, the slope is 0.05825 dollars per kWh or 5.825 cents per kWh. The slope tells us that PGE charges this customer 5.825 cents per kilowatt-hour.

d. The vertical intercept is (0 kWh, $5.50). This represents the monthly customer charge.

Putting this information together with part c, we know that PGE charges this customer $5.50 per month plus 5.825 cents per kilowatt-hour.

Example 5

Pressure Change.   The following set of data relates water pressure and depth of the water. Determine if the data can be modeled with a linear function. If so, draw a graph, and write a linear equation that models the data.

Depth (ft)	0	33	66	132
Pressure (atm)	1	2	3	5

Solution   From the first three pairs of data we can see that for each 33-foot increase in depth, the water pressure increases by 1 atmosphere (atm). The change from the third data point to the fourth is a 2-atm increase for a 66-foot increase in depth, which is equivalent to 1 atm per 33 feet. Therefore, the given set of data can be modeled with a linear function because the rate of change is constant. The data points are shown in the following graph and are connected with a line because the relationship is linear.

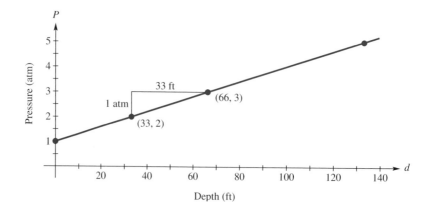

We can see from the graph and the table that the vertical intercept is the point $(0, 1)$ and

$$\text{slope} = \frac{\text{vertical change}}{\text{horizontal change}} = \frac{1 \text{ atm}}{33 \text{ feet}}$$

Therefore, the equation is $P = \frac{1}{33}d + 1$, where P is the pressure in atmospheres and d is the depth in feet.

Can every line be described by the equation $y = \text{slope} * x + \text{vertical intercept}$? If so, then every line has a defined slope and a vertical intercept. Are there any lines that do not have a slope and/or a vertical intercept? If so, then this equation will not work for those lines. Consider the following vertical line, which crosses the horizontal axis at $(5, 0)$. What is the slope of a vertical line? Does a vertical line always have a vertical intercept?

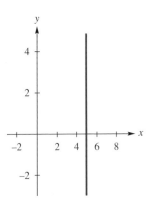

Let's pick two points on the line, say $(5, 2)$ and $(5, 4)$.

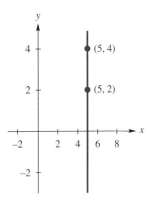

The

$$\text{slope} = \frac{\text{vertical change}}{\text{horizontal change}} = \frac{2}{0}$$

which is undefined, because division by zero is undefined. There is also no vertical intercept. So we need a different method for writing an equation. What is true about every point on this line? Is the horizontal coordinate always 5? What about the vertical coordinate? Because the horizontal coordinate is always 5 and the vertical coordinate runs through all real values, there is no restriction on y. Therefore, the equation of this vertical line is $x = 5$. Does this vertical line represent a function? Why or why not?

We have just seen that the equation $y = mx + b$ does not describe all lines. The equation $Ax + By = C$, where A, B, and C are constants, such that A and B are not both zero, is more general because this equation provides an algebraic description of *all* lines. Notice that an equation in slope–intercept form can also be written in general form. If $B = 0$ the equation $Ax + By = C$ describes a vertical line.

Because all linear equations can be written in the form $y = mx + b$ or $Ax + By = C$, we know that any equation that does not fit these two forms does not graph as a line. This means that if the variables in an equation are raised to any power other than one, then the equation is not linear. If the variable occurs in the denominator or under a radical sign, the equation is not linear. However, the numerical coefficients may include powers, roots, or fractions.

In this section, we introduced **linear functions.** At this point, we can recognize a linear function presented in any of the three models: numerical, graphical, or algebraic. The equation of a line is usually presented in **slope–intercept form,** $y = $ slope $* x + $ vertical intercept or $y = mx + b$, or in **general form,** $Ax + By = C$, where A, B, and C are real numbers and not both A and B are zero. From the equation of a linear function we can determine the **slope, horizontal intercept,** and **vertical intercept.** From a scaled graph we can determine, at least approximately, these same three characteristics of the line.

We can write the equation of a line if we are able to determine the slope and vertical intercept. Given an application for which a linear model is appropriate, we can present the model in all three forms: numerical, graphical or algebraic. Depending on what information is given in the problem and its context, any one of the three models may be the easiest entry point into the problem. As you get more practice, you will be able to pick the easiest place for you to start in a given situation. Until then, some trial and error will be involved. Try all three approaches before you ask for help.

Problem
Set
2.2

1. Solve the following equations. Check your solution.

   a. $0.28(5x - 100) + 100 = 877.7$

   b. $\dfrac{42}{x} = 25 + \dfrac{17}{x}$

2. Solve each literal equation for the indicated variable. Verify your solutions.

   a. $5(2x + a) = bx - c$ for x

   b. $A = P(1 + rt)$ for r

   c. $\dfrac{1}{P} + \dfrac{1}{Q} = \dfrac{1}{F}$ for F

   d. $I = \dfrac{nE}{nr + R}$ for n

   e. $F = \dfrac{mv^2}{r}$ for v

3. Select all of the following equations that are equivalent to $y = \dfrac{x + 6}{2}$.

   a. $y = x + 3$

   b. $y = \dfrac{1}{2}x + 3$

   c. $y = \dfrac{x}{2} + 3$

   d. $y = \dfrac{x}{2} + 6$

   e. $y = 2x + 12$

4. For each of the following graphs, write an equation of the line and check your equation.

a.

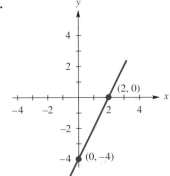

b.

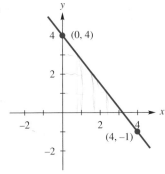

c.

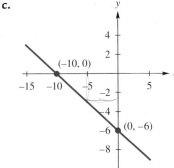

d.

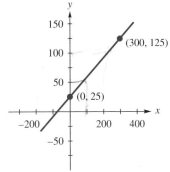

5. a. Rewrite $4x + 3y = 16$ in slope–intercept form. Identify the slope and vertical intercept.

   b. Rewrite $2x - 5y = 24$ in slope–intercept form. Identify the slope and vertical intercept.

6. a. Rewrite $^-2x + 5y = 12$ in slope–intercept form. Identify the slope and vertical intercept.

   b. Rewrite $^-9x - 20y = 120$ in slope–intercept form. Identify the slope and vertical intercept.

7. For each of the following graphs do the following.

   i. Write the equation of the line and check your equation.

   ii. Use your equation to determine the exact coordinates of the horizontal intercept.

a.

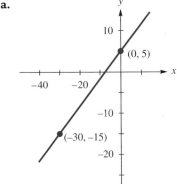

b.

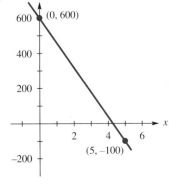

8. For each of the following linear functions, identify the slope and the vertical intercept. Use these to draw a graph.

a. $y = 3x - 2$

b. $y = 10 - 5x$

c. $y = -\dfrac{3}{4}x + 24$

d. $y = \dfrac{x}{2} - 10$

e. $y = \dfrac{15 - 4x}{5}$

f. $y = x$

9. For each linear function, algebraically determine the vertical and horizontal intercepts.

a. $y = 5.25x - 13.65$ **b.** $0.25x - 0.62y = 62$ **c.** $y = -\dfrac{5}{3}x + 25$

10. Draw a graph that clearly shows both intercepts of each linear function. Use the method of your choice. Label at least two points on your line.

a. $18 = y + 2x$

b. $y = {}^-4.2x + 1.5$

c. $2x + 15y = 50$

d. $y = 10.25x - 2500$

e. $2.6x + 0.25y = 24$

f. $\dfrac{x}{3} - y = 14$

g. $3x = 7y - 8$

11. a. Write the equation for the line shown in the graph.

b. Based on your results from part a, what is the slope of any horizontal line? What is the equation of any horizontal line?

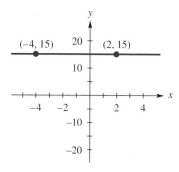

12. In Problem 11, you determined the equation of any horizontal line. A vertical line is similar to a horizontal line. What does the equation of any vertical line look like? Is a vertical line a function?

13. Without graphing, identify all of the following equations that graph as lines. Assume x is the independent variable and y is the dependent variable. For each equation that graphs as a line, determine the slope.

a. $y + 7x = 10$

b. $y = \sqrt{2}\,x$

c. $y = \sqrt{3x}$

d. $5x = 2y - 20$

e. $x = 12$

f. $y = \dfrac{5}{x}$

g. $y = \dfrac{x}{2}$

h. $\dfrac{5x}{4} = y$

14. Without graphing, identify all of the following equations that graph as lines. Assume x is the independent variable and y is the dependent variable. For each equation that graphs as a line, determine the slope.

a. $y + \dfrac{x}{2} = \dfrac{2}{3}$

d. $y = \sqrt[3]{9}$

g. $y = \dfrac{3}{5}x$

b. $\sqrt{7x} = 2y$

e. $xy = 25$

h. $\dfrac{2 + 3x}{5} = y$

c. $y = x$

f. $y = \dfrac{3}{5x}$

15. Graph the linear function $y = x$ in each of the following windows using your graphing calculator. (*Note:* The window settings are given in the form [xmin, xmax] by [ymin, ymax].)

a. $[^-10, 10]$ by $[^-10, 10]$

b. $[^-40, 40]$ by $[^-10, 10]$

c. $[^-10, 10]$ by $[^-40, 40]$

d. Do the slopes of the lines in parts a–c *appear* similar or different? Explain your response.

e. What is the slope of the line $y = x$? Explain why changing the window can make the slope of $y = x$ appear to be something other than 1.

16. Without graphing, determine which of the following *can* be the graph of the linear function $y = 30x + 1.2$. Explain how you made your decision.

a.

b.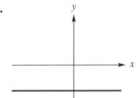

c.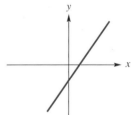

d.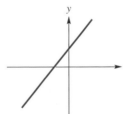

e.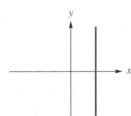

f.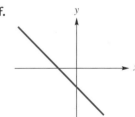

17. **Temperature Conversion.**   We know that water freezes at 0°C or 32°F and that water boils at 100°C or 212°F. Assume that the relationship between temperature on the Celsius scale and on the Fahrenheit scale is a linear relationship.

a. Graph the data with Celsius on the horizontal axis and Fahrenheit on the vertical axis.

b. Write an equation for the Fahrenheit temperature in terms of the Celsius temperature using a linear function.

c. Using your equation, predict the Celsius temperature if it is 82°F. Does your answer agree with your graph?

d. Using your equation, predict the Fahrenheit temperature if it is 16°C. Does your answer agree with your graph?

e. Interpret the meaning of the vertical intercept, the horizontal intercept, and the slope in the context of the problem.

18. Quick Convert.   A quick way to convert from Celsius to approximate Fahrenheit temperature is to double the Celsius temperature and add 30°.

  a. Write an equation for this quick conversion.

  b. Graph your quick conversion equation and the equation from Problem 17 on the same set of axes. (When you are asked to graph two equations on the same set of axes, draw one coordinate system and graph both of your equations on that one coordinate system.)

  c. For what temperature(s) do the two models match?

  d. For what temperatures does the quick conversion model give you a higher Fahrenheit temperature than is correct?

19. Comparing Wages.   A new employee at a sales company is offered one of three different possible salary packages: A, B, or C. The new employee works approximately 15 hours each week.

  Package A: $105 for the 15 hours of work each week

  Package B: $45 per week plus a commission of 8% of all sales during the week

  Package C: 12% commission of all sales during the week

  a. If a new employee expects to sell $1000 in merchandise, what salary package should the person choose?

  b. For what sales is salary package A the best deal? B? C?

  c. If you were the new employee, what package would you choose and why?

20. The Terrific Gadget Company.   The Terrific Gadget Company reports that when they produce and sell n Wazoos in a week, the cost is represented by the model $C = 175 + 5.1n$ and the revenue is represented by the model $R = 7.6n$, where both C and R are in dollars.

  a. Graph the two models together on the same set of axes. Scale your axes so that all of the "important features" are represented. (The vertical axis in this situation represents an amount of money in dollars.)

  b. What is represented by the intersection of the two models in the context of the problem?

  c. What is the slope, including units, of the revenue equation? What does the slope mean? What is the slope, including units, of the cost equation? What does the slope mean?

  d. Write an equation for the profit from producing and selling n Wazoos. Graph your profit equation on your axes from part a. (Profit = revenue − cost).

  NOTE: In English, we usually think of profit as positive, but in mathematics profit is revenue minus cost. This means if the cost is larger than the revenue, then our profit is negative.

  e. Identify the vertical intercept for your profit equation. What does this tell you about the problem situation?

  f. Reviewing the company records, you see that the company produced and sold 55 Wazoos last week. What would you advise them to do?

21. Pressure Changes in Water and Air.   The following tables show the pressure changes in water as the depth changes and in air as the altitude changes.

  a. Does either table show a linear relationship? Explain how you decided.

  b. For the relationship(s) that you selected as linear, calculate the slope using three different pairs of points. Did you expect the slope to come out exactly the same for each pair? If they aren't exactly the same, explain why.

Depth (ft)	Pressure (atm)
0	1
33	2
67	3
100	4
133	5
167	6
200	7
300	10
400	13
500	16

Altitude (1000s of ft)	Pressure (atm)
0	1.00
10	0.69
20	0.46
30	0.30
40	0.19
50	0.11
60	0.07

22. Use the following table to answer the questions.

  a. Determine the horizontal and vertical intercepts.

  b. Is the relation a function? Explain.

  c. Does the relation graph as a line? If so, write an equation for the line.

t	D
−8	24
−6	20
−4	16
−2	12
0	8
2	4
4	0
20	−32
30	−52

23. The following table of values represents a linear function. Use the table to answer the questions.

   a. Estimate the horizontal intercept using the table.

   b. Determine the vertical intercept using the table.

   c. Calculate the slope to the nearest hundredth.

   d. Write the approximate equation for the linear function.

x	y
−2.00	−0.17
−1.75	0.04
−1.50	0.25
−1.25	0.46
−1.00	0.67
−0.75	0.88
−0.50	1.08
−0.25	1.29
0.00	1.50
0.25	1.71
0.50	1.92
0.75	2.13
1.00	2.33

2.3 Equations of Linear Functions

Activity Set 2.3

1. a. Each of the following tables represents a linear function. Write an equation for each function. Check your equation.

  b. Graph each line by plotting the points from each table.

  c. Based on your equations, what are the vertical intercepts of the lines? Do your graphs confirm these?

LINE A		LINE B	
x	*y*	*x*	*y*
10	35	25	150
11	38	26	160
12	41	27	170

2. For each of the following graphs, write an equation for the line. Check your equation.

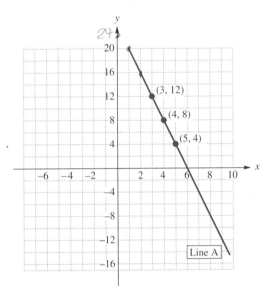

3. Use the following graph to answer parts a–d.

  a. Determine the slope of the line.

  b. Estimate the vertical intercept.

  c. Use the results of parts a and b to write an equation of the line.

  d. Check both of the known points in the equation you wrote in part c. Did your equation check exactly?

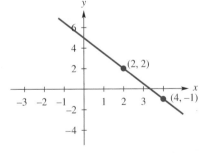

4. Use the following graph to answer parts a–d.

  a. Determine the slope of the line.

  b. Estimate the vertical intercept.

  c. Use the results of parts a and b to write an equation of the line.

  d. Check both of the known points in the equation you wrote in part c. Did your equation check exactly? Explain why or why not.

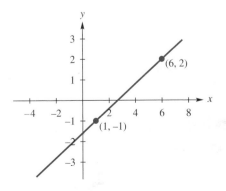

5. Write an equation for the following line. Check your equation. (*Note:* From now on, when you are asked to write the equation of a line, you need to determine the vertical intercept accurately and should not merely estimate its value from a graph. However, an estimate from the graph is a quick verification of your other calculations.)

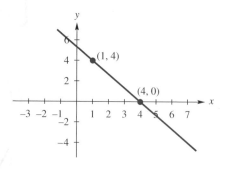

6. On each of the following graphs two lines are drawn.

   a. Determine the slope of lines L and M. What relationship do you observe between the lines and the slopes of the lines?

   b. Determine the slope of lines P and Q. What relationship do you observe between the lines and the slopes of the lines?

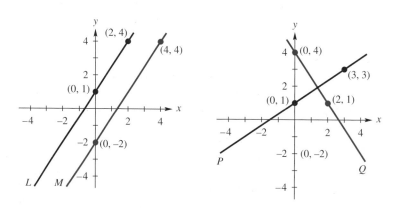

7. For each part carefully *hand draw* the graphs of both lines on the same set of axes, using the same scale for both the vertical and horizontal axes.

Based on your graphs, what can you say about the equations of parallel lines? What can you say about the equations of perpendicular lines?

a. $\begin{cases} y = 2x - 1 \\ y = 2x + 3 \end{cases}$

c. $\begin{cases} y = 2x + 1 \\ y = -\dfrac{1}{2}x - 3 \end{cases}$

b. $\begin{cases} y = \dfrac{2}{3}x - 2 \\ y = \dfrac{3}{2}x - 4 \end{cases}$

d. $\begin{cases} y = \dfrac{2}{3}x - 2 \\ y = -\dfrac{3}{2}x + 4 \end{cases}$

8. Given the equations of two lines, can you tell if the lines are parallel, perpendicular, or neither? If so, explain how you made your decision.

9. Redo parts c and d in Activity 7 using your calculator. Set the calculator to its standard graphing window. What do you observe? Explain why these lines no longer appear perpendicular.

Discussion
2.3

In the previous two sections, we looked at linear functions and in the case of vertical lines, linear relations. We know that any linear function can be written in the form $y = mx + b$, where m is the slope and $(0, b)$ is the vertical intercept.

In Activity 1, we were given a table of values for two different linear functions.

LINE A	
x	**y**
10	35
11	38
12	41

From the table for line A, we can see that the output increases by 3 each time the input increasesby 1. Therefore, the slope of line A is $\frac{3}{1} = 3$. With this information we might guess that the outputis $3x$. If we check this, we see that $3 * 10$ is 30, but the output is supposed to be 35. Therefore, we need to increase our guess by 5. Our new guess is $3x + 5$. Let's check this.

CHECK   When $x = 10$,

$$3 * 10 + 5 = 35 \qquad ✔$$

When $x = 11$,

$$3 * 11 + 5 = 38 \qquad ✔$$

When $x = 12$,

$$3 * 12 + 5 = 41 \qquad ✔$$

Because our expression checks, an equation for line A is $y = 3x + 5$. To determine the equation for line A, we used a combination of the knowledge that we have about lines together with guess and check. Let's formalize this process.

We know that any linear function can be written in the form $y = mx + b$; therefore, if we can determine the slope and vertical intercept of a line, we can write its equation. For line A, we know that the slope is 3. This means that the equation for line A must be $y = 3x + b$, where we still do not know b. In our guess-and-check method, we substituted the input 10 into our guess to see if we obtained the correct output of 35. Because the input is represented by x and the output is represented by y, we can substitute $x = 10$ and $y = 35$ into our equation $y = 3x + b$. This allows us to solve for b.

$$35 = 3 * 10 + b$$
$$35 = 30 + b$$
$$5 = b$$

Because $b = 5$, our equation for the line is $y = 3x + 5$. To check our equation, we substitute another point on the line to see if it satisfies the equation. Because we did this previously, we can omit it here.

To summarize, we find the slope of our line and substitute this value for m in the slope–intercept equation. However, the equation is not complete because we do not know the value of b. To find the value of b, we choose a known point on the line and substitute the corresponding values for x and y into our equation. This results in an equation that we can use to solve for b, the vertical coordinate of the vertical intercept. With the slope and vertical intercept we finally are able to write the equation of the line.

> *Writing an Equation of a Line Given Two Points*
>
> 1. Find the slope of the line using the two points given.
> 2. Substitute the value for the slope in the slope–intercept equation.
> 3. Find the value of b by substituting the coordinates of one of the given points into the slope–intercept equation. Solve the resulting equation for b.
> 4. Write the equation of the line by substituting the values for m and b into the slope–intercept equation.
> 5. Check the equation by substituting the coordinates of the other point into the equation.

To *write the equation of a line given the slope and one point,* you follow the same process as shown in the strategy except you can skip step 1 because you already know the slope.

Example 1

Write an equation for the following line. Check your equation.

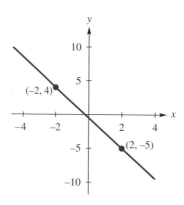

Solution   First we need to determine the slope. Going from the point $(^-2, 4)$ to $(2, ^-5)$ we go down 9 units and right 4 units.

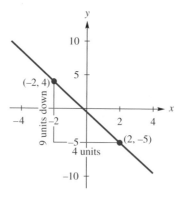

Therefore, the slope is $\frac{^-9}{4}$, or $^-2.25$. We can substitute this into the slope–intercept equation to obtain

$$y = ^-2.25x + b$$

Next, we substitute either of the points given into our equation, $y = {}^-2.25x + b$, to obtain a value for b. Let's substitute $({}^-2, 4)$.

$$4 = {}^-2.25 * {}^-2 + b$$
$$4 = 4.5 + b$$
$${}^-0.5 = b$$

Because b is ${}^-0.5$, this means that the line crosses the vertical axis at ${}^-0.5$. Based on the graph given, is this reasonable?

Now that we have determined the slope and the vertical intercept we can write the equation of the line as $y = {}^-2.25x - 0.5$. To check our equation, let's substitute the *other* known point. Because we used the point $({}^-2, 4)$ to write the equation of this line, we know it works in the equation. Therefore, we must substitute the point $(2, {}^-5)$ to see if our equation is correct.

CHECK

$$y = {}^-2.25x - 0.5$$
$${}^-5 \stackrel{?}{=} {}^-2.25 * 2 - 0.5$$
$${}^-5 = {}^-5 \quad ✔$$

Therefore, the equation of the line is $y = {}^-2.25x - 0.5$.

Example 2

The Blue Book Value.   Suppose you own a car that is presently two years old and the *Kelley Blue Book* trade-in value is $6270. In an old *Blue Book* you found that the trade-in value of your car one year ago was $7800. Assume that the trade-in value of your car decreases linearly with time.

a. Write an equation for the trade-in value of your car in terms of its age.

b. When should you sell your car if you want to sell it before the trade-in value drops below $2000?

c. What are the units associated with the slope? What does the slope tell you about the problem situation?

d. Interpret the meaning of the intercepts in the context of the problem.

Solution

a. First let's sketch a graph. To draw a graph we must first choose the independent and dependent variables. The trade-in value of a car depends on its age. Therefore, age is the independent variable and is graphed on the horizontal axis. The trade-in value is the dependent variable and is graphed in the vertical axis.

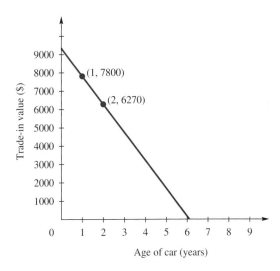

To write the equation of a line, we need to determine the slope. From the graph, the slope is

$$\frac{\text{vertical change}}{\text{horizontal change}} = \frac{^-\$1530}{1 \text{ year}}$$

If T represents the trade-in value in dollars and g represents the age of the car in years, we know that

$$T = {}^-1530g + b$$

Next, we need to find the value for b. To do this, we can substitute one of the known points into the equation and solve for b. Let's use $(1, 7800)$.

$$7800 = {}^-1530 * 1 + b$$
$$7800 = {}^-1530 + b$$
$$9330 = b$$

Because $b = 9330$, we know that the vertical intercept is $(0, 9330)$. Is this conclusion reasonable based on our graph?

Now that we know the slope and vertical intercept we can write the equation.

$$T = {}^-1530g + 9330$$

To check our equation, we need to substitute the other known point.

CHECK   Using the point $(2, 6270)$,

$$6270 \overset{?}{=} {}^-1530 * 2 + 9300$$
$$6270 = 6270 \qquad ✔$$

Our equation checks.

b. We can solve this problem in two different ways: we can approximate the solution using the graph, and we can use the equation that we wrote in part a.

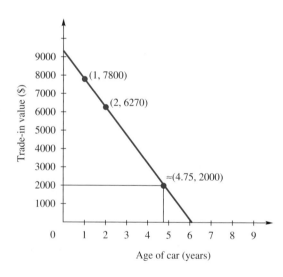

From the graph it appears that the car will be worth $2000 when it is about 4 years and 9 months old.

Using the linear function, $T = {}^-1530g + 9300$ from part a, we need to solve the equation $2000 = {}^-1530g + 9330$ for g. In doing this you will find that $g \approx 4.79$. Because we want to sell the car before the trade-in value drops below $2000 we should sell it by the time it is 4 years and 9 months old.

In this problem we made many simplifying assumptions. For example, we assumed that the data is linear. In reality, many factors besides the age of the car affect its trade-in value. Therefore, whether we use the graph to answer the question or the equation we only obtain an estimate. However, be aware that in other problems you may encounter you will need to be more concerned about the accuracy of your solution. In these situations, you need to choose a solution process that provides you with the appropriate accuracy.

c. Because the vertical change (rise) is in dollars and the horizontal change (run) is in years, the units associated with the slope are dollars/year, or dollars per year. Therefore, the slope indicates the car depreciates in value by $1530 per year.

d. The vertical intercept is the point $(0, 9330)$. This tells us that the initial value of the car was $9330. From the graph the horizontal intercept appears to be the point $\approx (6.1, 0)$. How can we find the horizontal intercept algebraically? Algebraically, we can set T equal to 0 and solve for g.

$$0 = {}^-1530g + 9300$$
$$1530g = 9300$$
$$g = \frac{9300}{1530} \approx 6.1$$

This confirms our original estimate that the horizontal intercept is the point $\approx (6.1, 0)$.

The horizontal intercept represents the age when the trade-in value of the car is 0. Therefore, in about six years the trade-in value of the car will be 0.

As you found in Activities 6–9, parallel lines have the same slope, and perpendicular lines have opposite reciprocal slopes. We can use this information to analyze many geometric shapes.

Definition _____

Two lines are **parallel** if and only if they have the same slope.

Definition _____

Two nonhorizontal/nonvertical lines are **perpendicular** if and only if their slopes are opposite reciprocals. That is, if L is a line whose slope is m (where $m \neq 0$), then any line perpendicular to line L has slope $-\frac{1}{m}$.

Example 3   **The Tangent Line.**   Draw a circle with radius 5, centered at the origin. Is the point $(3, 4)$ on the circle? Write the equation of the line tangent to the circle at the point $(3, 4)$.

Solution   First we draw the circle with radius 5, centered at the origin.

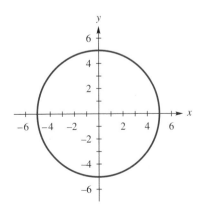

We must decide if the point $(3,4)$ lies on the circle. If the point $(3,4)$ is on the circle, then the length of line segment $\overline{AC}$ in the following figure must be 5.

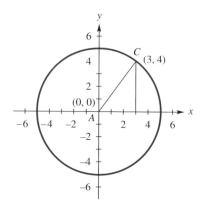

We can use the Pythagorean theorem to conclude that the point $(3,4)$ lies on the circle because $3^2 + 4^2 = 5^2$.

To determine the equation of the tangent line we must recall some geometry. The line joining the center of a circle to any point on the circle is perpendicular to the tangent line at that point. In this situation, that means that the line joining A and C is perpendicular to the tangent line at $(3,4)$. Because the slope of the line joining A and C is $\frac{4}{3}$, the slope of the tangent line at $(3,4)$ is $-\frac{3}{4}$. We can use this slope, together with the point $(3,4)$, to determine the equation of the tangent line.

$$y = -\frac{3}{4}x + b$$
$$4 = -\frac{3}{4} * 3 + b$$
$$4 + \frac{3}{4} * 3 = b$$
$$6\frac{1}{4} = b$$

This gives us the equation $y = -\frac{3}{4}x + 6\frac{1}{4}$. To verify this we can sketch the graph of the line $y = -\frac{3}{4}x + 6\frac{1}{4}$ together with the circle.

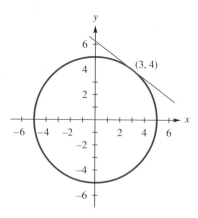

The graph of the line $y = -\frac{3}{4}x + \frac{25}{4}$ appears to be tangent to the circle at $(3,4)$ as desired. We conclude that the equation of the line tangent to the circle is $y = -\frac{3}{4}x + \frac{25}{4}$.

Example 4

The Hershey Bar.   The following table shows the cost of a Hershey bar for several years between 1969 and 1998. Determine an approximate linear model for the data. Then use your linear model to predict the cost of a Hershey bar in the year 2002 and predict the year when the cost of a Hershey bar will reach $1.

Year	Cost (¢)
1969	10
1974	15
1975	20
1978	25
1982	30
1983	35
1991	45
1998	55

Solution   To begin, we graph the data. When we are modeling problems that involve years it is often convenient to look at the number of years past a certain date instead of the actual year itself. For this problem we let the independent variable be the number of years after 1900. With this we obtain the following graph.

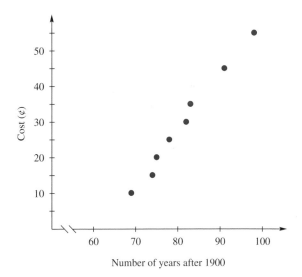

We can see that the data plotted can be approximated by a linear model. This is not a perfect fit but approximates the trend and can be used for making rough predictions. First we need to draw a line that we feel "fits" the data. That is, we "eyeball" it to determine a good fit.

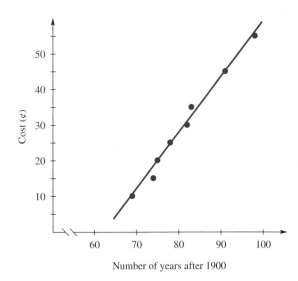

Notice that the line drawn appears to pass through several of the data points. To write an equation for this line we can pick any two of the points that the line appears to pass through.

In this case, let's assume the line passes through $(78, 25)$ and $(91, 45)$.

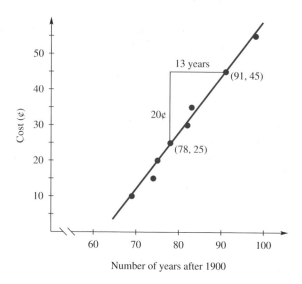

Number of years after 1900

From the graph we can see that the slope of the "eyeball fit" line is $\frac{20¢}{13\ \text{yr}} \approx 1.54¢$ per year. Because our line does not model the data exactly, we do not need to keep the slope exact. Therefore, we use 1.54 as the slope of our line. We can use this together with one of the data points that the line passes through to determine the vertical intercept, and thus the equation of the "eyeball fit" line. Let C be the cost in cents of a Hershey bar and t be the number of years after 1900. Then

$$C \approx 1.54t + b$$
$$25 \approx 1.54 * 78 + b$$
$$b \approx {}^{-}95.1$$

Because we rounded the slope to three significant digits, we also round b to three significant digits.

Therefore, the equation of our "eyeball fit" line is

$$C \approx 1.54t - 95.1$$

Notice, from the equation, the vertical intercept is $\approx (0, {}^{-}95.1)$. However, this intercept cannot be seen on our graph because the line cannot be extended across the break.

To predict the cost in 2002 and to determine when the price will reach \$1, we assume that the linear trend in the data points continues. Do you think this is a reasonable assumption?

To predict the cost of a Hershey bar in 2002 we substitute 102 for t.

$$C \approx 1.54 * 102 - 95.1$$
$$C \approx 62$$

To predict when the cost of a Hershey bar will reach \$1 we substitute 100 for C and solve for t.

$$100 \approx 1.54t - 95.1$$
$$t = 127$$

This is 127 years after 1900, which would be the year 2027.

In conclusion, we predict that a Hershey bar will cost 62¢ in the year 2002 and reach \$1 in the year 2027.

For what years do you think the eyeball fit linear model in the Hershey bar example is valid? Is there only one eyeball fit linear model for the data?

In Section 2.2, we were able to write the equation for a linear function only if we were given the vertical intercept. In this section, we extended that knowledge to write the equation given any two points or the slope and any one point. This information might be given to us directly, through a graph, or in a numerical table.

We also found that parallel lines have equal slopes and that perpendicular lines have opposite reciprocal slopes. These facts were used directly and in application situations. Eyeball fit models for real data were introduced. If data appear to be linearly related when we graph them, we can come up with an approximate linear function to model the data.

Problem Set 2.3

1. Solve each of the following equations, and check your solutions.

   a. $2.4(2k - 8) + 17 = 27k$

   b. $\dfrac{x - 3}{5} + \dfrac{3x}{2} = 4 - x$

   c. $14 - (3m - 30) = 15m - 10$

   d. $\dfrac{5 + x}{6} - \dfrac{10 - x}{3} = 1$

2. Write the equation for each line and check your equation.

   a.

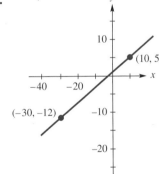

   b.

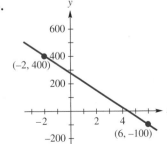

3. Write the equation for each line and check your equation.

a.

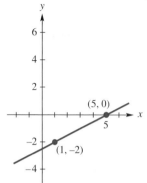

b.

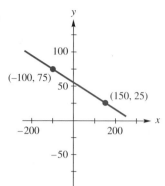

c.

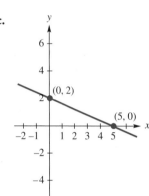

d.

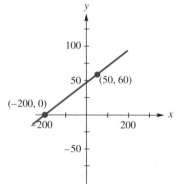

4. Write the equation for the line and check your equation.

a.

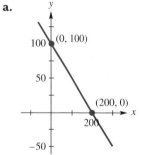

b.

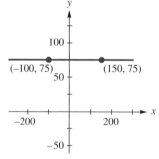

c.

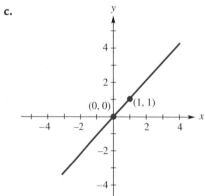

d.

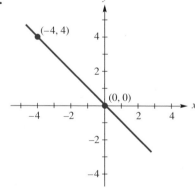

5. For each part, sketch the graph, and write the equation of the line satisfying the given conditions. Label two points on the line.

   a. With a slope of $-\frac{3}{2}$ and going through the point $(^-1, 4)$

   b. With a slope of 4 and going through the point $(6, 0)$

   c. Passing through the points $(3, ^-2)$ and $(^-2, 3)$

   d. Passing through the points $(^-5.8, 3.6)$ and $(4.4, 18.9)$

6. For each part, sketch the graph, and write the equation of the line satisfying the given conditions. Label two points on the line.

   a. Parallel to $y = 4x + 1$ and passing through the point $(0, ^-3)$

   b. Parallel to the line $2x - 3y = 5$ and passing through the point $(0, 5)$

   c. Perpendicular to $y = 5x + 2$ and passing through the point $(0, ^-4)$

   d. Perpendicular to $y = 0.25x - 10$ and passing through the point $(2, 5)$

7. For each part, sketch the graph, and write the equation of the line satisfying the given conditions. Label two points on the line.

   a. Parallel to the line $y = \frac{2}{5}x - 5$ and passing through the point $(^-3, 1)$

   b. Parallel to the line $x - 3y = 16$ and passing through the point $(^-4, 7)$

   c. Perpendicular to $y = -\frac{5}{4}x + 8$ and passing through the point $(^-3, ^-2)$

   d. Perpendicular to $x = 3$ and passing through the point $(2, 7)$

8. Draw a graph of each linear function, using the method of your choice. Label at least two points on your line. Your graph should clearly show both intercepts.

   a. $y = 20 - 4x$

   b. $y = 65x - 2500$

   c. $3.2x + 0.25y = 24$

9. In each of the following problems many equations are possible.

   a. Write the equations of three different lines that have a slope of $^-2$.

   b. Write the equations of three different lines that pass through the point $(4, 5)$.

   c. Write the equations of three different lines that are parallel to $y = 5x - 6$.

10. In each of the following problems many equations are possible.

    a. Write the equations of three different lines that have a vertical intercept of $(0, ^-4)$.

    b. Write the equations of three different lines that have a horizontal intercept of $(3, 0)$.

    c. Write the equations of three different lines that are perpendicular to $y = \frac{1}{5}x - 10$

11. In each of the following problems many equations are possible.

    a. Write the equations of three different lines that pass through the origin.

    b. Write the equations of three different lines that do not pass through quadrant III.

    c. Write the equations of three different lines that are perpendicular to $y = ^-4x - 3$.

12. Make up an equation of a linear function that is increasing for price P in terms of time t.

13. One member of your team tells you that $x = 5$ is a point. Another claims it is the equation of a line. A third claims that $x = 5$ is a function. Which members of your team are correct? Explain.

14. Hay for the Winter.   Al has stored bales of hay in his barn loft to feed his horses from December through May. Four weeks after he began using the hay from the loft, he took inventory and had 230 bales. Ten weeks later he took inventory again and had 125 bales.

    a. Assuming the hay usage is linear, write an equation for the number of bales of hay in the loft in terms of the number of weeks after he began using the stored hay.

    b. How many bales of hay did Al originally store in the loft?

    c. Will the hay supply last through the 25 weeks that he anticipates needing it?

15. Pollution.   During the summer in a large West Coast city, the pollution index increases in an approximately linear fashion from 8 A.M. to around 4 P.M. On an average summer day the reading at 10 A.M. is around 43.0 parts per million (ppm) and at 1 P.M., 80.5 ppm.

    a. Write an equation to express the pollution index in terms of the *hours since 8 A.M.*

    b. Graph your equation for the appropriate time span.

    c. What units are associated with the slope? What does the slope tell you about the problem situation?

    d. Identify the vertical intercept. What does the vertical intercept tell you about the problem situation?

    e. What does your equation predict the pollution index will be at 11:30 A.M.? At 11:30 P.M.? Are these answers reasonable? Why or why not?

    f. Bob has been advised by his doctor not to go outdoors when the pollution index is above 100 ppm. If we assume our model is a fairly safe indicator, what would your advice to Bob be in terms of the time of day he can be outdoors?

16. Penny's Phone Bill.   On Penny's long-distance bill for last month were several calls from Vancouver, WA, to Olympia, WA, where one of her sons attends college. Among those calls were evening calls for 1 minute for $0.29, 4 minutes for $0.74, 6 minutes for $1.05, and 11 minutes for $1.80.

    a. Is this a linear relationship? Explain how you decided.

    b. Graph these data. What makes most sense as the dependent variable? Why?

    c. Assuming we can model this relationship with a linear function, use two of the phone calls to write an equation. Check the other two phone calls in your model.

    d. What are the units associated with the slope? What does the slope tell about the problem situation?

    e. Interpret the meaning of the vertical intercept in the context of the problem.

    f. According to your equation, what is the cost of a 20-minute phone call?

    g. How long can an evening call be and still keep the cost under $5?

17. Supermarket.   Go to your local supermarket and record the total cost in terms of the weight for at least six whole chickens of one particular brand (your choice).

    a. Plot your data.

    b. If the data appear to be linear, write an equation.

    c. Is this what you expected? Why? Consider the meaning of the slope including its units.

18. Suburban Population. The following charts show the populations for two suburbs from 1980 to 1990.

Year	Suburb 1
1980	12,690
1982	13,071
1984	13,463
1986	13,867
1988	14,283
1990	14,711

Year	Suburb 2
1980	23,112
1982	23,693
1984	24,274
1986	24,855
1988	25,436
1990	26,017

   a. Analyze the data, and decide which suburb's population growth can be modeled by a linear function. Explain how you decided.

   b. Write an equation for the one that is linear.

   c. What does your equation predict the population will be in 2010 if the current trends continue?

   d. When will the population be above 30,000?

19. a. Determine the area of the quadrilateral in the figure.

   b. Is the quadrilateral a parallelogram? Explain how you made your decision.

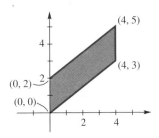

20. a. The formula for the midpoint of a line segment is shown in the following box. Determine the midpoint M of side AC in $\triangle ABC$. Determine the midpoint N of side BC in $\triangle ABC$.

   b. Determine the slope of the line MN and the slope of line AB. What can you say about these two lines?

> *Midpoint Formula*
>
> For a line segment drawn between (x_1, y_1) and (x_2, y_2), the **midpoint** is
>
> $$\left(\frac{x_1 + x_2}{2}, \frac{y_1 + y_2}{2}\right)$$

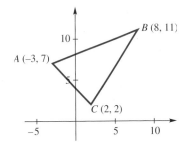

21. Draw a circle with radius 13, centered at the origin. Does the point $(^-12, ^-5)$ lie on the circle? Write the equation of the line tangent to the circle at the point $(^-12, ^-5)$.

22. Chirping Crickets.   It is surprising but true that the temperature can be determined fairly accurately by counting the number of times a cricket chirps in a minute. Your lab team has supplied you with the following data.

Number of chirps per minute	40	60	100	140
Temperature (°F)	50	55	65	75

a. Graph the temperature in terms of the number of chirps per minute.

b. Using your graph, predict the temperature if the number of chirps per minute is 80.

c. Using your graph, predict the number of chirps per minute if the temperature is 70°F.

d. Do the given data show a linear relationship?

e. Write an equation for the linear function that gives the temperature in terms of the number of chirps per minute.

f. Use your equation to find the temperature if the number of chirps per minute is 170.

g. Use your equation to find the number of chirps per minute if the temperature is 25°C. (*Hint:* Read carefully!)

h. At what temperature do the crickets stop chirping?

i. What is the number of chirps per minute if the cricket is in water at 212°F? Is it reasonable to use your model to answer this question?

23. Heating and Temperature.   In moderate and colder climates, the daily temperature affects the amount of electricity used in homes that heat electrically. The following table shows this relationship for one household.

a. Plot the points given in the chart.

b. Draw an eyeball fit line through the data. (*Hint:* Your next task will be easier if the line goes through two of the data points.)

c. Write an equation for the line you drew.

d. Use your equation to predict the amount of kilowatt hours this household might use if the average daily temperature for March is 45°F. If the cost for electricity is $0.05825 per kilowatt hour plus a monthly fee of $5.50, how much will the bill be for March?

Average Daily Temperature for Each Month (°F)	Number of Kilowatt-Hours (kWh) per Month (in thousands)
37	6.0
44	5.1
46	3.9
55	2.3
62	1.8
63	0.9
64	1.1
63	1.0
63	1.1
60	1.8
49	3.7
41	5.2

24. Scatter plots display the relation between two variables. Some scatter plots exhibit a clear pattern between two variables, and others show no clear pattern. Use graphs A–D to answer the following questions.

   a. Which graph appears to show a linear relation with a negative slope?

   b. Which graph appears to show a linear relation with a positive slope?

   c. For which graph is a line of best fit most appropriate at predicting the relationship between the two variables?

   d. Which graph exhibits a nonlinear relationship?

   e. Which graph shows no relationship?

Graph A

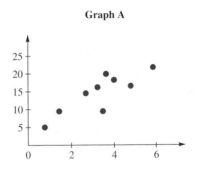

Graph B

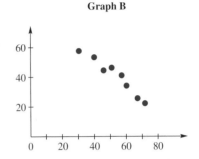

Graph C

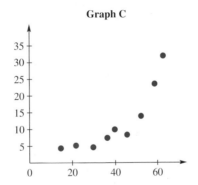

Graph D

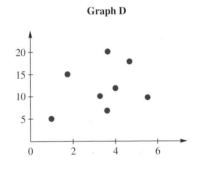

25. Population Growth in the United States. The following table shows the population in the United States at the beginning of each decade.

   a. Plot the data given in the chart.

   b. Draw an eyeball fit line through the data.

   c. Write an equation for the line you have drawn.

   d. What are the units on the slope? What does the slope mean in the context of this problem?

   e. Assuming the trend continues, what will the population be in the United States in the year 2005?

Years after 1900	U.S. Population (in millions)
0	76.2
10	92.3
20	106.0
30	123.2
40	132.2
50	151.3
60	179.3
70	203.3
80	226.5
90	249.6

26. World Record Times.   The following graph shows the improvement in world-record times for the men's 1500-meter running event. (*Note:* The units on the vertical axis are (min:sec), for example, 3:15.0 means 3 minutes and 15.0 seconds.)

a. Draw an eyeball fit line through the data. (*Hint:* Your next task will be easier if the line goes through two of the data points.)

b. Write an equation for the line you drew.

c. What are the units on the slope? What does the slope mean in the context of this problem?

d. Assuming this trend continues, use your equation to predict the finish time for the 1500-meter event at the 2004 Olympics in Athens, Greece.

MEN'S 1500-METER RECORD TIMES

Runner	Country	Year (after 1950)	Time (in min:sec)
Herbert Elliott	Austria	8	3:36.0
Elliott	Austria	10	3:35.6
James Ryan	USA	17	3:33.1
Filbert Bayi	Tanzania	24	3:32.2
Sebastian Coe	Great Britain	29	3:32.03
Steve Ovett	Great Britain	30	3:31.36
Sydney Maree	USA	33	3:31.24
Steve Ovett	Great Britain	33	3:30.77
Steve Cram	Great Britain	35	3:29.67
Said Aouita	Morocco	35	3:29.46
Noureddine Morceli	Algeria	42	3:28.86
Morceli	Algeria	45	3:27.38
Guerrouj	Morocco	48	3:26.00

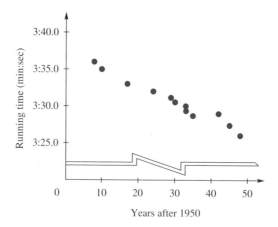

Years after 1950

27. Fast Food Facts.   The following table lists the calories and fat content of fast food items at four popular fast food restaurants.

   a. Make a scatter plot of the data.

   b. Based on the scatter plot, is the fat content and the number of calories in the selected food items linearly related?

   c. Draw an eyeball fit line through the data, and write the equation for this line.

	Calories	Fat (g)
Burger King		
Double Cheeseburger	600	36
Chicken Salad	710	43
Chunky Chicken Salad	142	4
Wendy's		
Jr. Bacon Cheeseburger Dlx	440	25
Grilled Chicken Sandwich	290	7
Breaded Chicken Sandwich	450	20
Chili	210	7
McDonald's		
Hamburger	260	9
Cheeseburger	320	13
Big Mac	560	31
Taco Bell		
Taco	170	10
Taco Supreme	220	13
Taco Salad with Salsa	840	52

28. Increasing College Tuition.   College tuition is constantly going up, but so is inflation. Can we use the rate of inflation to predict the increase in college costs? Use the following data for 1987 through 1994 to answer this question. Explain how you used the data to answer the question.

AVERAGE ANNUAL PERCENTAGE INCREASES FROM 1987 TO 1994							
Consumer Price Index (%)	6.0	4.5	7.0	7.1	12.0	9.9	8.0	6.0
Public 4-Year Colleges (%)	7.8	9.1	8.9	7.8	7.0	6.9	6.1	6.0

2.4 Systems of Linear Equations

1. Determine the solution to the equation $x + 3 = 5$.

2. List three solutions to the equation $y = x + 3$.

3. *Use the following graph* to answer parts a–c. (Many people find it helpful to color the graphs and their corresponding equations in different colors to help them see the relationships.)

   a. Identify two different solutions to the equation $y = {}^{-}2.5x + 9.5$.

   b. Identify two different solutions to the equation $y = \frac{2}{9}x + \frac{4}{3}$.

   c. Identify the solution to the equation ${}^{-}2.5x + 9.5 = \frac{2}{9}x + \frac{4}{3}$.

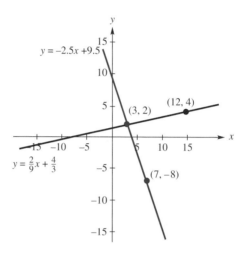

4. The following equations constitute a two-by-two linear system. To solve such a system means to find a solution that satisfies both equations. Use the graph to determine the solution to the system. How would you check your solution?

$$\begin{cases} y = -2.5x + 9.5 \\ y = \frac{2}{9}x + \frac{4}{3} \end{cases}$$

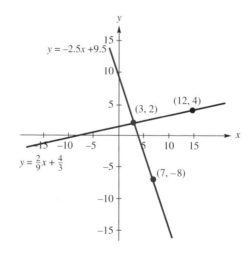

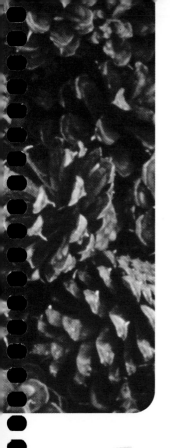

5. Look at the two balances in the following figure. Each brick weighs the same, and each prism weighs the same. When equal weights are on each side of a balance, the top bar is horizontal. For example, the first balance below shows that five bricks and six prisms weigh the same as 23 pounds.

Use the two scales to determine how much each brick and each prism weigh. Explain your process. (*Hint:* You might try *removing* objects from both sides of a scale. For example, if you removed one brick and two prisms from the left side of scale 1, what would you need to remove from the right-hand side to make it balance?

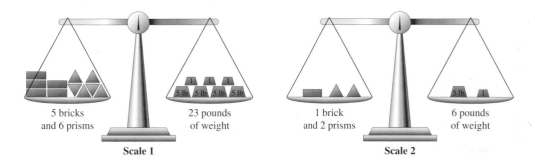

| 5 bricks and 6 prisms | 23 pounds of weight | 1 brick and 2 prisms | 6 pounds of weight |

Scale 1   **Scale 2**

Discussion 2.4

Graphical Solutions to Systems of Equations

As we saw in the activities, a linear equation in two variables does not have a single solution. An infinite number of possible ordered pairs satisfy the equation, and these ordered pairs lie on the line that is the graph of the equation. Furthermore, we saw in the activities that if we have two linear equations in two variables, they intersect in one point (assuming the graphs of these equations are not parallel). Because these *two linear equations* are in *two variables,* this is called a **2 × 2 linear system of equations.** To solve a system of equations, we must find a value for each of the two variables that makes *both* of the equations true. Graphically, the solution corresponds to the point of intersection of the graphs of the two equations. Because the solution to a 2 × 2 system must satisfy both equations, we can check our solution by substituting the values for the variables into both of the equations in the system.

Example 1

Solve the system of equations graphically. Check your solution.

$$\begin{cases} y = \frac{-1}{4}x + 12 \\ y = \frac{2}{3}x - 10 \end{cases}$$

Solution   First, looking at the equations we see that the equation $y = \frac{-1}{4}x + 12$ graphs as a line with a vertical intercept at $(0, 12)$ and a slope of $\frac{-1}{4}$. The equation $y = \frac{2}{3}x - 10$ graphs as line with a vertical intercept at $(0, {}^-10)$ and a slope of $\frac{2}{3}$. Because the vertical intercepts are at $(0, 12)$ and $(0, {}^-10)$, we want a slightly larger window than the standard window. We let the horizontal and vertical axes range from $^-20$ to 20.

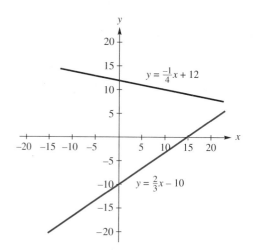

We can see that these two lines intersect if we extend the graph just a little to the right. So we extend the positive *x*-axis to 35.

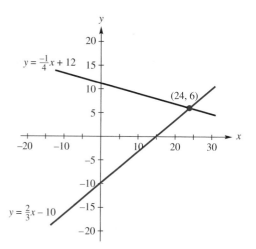

The two lines intersect at the point $(24, 6)$. So, our solution appears to be the point $(24, 6)$ or $x = 24$ and $y = 6$. To be a solution to a system means that the point must be a solution for both of the equations. Therefore, we can numerically check this solution by substituting these values into *both* of the original equations.

CHECK:

$$y = \frac{-1}{4}x + 12 \qquad\qquad y = \frac{2}{3}x - 10$$

$$6 \overset{?}{=} \frac{-1}{4} * 24 + 12 \qquad 6 \overset{?}{=} \frac{2}{3} * 24 - 10$$

$$6 = 6 \quad \checkmark \qquad\qquad 6 = 6 \quad \checkmark$$

The ordered pair $(24, 6)$ satisfies both equations, so our solution is $(24, 6)$. We can also write this solution as $x = 24$ and $y = 6$.

In Example 1, we solved a system of linear equations graphically. We will now look at two different methods for solving systems of equations algebraically.

Elimination Method to Solve Linear Systems of Equations

In the activities, you solved the problem of how much each prism and brick weighed by imagining removing equal amounts from both sides of scale 1. Using the information in scale 2, we know that two prisms and one brick together weigh 6 pounds. Therefore, we can remove two prisms and one brick from the left side of scale 1 as long as we remove 6 pounds from the right side of the scale.

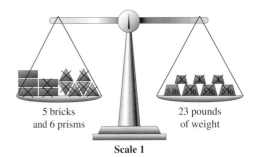

5 bricks
and 6 prisms

23 pounds
of weight

Scale 1

By repeating this process three times, it was possible to get the balance down to just 2 bricks on the left and 5 pounds on the right. Then, two bricks are equal in weight to 5 pounds, so each brick must weigh 2.5 pounds. Knowing that one brick weighs 2.5 pounds, it is possible to determine the weight of the prisms. Let's see how this process can be done algebraically.

We begin by writing an equation to model each of the balance scales. First define the variables.

Let p = weight of one prism in pounds and
b = weight of one brick in pounds

Then, from the scales we know that

Ⓐ $\begin{cases} 5b + 6p = 23 \\ b + 2p = 6 \end{cases}$
Ⓑ

Notice that we have added labels Ⓐ and Ⓑ to the equations. This makes it easier to refer to each of the equations as we work through the solution process.

In the activities, you took away the amount of the second scale three times. We can model this by multiplying equation Ⓑ by $^-3$ and adding the result to equation Ⓐ. (*Note:* Adding $^-3$ times a quantity is equivalent to subtracting 3 times a quantity and actually makes the arithmetic easier.)

The process is modeled below.

$$\begin{array}{rl} \text{Ⓐ} & 5b + 6p = 23 \\ ^-3 \ \text{Ⓑ} & \underline{-3b - 6p = {}^-18} \\ & 2b \quad\ \ = 5 \end{array}$$

Multiply *both sides* of equation Ⓑ by $^-3$.
Add the two equations.

We have now *eliminated* the variable p and have one equation containing only the variable b. Now solve this equation.

$$2b = 5$$
$$\frac{2b}{2} = \frac{5}{2}$$
$$b = 2.5$$

Remember that to solve this problem we need to know how much each shape weighs. To solve for p, substitute $b = 2.5$ into any equation that contains both of the variables. We choose equation Ⓑ.

Ⓑ $\quad b + 2p = 6$
$\quad 2.5 + 2p = 6$ Substitute 2.5 for b.
$\qquad\quad 2p = 3.5$
$\qquad\quad \dfrac{2p}{2} = \dfrac{3.5}{2}$
$\qquad\qquad p = 1.75$

The solution appears to be $b = 2.5$ and $p = 1.75$. To check the solution, substitute the values for b and p into *both of the original equations.*

CHECK

Ⓐ
$$5b + 6p = 23$$
$$5 * 2.5 + 6 * 1.75 \overset{?}{=} 23$$
$$23 = 23 \quad ✔$$

Ⓑ
$$b + 2p = 6$$
$$2.5 + 2 * 1.75 \overset{?}{=} 6$$
$$6 = 6 \quad ✔$$

Therefore one brick weighs 2.5 pounds, and one prism weighs 1.75 pounds.

The algebraic method we just demonstrated is called the **elimination method.** Multiplying the first equation by $^-3$ and adding it to the second equation eliminated the variable p. The result was an equation in one variable that we could then solve. The first step in the elimination method is determining a numerical factor, or sometimes two factors, that can be used to eliminate one of the variables.

Example 2

Solve the following system of equations using the elimination method. Check your solution.

Ⓐ $\begin{cases} 5x + 3y = 25 \\ 8x + 10y = 118 \end{cases}$
Ⓑ

Solution

First we need to find a factor that will eliminate one of the variables. You can't easily multiply 5 by anything and get 8, or 3 by anything and get 10. In this case, it is easier to multiply both of the equations by factors that result in the elimination of one of the variables. Suppose we choose to eliminate the terms containing y. By multiplying equation Ⓐ by 10 and equation Ⓑ by $^-3$, the coefficients of y are 30 and $^-30$. Adding these together eliminates the variable y.

$$\begin{array}{rr} 10 \, Ⓐ & \{ \quad 50x + 30y = 250 \\ ^-3 \, Ⓑ & \{ \underline{^-24x - 30y = ^-354} \\ & 26x \qquad\quad = ^-104 \end{array}$$

Next, we solve this equation for x.

$$26x = ^-104$$
$$\frac{26x}{26} = \frac{^-104}{26}$$
$$x = ^-4$$

To find y, we substitute $x = ^-4$ into any equation that contains both variables. Let's choose equation Ⓐ.

Ⓐ
$$5x + 3y = 25$$
$$5 * {}^-4 + 3y = 25$$
$$^-20 + 3y = 25$$
$$3y = 45$$
$$\frac{3y}{3} = \frac{45}{3}$$
$$y = 15$$

To check the solution, substitute these values for x and y into both of the original equations.

CHECK

Ⓐ $5x + 3y = 25$ Ⓑ $8x + 10y = 118$

$5 * {}^-4 + 3 * 15 \overset{?}{=} 25$ $8 * {}^-4 + 10 * 15 \overset{?}{=} 118$

$25 = 25$ ✔ $118 = 118$ ✔

So, our solution is $x = {}^-4$ and $y = 15$. This could also be written as the ordered pair $({}^-4, 15)$.

In this example, we chose to eliminate y as the first step. This resulted in our selection of 10 and $^-3$ as factors to multiply each of the equations. We could have decided to eliminate x as the first step. What factors could we use in this case?

Substitution Method to Solve Linear Systems of Equations

We looked at solving systems of linear equations graphically and algebraically using the elimination method. The elimination method always works to algebraically solve linear systems of equations, although at times this method can be cumbersome. In later sections we will be solving nonlinear systems of equations. The elimination method does not always work with these systems.

Now, we want to introduce you to the **substitution method.** Both the elimination method and the substitution method are algebraic methods for solving a system. When you are asked to solve a linear system algebraically, you can choose either method. Suppose we want to solve the following system of equations using the substitution method.

Ⓐ $\begin{cases} 5x + 3y = 37 \\ y = 4x - 75 \end{cases}$
Ⓑ

The first step in solving a system using the substitution method is to solve one of the equations for one of the variables. In this system, equation Ⓑ is already solved for y. The next step is to substitute the expression $(4x - 75)$ for y in equation Ⓐ. This step is how the substitution method got its name. It is important when using the substitution method that you solve one of the equations for a variable and then substitute the resulting expression into *the other* equation (that is, the equation that you did *not* use to solve for the variable).

Ⓐ $5x + 3(y) = 37$

Ⓐ $5x + 3(4x - 75) = 37$ Substitute $4x - 75$ for y in equation Ⓐ.

Notice that this equation now contains only one variable, x. We can now solve this equation.

$$5x + 3(4x - 75) = 37$$
$$5x + 12x - 225 = 37$$
$$17x - 225 = 37$$
$$17x = 262$$
$$\frac{17x}{17} = \frac{262}{17}$$
$$x \approx 15.4$$

We are not finished yet; we need to find y. To find y, we substitute 15.4 for x in any of the equations containing both x and y. Because the equation $y = 4x - 75$ is already solved for y, selecting this equation makes our next step easier.

$$y = 4x - 75$$
$$y \approx 4 * 15.4 - 75$$
$$y \approx {}^-13.4$$

Finally, we check our solution using both of the original equations.

CHECK

Ⓐ $$5x + 3y = 37$$
$$5 * 15.4 + 3 * {}^-13.4 \overset{?}{=} 37$$
$$36.8 \approx 37 \quad ✔$$

Ⓑ $$y = 4x - 75$$
$${}^-13.4 \overset{?}{=} 4 * 15.4 - 75$$
$${}^-13.4 = {}^-13.4 \quad ✔$$

Therefore, the solution to the system is $x \approx 15.4$ and $y \approx {}^-13.4$, which could also be written as $\approx (15.4, {}^-13.4)$.

Example 3

Creative Computing Company's Clutter. In the clutter of everyday work, Creative Computing Company's secretary has lost the price list for paper for their two fax machines. One of the machines uses plain paper and the other uses fax paper. The secretary did find the notes for the last two orders. Creative Computing ordered seven boxes of plain paper and ten boxes of fax paper at a total cost of $250.65 and six boxes of plain paper and one box of fax paper at a total cost of $120.20.

Write a 2×2 linear system of equations for this problem situation and solve the system algebraically.

Solution   First we must write two equations to model the last two paper orders. Let P represent the price per box for plain paper and F represent the price per box for fax paper.

Ⓐ $\begin{cases} 7P + 10F = 250.65 \\ 6P + F = 120.20 \end{cases}$
Ⓑ

We have two methods that can be used to solve this system algebraically: the elimination method and the substitution method.

If we choose the elimination method, our first step is to decide which variable to eliminate. By multiplying equation Ⓑ by $^-10$, we eliminate the variable F.

To use the substitution method, our first step is to solve one of the equations for one of the variables. Variable F has a numerical coefficient of 1 in equation Ⓑ. This is the easiest variable to solve for. We complete the solution using the substitution method.

Ⓑ $\quad 6P + F = 120.20$
Ⓑ $\qquad F = 120.20 - 6P \qquad$ Solve for F.

Next substitute the expression, $120.20 - 6P$, for F into equation Ⓐ.

Ⓐ $$7P + 10(F) = 250.65$$
$$7P + 10(120.20 - 6P) = 250.65$$
$$7P + 1202.0 - 60P = 250.65$$
$${}^-53P + 1202.0 = 250.65$$
$${}^-53P = {}^-951.35$$
$$\frac{{}^-53P}{{}^-53} = \frac{{}^-951.35}{{}^-53}$$
$$P = 17.95$$

To find F, we need to substitute 17.95 for P into any equation containing both variables and solve for F. Our next step is made easier by selecting the equation $F = 120.20 - 6P$.

$$F = 120.20 - 6P$$
$$F = 120.20 - 6 * 17.95$$
$$F = 12.5$$

The solution appears to be $(P, F) = (17.95, 12.50)$.

We now need to check if this answer works in the problem situation, not just the equations we wrote, because we may have set up our equations incorrectly.

The first order was for seven boxes of plain paper and ten boxes of fax paper for a total of $250.65. The second order was for six boxes of plain paper and one box of fax paper for a total of $120.20

CHECK

First order:	7 boxes * \$17.95 per box =	\$125.65
	10 boxes * \$12.50 per box =	125.00
Total order is	\$125.65 + \$125.00 =	250.65 ✔
Second order:	6 boxes * \$17.95 per box =	\$107.70
	1 box * \$12.50 per box =	12.50
Total order is	\$107.70 + \$12.50 =	120.20 ✔

We can now say that the cost of the plain paper is $17.95 per box, and the cost of the fax paper is $12.50 per box.

Although it may seem confusing to introduce two methods for solving a system algebraically, you may find one method easier to use on a given problem.

The Number of Solutions to a System of Linear Equations

Example 4

Solve each of the following linear systems.

a. $\begin{cases} y = ^-5x + 2 \\ 1.25x + 0.25y = 0 \end{cases}$
   b. $\begin{cases} 3x - 12y = 4 \\ ^-6x + 24y + 8 = 0 \end{cases}$

Solution a. Because the first equation is already solved for y, let's solve the system in part a by substitution. We can substitute the expression $(^-5x + 2)$ for y in the second equation.

$$1.25x + 0.25(y) = 0$$
$$1.25x + 0.25(^-5x + 2) = 0$$
$$1.25x - 1.25x + 0.5 = 0$$
$$0.5 = 0$$

This statement is clearly not true! But what does it tell us about the solution to the system of equations in part a? Let's try looking at the graphs. Rewrite each equation in slope–intercept form. The equation $1.25x + 0.25y = 0$ becomes $y = ^-5x$. What do you notice about the two equations?

a. $\begin{cases} y = ^-5x + 2 \\ y = ^-5x \end{cases}$

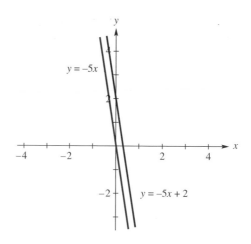

The graph shows what we expected, that these lines are parallel. The solution to a system of linear equations is the point of intersection of the two lines. Because parallel lines never intersect, this system has no solution. This is why our algebraic solutions resulted in a **false statement,** a statement that is always false regardless of the input. Our final answer must be that the system of equations in part a has no solutions.

b. Solving for x or y in either of the equations in the system in part b results in fractional coefficients. Therefore, the method of elimination looks like the best algebraic method for solving this system. To use the elimination method, the variables and constants must "line up." So, the first step is to rewrite the second equation with variables to one side of the equal sign and constants on the other. We rewrite $^-6x + 24y + 8 = 0$ as $^-6x + 24y = ^-8$.

$$\text{Ⓐ} \begin{cases} 3x - 12y = 4 \\ ^-6x + 24y = ^-8 \end{cases} \text{Ⓑ}$$

We can see that by multiplying equation Ⓐ by 2 and adding it to equation Ⓑ, x is eliminated.

$$\begin{array}{r} 2\text{ Ⓐ} \begin{cases} 6x - 24y = 8 \\ ^-6x + 24y = ^-8 \end{cases} \text{Ⓑ} \\ \hline 0 = 0 \end{array}$$

This equation is definitely true, but does not seem to help solve the system. Again, we explore the possible solution by looking at the graphs of the two lines. To graph the equations we must first solve each equation for y. We can rewrite $3x - 12y = 4$ as $y = \frac{1}{4}x - \frac{1}{3}$ and $^-6x + 24y + 8 = 0$ as $y = \frac{1}{4}x - \frac{1}{3}$. What do you observe about these two equations?

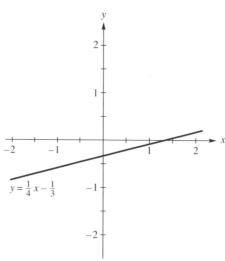

The two equations are identical and, therefore, graph as the same line. Two lines that have the same graph and the same set of solutions are called **coincidental.** The solutions for x and y are the coordinates of the points of intersection of the lines. Therefore all points lying on the line $y = \frac{1}{4}x - \frac{1}{3}$ are solutions to this system. Note that the solution is not "all real numbers." Every point in the Cartesian coordinate system does not satisfy these two equations, only those that lie on the line. The solution is an infinite number of points. This means we cannot list all possible solutions, and, therefore, we describe the solution by writing the equation of the line. Any equivalent form of the equation can be used in the statement of the solution. For example, the solution could have been written as, "all points lying on the line $3x - 12y = 4$."

Examples 3 and 4 are special cases. In most applications the lines are nonparallel and noncoincidental. In other words, they intersect in just one point. When lines are nonparallel and noncoincidental, exactly one point of intersection occurs, and, therefore, one ordered pair is the solution to the system of equations.

3 × 3 Linear System of Equations

So far, we looked at linear systems of equations involving two variables and two equations. In some problem situations three or more variables may be required. In those cases we also need to have the same number of equations. Here we look at solving a 3 × 3 linear system of equations. We learned three techniques for solving 2 × 2 linear systems: graphically, and algebraically using elimination or substitution. In the case of a 3 × 3 system, we do not consider a graphical solution because it involves graphing on a system with three axes. This means we solve a 3 × 3 system algebraically using elimination or substitution. Both of these techniques can be useful, but usually the elimination method is the most efficient. Therefore, this is the method we will demonstrate in the next example.

Example 5

Solve the following 3 × 3 linear system of equations using the elimination method.

$$\begin{cases} 2x + 5y + z = {}^-1.5 \\ {}^-3x + 2y + 2z = {}^-4.4 \\ 10x + 4y - 3z = 5.2 \end{cases}$$

Solution   To use the elimination method on a 3 × 3 linear system, we start by choosing a variable to eliminate. It does not matter which variable. In this example, let's start by eliminating the variable z, because the coefficient of z in the first equation is 1. We need to choose two equations to use (it does not matter which two). Assume the equations are labeled Ⓐ, Ⓑ, and Ⓒ from top to bottom. Let's eliminate z by using equations Ⓐ and Ⓑ. To do this we need to multiply equation Ⓐ by $^-2$ and then add this to equation Ⓑ. We label the resulting equation with the letter Ⓓ.

$$\begin{array}{rl} {}^-2 \text{ Ⓐ} & \begin{cases} {}^-4x - 10y - 2z = 3 \\ {}^-3x + 2y + 2z = {}^-4.4 \end{cases} \\ \text{Ⓓ} & {}^-7x - 8y \qquad\;\; = {}^-1.4 \end{array}$$

Equation Ⓓ contains two variables. If we can create a second equation with the same two variables, we will have a 2 × 2 system that we know how to solve. To create this second equation, we must again eliminate the variable z but use a different pair of equations. We can either choose the pair Ⓐ and Ⓒ or the pair Ⓑ and Ⓒ. Let's use Ⓐ and Ⓒ. We multiply equation Ⓐ by 3 and add it to equation Ⓒ. We label the resulting equation Ⓔ.

$$\begin{array}{rl} 3 \text{ Ⓐ} & \begin{cases} 6x + 15y + 3z = {}^-4.5 \\ 10x + 4y - 3z = 5.2 \end{cases} \\ \text{Ⓔ} & 16x + 19y \qquad\; = 0.7 \end{array}$$

Equations Ⓓ and Ⓔ together make a 2×2 linear system. We can solve this system using the elimination method as well.

$$\begin{array}{l} Ⓓ \\ Ⓔ \end{array} \left\{ \begin{array}{l} {}^-7x - 8y = {}^-1.4 \\ 16x + 19y = 0.7 \end{array} \right.$$

Let's eliminate the variable x.

$$\begin{array}{l} 16\ Ⓓ \\ 7\ Ⓔ \end{array} \left\{ \begin{array}{l} {}^-112x - 128y = {}^-22.4 \\ \underline{112x + 133y = 4.9} \\ 5y = {}^-17.5 \end{array} \right.$$

We can solve this equation for y.

$$5y = {}^-17.5$$
$$y = {}^-3.5$$

To finish solving the 2×2 system we need to find the value for x. We can use either equation Ⓓ or Ⓔ.

$$Ⓓ \quad {}^-7x - 8 * {}^-3.5 = {}^-1.4$$
$$-7x + 28 = {}^-1.4$$
$$-7x = {}^-29.4$$
$$x = 4.2$$

This means the solution to the 2×2 system consisting of equations Ⓓ and Ⓔ is $x = 4.2$ and $y = {}^-3.5$. But our goal was to solve the 3×3 system. To do this we need to find the value for z. Because we have values for x and y, we can substitute these values into any of the equations that contain all three variables. Let's use equation Ⓐ.

$$2 * 4.2 + 5 * {}^-3.5 + z = {}^-1.5$$
$$8.4 - 17.5 + z = {}^-1.5$$
$$-9.1 + z = {}^-1.5$$
$$z = 7.6$$

We found that $x = 4.2$, $y = {}^-3.5$, and $z = 7.6$. Now we must check our solution. The solution to a 3×3 system must satisfy all three of the original equations. Therefore, to check this solution we must substitute our values into all three equations.

CHECK

$$Ⓐ \quad 2 * 4.2 + 5 * {}^-3.5 + 7.6 = {}^-1.5$$
$$-1.5 = {}^-1.5 \quad ✔$$

$$Ⓑ \quad {}^-3 * 4.2 + 2 * {}^-3.5 + 2 * 7.6 = {}^-4.4$$
$$-4.4 = {}^-4.4 \quad ✔$$

$$Ⓒ \quad 10 * 4.2 + 4 * {}^-3.5 - 3 * 7.6 = 5.2$$
$$5.2 = 5.2 \quad ✔$$

The values satisfy all three equations, so we know that $x = 4.2$, $y = {}^-3.5$, and $z = 7.6$ is the solution to this system. This can also be written as an ordered triple, $(4.2, {}^-3.5, 7.6)$.

As you might imagine, the technique used in Example 5 can be used to solve 4×4 linear systems and higher. On a 4×4 linear system we start by eliminating one variable using three different pairs of equations to create a 3×3 system. Then choose two pairs of equations, and eliminate a second variable to create a 2×2 system, and so on.

We learned three techniques for solving **2 × 2 linear systems of equations:** one graphical method and two algebraic methods. **Solving a system of equations** means finding values of the variables that simultaneously satisfy all equations in the system. The **graphical method** involves graphing the two linear equations and looking for the point of intersection. The other two methods we explored were algebraic. Both the **elimination method** and the **substitution method** give us techniques to reduce the two equations in two variables to a single equation in one variable that we can easily solve. It is important to remember that *checking* a system of equations requires that we substitute the values for the variables into both equations. An ordered pair that works in one equation and not the other is not a solution to the system. How would such an ordered pair show up in a graphical representation of the system?

We also used the elimination method to solve a 3 × 3 linear system. After reading through and doing Example 5, you might imagine that in the process of solving a 3 × 3 system many arithmetic errors are possible. For this reason, it is a good idea to proceed slowly, write down all of your steps carefully, and don't confuse the letter z with the number 2. It can be helpful to put a line through the middle of your z's to distinguish them from your 2's.

Problem Set 2.4

1. For each part, check to see whether $x = {}^-2$ and $y = 5$ is a solution.

   a. $\begin{cases} 2x - 3y = {}^-19 \\ 5x + 2y = 0 \end{cases}$

   b. $\begin{cases} y = {}^-3x - 1 \\ 5x - 2y = 11 \end{cases}$

2. For each part, check to see whether $x = 3$ and $y = {}^-2.1$ is a solution.

   a. $\begin{cases} x + y = 5.1 \\ 4x - 3y = 18.3 \end{cases}$

   b. $\begin{cases} y = 3x - 11.1 \\ x = 2.1y + 7.41 \end{cases}$

3. Solve each 2 × 2 linear system graphically. Sketch the graph used to solve the system.

   a. $\begin{cases} y = 2x - 7 \\ x + y = 8 \end{cases}$

   c. $\begin{cases} 1.5x - 3y = 45 \\ y = 1.5x - 25 \end{cases}$

   b. $\begin{cases} 0.8x + y = 4 \\ y = 2x - 3 \end{cases}$

   d. $\begin{cases} 9x - y = 21 \\ 3x + 4y = 150 \end{cases}$

4. Solve each 2 × 2 linear system algebraically using either the elimination or the substitution method. Numerically check all solutions.

   a. $\begin{cases} y = 2x - 5 \\ x + y = 7 \end{cases}$

   d. $\begin{cases} 1.5x - 3y = 45 \\ y = 1.5x - 25 \end{cases}$

   b. $\begin{cases} 4x - 5y = 9 \\ 2x + 3y = {}^-12 \end{cases}$

   e. $\begin{cases} y = 2.5x + 7.5x \\ y = 6.5x + 30 \end{cases}$

   c. $\begin{cases} 9x - y = 21 \\ 3x + 4y = 150 \end{cases}$

   f. $\begin{cases} 1.5x - y = 45 \\ y = 1.5x + 25 \end{cases}$

5. Which of the following graphs can be a graph of the system shown? Explain how you arrived at your decision.

$$\begin{cases} y = \frac{3}{5}x - 27 \\ y = \frac{3}{5}x - 52 \end{cases}$$

a. 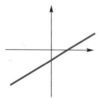   **b.**    **c.**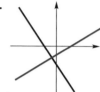

6. Suppose you are going to solve each of the following systems algebraically. Decide whether to solve the system using the elimination method or substitution method. Explain your reasoning. You do *not* need to solve these systems.

a. $\begin{cases} 3x + y = 12 \\ 5x - 2y = 20 \end{cases}$ **c.** $\begin{cases} -3x + 10y = 56 \\ 4x + 2y = 2 \end{cases}$

b. $\begin{cases} 4.5x + 6.5y = 2902 \\ y = 500 - x \end{cases}$ **d.** $\begin{cases} -x + 5y = 36.25 \\ x = 16.25 - 2y \end{cases}$

7. In Problem 3, you solved each system graphically. In Problem 4, you solved systems algebraically. In this problem you are to solve each 2 x 2 linear system by choosing one method, either graphical or one of the two algebraic methods. Your experience in Problems 3 and 4 may help you decide when to choose a particular method. Regardless of what method you choose, check all solutions. Round any approximate solutions to the nearest hundredth.

a. $\begin{cases} 2x + y = 6 \\ x - 3y = 10 \end{cases}$ **f.** $\begin{cases} x = 3x - 2y + 25 \\ y = x - 3y + 14 \end{cases}$

b. $\begin{cases} 7x - 3y = 4 \\ 5x + 4y = 52 \end{cases}$ **g.** $\begin{cases} 2x - 15y = 180 \\ y - 6x = 125 \end{cases}$

c. $\begin{cases} 2x - 105 = y \\ x = 6 - 0.5y \end{cases}$ **h.** $\begin{cases} 0.25x + 1.65y = 42 \\ x + y = 200 \end{cases}$

d. $\begin{cases} -x + 5y = 19 \\ x = 2.1y + 11.45 \end{cases}$ **i.** $\begin{cases} 3.5x - 4.4y = 8.6 \\ 7.2x - 2.9y = 17.0 \end{cases}$

e. $\begin{cases} x + y = 20 \\ 0.25x + 0.75y = 7 \end{cases}$

8. Solve the following systems of equations using either the graphical method or one of the two algebraic methods. Round approximate solutions to the hundredths place. Numerically check all solutions.

a. $\begin{cases} 0.75x + 0.25y = 293.75 \\ 0.80x + 0.20y = 280 \end{cases}$ **c.** $\begin{cases} A = 1.25B + 32 \\ A = 0.8B - 25.6 \end{cases}$

b. $\begin{cases} y = 3x + 50 \\ 2x + 3y = 105 \end{cases}$

9. Solutions to systems of three equations in three variables can require several steps. Follow the steps in a through f to determine an ordered triple (x, y, z) that is a solution to the following system.

$$\begin{cases} x + 2y - 3z = 8 \\ 3x - 4y + z = {}^-1 \\ {}^-x + 10y + 4z = 9 \end{cases}$$

a. Using the first two equations, use the method of elimination to eliminate the terms containing the variable x.

b. Choose a different pair of equations from the original set of three, and use the method of elimination to eliminate the terms containing the variable x.

c. Using the method of elimination or substitution, solve the 2×2 system formed from the resulting equations in part a and part b. Solve this system of two equations to find values for y and z.

d. Substitute your solutions for y and z, into any *one* of the original equations and then solve for x.

e. Check your solution in all three of the original equations.

f. Could you have solved this system of equations by eliminating y or z in the first two steps?

10. Solve each 3×3 linear system algebraically. Check your solution.

a. $\begin{cases} x - 2y + z = 8 \\ 3x + y - 2z = 5 \\ 2x - y + 3z = 11 \end{cases}$ b. $\begin{cases} 3x - y + 2z = 5 \\ 2x + 3y + z = 1 \\ 5x + y + 4z = 8 \end{cases}$ c. $\begin{cases} 3x + y = 16 \\ 2x + z = 15 \\ x - 2y + 2z = 16 \end{cases}$

11. The Paint Sprayer.   The Radical Rental Emporium charges $10.00 plus 75¢ per hour to rent a paint sprayer. A competitor, Truly Temporary Equipment Service, rents the same sprayer for $2.95 plus $1.22 per hour.

a. Write an equation for the cost in terms of the number of rental hours for each business.

b. Determine algebraically and graphically the number of hours for which the cost of renting a paint sprayer is the same from Radical Rental or Truly Temporary.

c. If you need to rent a paint sprayer how will you decide which company to rent from?

12. The Investment.   Cathy just won $10,000 in the Oregon state lottery. She wants to put some of the money in an account where it is accessible and to commit the rest to a long-term investment. She has been considering a savings account that has been paying an annual percentage rate (APR) of 2.64% and a mutual fund that has been paying an APR of 8.9%. At the end of one year, Cathy hopes to earn $750. How much money must she invest in each type of account?

a. Model the situation algebraically.

b. Determine the solution algebraically and graphically.

c. Is there any way that Cathy can invest the money to earn $1000 in interest in one year? Explain how you arrived at your conclusion.

13. TGIF.   Chris and Terry have been trying to live within their budget but miss going out to dinner on Friday nights. They decide not to spend any nickels or quarters they receive as change for a month and save these in a jar. At the end of the second week they have a total of 90 coins in the jar. The value of the coins totals $15.10. How many nickels and how many quarters are in the jar?

14. The Boat Collision. A sailboat and a speedboat collided recently. The speedboat was returning from a picnic on an island. The island is located 5.2 miles east and 12.6 miles north of the dock. The speedboat was heading from the island toward a point that is 3.0 miles west of the dock. At the same time the sailboat left the dock heading toward a buoy that is 5.0 miles west and 7.5 miles north of the dock.

   a. Using the dock as the origin, sketch this situation.

   b. Write an equation for the path of each boat.

   c. Determine the point of collision algebraically and graphically.

15. Chance Meeting. Jody and Cindy went on vacation near a resort. Jody is staying in the lodge at the resort. Cindy is camping at a site that is located 4 miles east and 15 miles north of the lodge. Early on the second morning of their vacation, Jody decided to go for a bike ride to a lake nearby. The lake is located 12 miles east and 18 miles north of the lodge. On that same morning, Cindy went for a run to the local store. The store is located 10 miles due east of the lodge. During the course of the morning Jody and Cindy ran into each other.

   a. Using the lodge as the origin, sketch this situation.

   b. Write an equation for the path of Jody's bike ride and Cindy's run.

   c. Determine the point where Jody and Cindy meet.

16. The Electron. An electric current is running through a wire along the line $y = -2x + 11$. An electron at $(-3, -10)$ is attracted toward the wire and moves toward the closest point on the wire (that is, it moves along a line perpendicular to the wire).

   a. Draw the line $y = -2x + 11$ that represents the wire.

   b. Write the equation of the line along which the electron travels.

   c. Determine the coordinates of the point where the electron hits the wire.

   d. What is the distance the electron travels (along the line) before hitting the wire?

17. Jazz Concert. Last spring, several of the Mt. Hood Community College (MHCC) jazz groups held a joint concert. Tickets sold for $5.50 to the general public and for $3.00 for students and staff. Ticket sales brought in $1607.50. If 390 people attended the concert, how many were MHCC students and staff, and how many were from the general public?

18. Chemistry Lab. The chemistry lab aide has been told that tomorrow's lab experiments will require 2 liters of 17% sulfuric acid. When she checks the storeroom she finds only 10% solution and 30% solution of sulfuric acid in stock. She realizes that she can mix these two to get what is needed. How much of the 10% acid and how much of the 30% acid should she mix together to end up with 2 liters of a 17% sulfuric acid solution?

19. Classified Ads. From the classified section of a newspaper, select one particular brand and model of used car that appears several times. Collect data on the asking price in terms of the age (number of years old) of the vehicle. You should have at least eight data points.

   a. Graph the asking price of the car in terms of the age of the car (Is the car 2 years old? 5 years old? and so on). Carefully label your graph.

   b. Draw an eyeball fit line through the data and write the equation for this line.

   c. What are the units associated with the slope of your line? What does the slope tell you about the problem situation?

   d. What does the vertical intercept tell you about the problem situation?

   e. Your data points probably do not lie in a straight line, but most, if not all, are relatively close to the line you drew. Are all of your data points relatively close to the line you drew, or do you have any data points that are extremely far from your line? If you have a data point that is extremely far from your line, explain why you think it is so far from your other points.

20. The following table shows some of the input–output pairs for a 2×2 system of linear functions. Use the table to answer the following questions about the two linear functions. You do not need to write the equations for the functions.

a. Is the ordered pair $(1, {}^-2)$ a solution to the first function?

b. Is the ordered pair $(1, {}^-2)$ a solution to the second function?

c. What is the solution to the system of equations?

Input	Function 1 Output	Function 2 Output
$^-2$	$^-11$	14
$^-1$	$^-8$	12
0	$^-5$	10
1	$^-2$	8
2	1	6
3	4	4
4	7	2
5	10	0
6	13	$^-2$

21. The following tables show some input–output pairs for a 2×2 system of linear functions. Use the tables to find the solution to the system . You do not need to write the equations for the functions.

FUNCTION 1		FUNCTION 2	
Input	Output	Input	Output
24.5	29	24.5	25.333
25	28	25	24.667
25.5	27	25.5	24.000
26	26	26	23.333
26.5	25	26.5	22.667
27	24	27	22.000
27.5	23	27.5	21.333
28	22	28	20.667
28.5	21	28.5	20.000
29	20	29	19.333
29.5	19	29.5	18.667
30	18	30	18.000
30.5	17	30.5	17.333
31	16	31	16.667
31.5	15	31.5	16.000
32	14	32	15.333
32.5	13	32.5	14.667

Chapter Two Summary

In this chapter we studied a specific class of equations in two variables called **linear equations.** We learned how to identify a linear relationship from a table of values and from an equation, we learned how to graph a linear equation, and we learned how to write equations for linear relationships.

From a table of values, we can determine whether a relationship graphs as a line by looking at the change in output and change in input. If the change in output is constant for a constant change in input, then the data graph as a line. Another way to look at this relationship is to look at the ratio

$$\frac{\text{change in output}}{\text{change in input}}$$

Linear Relationships—Numerically

A table of data graphs as a line if and only if the ratio

$$\frac{\text{change in output}}{\text{change in input}}$$

is constant for all pairs of data in the table.

Because this ratio is constant for any line, we call this ratio the **slope** of the line. On a graph, we refer to this ratio as

$$\frac{\text{vertical change}}{\text{horizontal change}} \quad \text{or} \quad \frac{\text{rise}}{\text{run}}$$

The slope, together with its units, can tell us important information about a problem situation.

Linear Relationships—Algebraically

Algebraically, we found that linear equations are seen in two forms:

• Slope–intercept form of a linear equation

$y = mx + b$

where $m = $ slope

$(0, b) = $ vertical intercept

• General form of a linear equation

$Ax + By = C$

where A and B cannot both be zero.

Because these are the only two simplified forms of a linear equation, we know that any equation that does not fit at least one of these two forms does not graph as a line. This means that if the variables in an equation are raised to any power other than 1, then the equation is not linear. If the variable occurs in the denominator or under a radical sign, the equation is not linear. However, the numerical coefficients may include powers, roots, or fractions.

With this information, we are able to inspect an equation to determine whether the equation graphs as a line. When we determine that an equation graphs as a line, we only need to plot two points, because two points determine a line.

When starting with an equation, we learned three options for determining two possible points through which to draw our line.

Option 1   Make a table of values. Again we only need two points if we know the graph is a line.

Option 2   Use the slope and intercept. Start by plotting the vertical intercept. Then, determine a second point by using the slope to move up or down and over the appropriate amount.

Option 3   Algebraically determine both the vertical intercept and the horizontal intercept. The vertical intercept can be found by substituting $x = 0$ into the equation and solving for y. Similarly, the horizontal intercept can be found by substituting $y = 0$ into the equation and solving for x.

The method we choose to graph a linear equation depends on the form of the equation, the magnitude of the numbers in the equation, and our preference on methods.

If the equation we are trying to graph is not linear, then we can plot several points to get an idea of how the graph looks. As you proceed in your mathematics courses, you will learn more about different classes of equations and their graphs.

In this chapter, we saw how to write the equation of a line if we know or can determine the vertical intercept and slope of a line. We can write its equation by substituting these values into the slope–intercept equation. The information we use to determine the vertical intercept and slope may be given to us in graphical form or in an application. If data is given in an application, it is sometimes helpful to graph the data or put the data in a table, before trying to write an equation.

In Section 2.3, we extended our knowledge of how to write linear equations so that we could write equations for lines for which we do not know the vertical intercept. In that section we learned how to write the equation of a line given any two points or given the slope and one point.

Strategy

> *To write the equation of a line given two points,*
>
> 1. Find the slope of the line using the two points given.
> 2. Substitute the value for the slope in the slope–intercept equation.
> 3. Find the value of b by substituting the coordinates of one of the given points into the slope–intercept equation. Solve the resulting equation for b.
> 4. Write the equation of the line by substituting the value for m and b into the slope–intercept equation.
> 5. Check the equation by substituting the coordinates of the other point into the equation.

To write the equation of a line given the slope and one point, you follow the same process as shown in the box except you can skip step 1 because you already know the slope.

Many situations are best modeled using two variables. If two variables are required, then we need two equations in order to solve the situation. In Section 2.4 we looked at 2×2 linear systems of equations. Three methods can be used to solve a 2×2 linear system. We can solve a 2×2 linear system graphically, or algebraically using either the elimination method or the substitution method.

To **solve a 2×2 linear system graphically,** we graph both of the equations on a calculator. It is helpful to consider the slope and vertical intercept of each equation to determine a reasonable window. The intersection of the two graphs is the solution to the system. We can check our solution by substituting the values for both variables into the *two original equations.*

To solve a 2×2 linear system using the **elimination method,** we choose factors to multiply one or both equations so that when the equations are added, one of the variables is eliminated. This produces an equation with only one variable. We solve this equation. Substitute the value for this variable into any of the equations containing two variables to determine the value of the other variable. Once we have the value of both variables we check our solution by substituting these values into the *two original equations.*

To solve a 2×2 linear system using the **substitution method,** we solve one of the equations for one of the variables. Then we substitute the resulting expression into the other equation. This produces an equation with only one variable. We solve this equation and proceed as we did in the elimination method.

Additionally, we applied the elimination method to solve a 3×3 linear system. Although this takes many more steps than needed for solving a 2×2 system, the basic processes remain the same.

Chapter Three

Integer Exponents and Probability

We know that exponents are used frequently in mathematics. In this chapter, we will look at integer exponents and how to use these exponents with signed number bases. The properties of exponents will be discussed and used to simplify expressions.

In addition, we will begin a discussion of probability and counting techniques. These can be used to indicate the likelihood of the occurrence of an event. We will see that exponents can play a role in some counting techniques.

3.1 Integer Exponents

Activity
Set
3.1

1. Simplify the following expressions. Verify your results.
   a. $k^3 k^4$
   b. $4x * 5x^3$

2. Simplify the following expressions. Verify your results.
   a. $R(R^2 - R^3)$
   b. $3w^4(8 - 5w^2) + w^6$

3. You may or may not have used properties (rules) of exponents to simplify the expressions in Activities 1 and 2. However, write the rule for each of the following expressions. If you are not sure of the rule, make up an example, and write out what it means.
   a. $x^m x^n$ (*Hint:* Try $x^2 x^3$ if you are not sure of the rule.)
   b. $\dfrac{x^m}{x^n}$
   c. $(xy)^m$
   d. $\left(\dfrac{x}{y}\right)^m$
   e. $(x^m)^n$

4. *Use the previous properties* to simplify each of the following expressions. Verify your results.
   a. $x^5(x^8)$
   b. $(k^{10})^5$
   c. $(5x)^3$
   d. $p^4 * p^4$
   e. $\left(\dfrac{a^2}{b}\right)^3$
   f. $(x + y)^2$
   g. $3x^5 * 7x^8$
   h. $2w^4 + (2w)^4$
   i. $\left(\dfrac{6m^3}{mn}\right)^2$

5. In biology we often study bacteria that reproduce by splitting. In ideal situations, a population can double over a specific period of time. Suppose we are studying a population of bacteria that can double every hour. It has been determined that Monday at 9:00 A.M. 800 bacteria are in the population.
   a. How many bacteria are in the population Monday at 10:00 A.M.? 11:00 A.M.? How many bacteria were in the population at 8:00 A.M.? 7:00 A.M.?
   b. Complete a table similar to the following one for this situation.

Number of Hours After 9:00 A.M.	-3	-2	-1	0	1	2	3	24
Number of Bacteria in Population				800				

   c. If n represents the number of hours after 9:00 A.M. and P represents the number of bacteria in the population, write an equation for P in terms of n.
   d. Check your equation for $n = 2, 3, -2,$ and -3.
   e. Substitute $n = 0$ into your equation in part c. What is the value of 2^0?
   f. What fraction is equivalent to 2^{-1}? 2^{-2}? 2^{-3}?
   g. Write a rule for x^{-n}, where x is a nonzero real number and n is a positive integer.

6. Simplify each expression using the properties of integer exponents. Verify by numerical substitution.

   a. $k^3 * k^8$ **b.** $m^7 * m^{-2}$

7. Simplify each expression using the properties of integer exponents. Verify your result.

   a. $\dfrac{x^8}{x^3}$ **b.** $\dfrac{p^{10}}{p^{-6}}$

8. Simplify each expression using the properties of integer exponents. Verify your result.

   a. $(x^3)^5$ **b.** $(k^{-2})^5$

9. Simplify each expression using the properties of integer exponents. Verify your result.

   a. $(2y^4)^3$ **b.** $(3m^4)^{-2}$

10. Use your calculator to evaluate each expression. Try to guess the value each time before entering it into the calculator.

    $$\frac{1}{2^{-2}} \quad \frac{1}{3^{-2}} \quad \frac{1}{4^{-2}} \quad \frac{1}{5^{-2}} \quad \frac{1}{2^{-1}} \quad \frac{1}{2^{-2}} \quad \frac{1}{2^{-3}} \quad \frac{1}{2^{-4}}$$

    Write a rule for $\dfrac{1}{x^{-n}}$, where x is a nonzero real number and n is a positive integer.

Discussion 3.1

In this section, we will be working with expressions involving integer exponents. Emphasis will be placed on reading the expressions in a way that allows us to interpret their meaning precisely. Additionally, we will review the properties of exponents. These rules allow us to simplify expressions with exponents. The knowledge you acquire in this section will be important later when we extend the definition of exponents to include fractional powers.

In the activities, you may have discovered how to interpret zero as an exponent. The expression $800 * 2^0$ represents the number of bacteria in the population 0 hours after 9:00 A.M., that is, at 9:00 A.M. We know that at 9:00 A.M. 800 bacteria were in the population, that means 2^0 must equal 1. Therefore, we have the following definition.

> **Definition** _____
>
> If x is a nonzero real number, then $x^0 = 1.$

You may have noticed that in the definition, x must be a nonzero number. What if we want to know what the expression 0^0 equals? Let's see if looking at some patterns can help us decide. Using the preceding definition, we can create a pattern of numbers raised to the zero power.

$$3^0 = 1$$
$$2^0 = 1$$
$$1^0 = 1$$

Following this pattern, we might think that $0^0 = 1$. But, suppose we look at a pattern of powers where the base is zero.

$$0^3 = 0$$
$$0^2 = 0$$
$$0^1 = 0$$

Following this pattern, we might conclude that $0^0 = 0$. Because both of these patterns are reasonable and yet the results are different, we say that 0^0 is undefined.

Definition

The expression $\mathbf{0^0}$ is undefined.

You may have also discovered the definition for negative exponents while working through the activities. Let's see how the properties of exponents can be used to arrive at the same definitions. Consider the expression $2^1 2^{-1}$. The properties of exponents should hold for negative exponents, so

$$2^1 * 2^{-1}$$
$$= 2^{1 + {}^-1} \qquad \text{When multiplying powers of 2, add the exponents.}$$
$$= 2^0$$
$$= 1$$

We see that $2^1 * 2^{-1} = 1$. This means that 2^{-1} must be equal to $\frac{1}{2}$, because $2 * \frac{1}{2} = 1$. We say that 2 and $\frac{1}{2}$ are multiplicative inverses because their product is 1. We conclude that $2^{-1} = \frac{1}{2}$.

Using this idea and the property, $(x^m)^n = x^{m*n}$, $2^{-3} = (2^{-1})^3 = \left(\frac{1}{2}\right)^3$. That is, the negative third power of two is the cube of the reciprocal of two. This motivates the following definition.

Definition

If n is a positive integer and x is a nonzero real number, then

$$x^{-n} = \left(\frac{1}{x}\right)^n$$

In words we would say that x^{-n} is the nth power of the reciprocal of x.

If we then apply the property

$$\left(\frac{x}{y}\right)^m = \frac{x^m}{y^m}$$

to this result we obtain

$$x^{-n} = \left(\frac{1}{x}\right)^n = \frac{1^n}{x^n} = \frac{1}{x^n}$$

That is, x^{-n} is also the reciprocal of the nth power of x.

Definition

The expression x^{-n} can be defined either as the nth power of the reciprocal of x or as the reciprocal of the nth power of x. That is, if x is a nonzero real number,

$$x^{-n} = \left(\frac{1}{x}\right)^n = \frac{1}{x^n},$$

where x is a nonzero real number.

Again, notice that the definition requires x to be a nonzero real number. If x were 0, then $\frac{1}{x}$ is undefined since this is division by zero. Now that we know what zero and negative exponents mean, we need to practice careful reading and evaluating of expressions. Remember that exponents apply only to what is to their immediate left in an expression.

Example 1

Evaluate each expression without using your calculator.

   a. 5^{-2} b. -5^2 c. $(-5)^{-2}$ d. -5^{-2}

Solution   a. The expression 5^{-2} can be read as *the negative second power of five*. The negative second power of 5 can be rewritten as the reciprocal of the second power of 5, $5^{-2} = \frac{1}{5^2}$. Then $\frac{1}{5^2} = \frac{1}{25}$. Therefore,

$$5^{-2}$$

$$= \frac{1}{5^2} \qquad \text{Rewrite the negative second power of 5 as the reciprocal of the second power of 5.}$$

$$= \frac{1}{25}$$

b. The expression -5^2 can be read as *the opposite of the second power of five*. Remember that the power does not apply to the negative sign if it is not included inside parentheses. The square of 5 is 25 and the opposite of the square of 5 is -25. Therefore,

$$-5^2 = -25$$

c. The expression $(-5)^{-2}$ can be read as *the negative second power of negative five*. The negative second power of -5 can be rewritten as the reciprocal of the second power of -5, $(-5)^{-2} = \frac{1}{(-5)^2}$. Then $\frac{1}{(-5)^2} = \frac{1}{25}$. Therefore,

$$(-5)^{-2}$$

$$= \frac{1}{(-5)^2} \qquad \text{Rewrite the negative second power of } {}^-5 \text{ as the reciprocal of the second power of } {}^-5.$$

$$= \frac{1}{25} \qquad \text{The square of } {}^-5 \text{ is 25.}$$

d. The expression -5^{-2} can be read as *the opposite of the negative second power of five*.

$$-5^{-2}$$

$$= -\frac{1}{5^2} \qquad \text{Notice that the power applies only to the number 5 and not to the negative sign.}$$

$$= -\frac{1}{25}$$

Example 2

a. Is the expression $8y^{-4}$ positive or negative when $y = {}^-10$? Explain.

b. Evaluate the expression $8y^{-4}$ when $y = {}^-10$. Show your substitution.

Solution

a. Rewriting the expression $8y^{-4}$ helps us decide whether it is positive or negative when $y = {}^-10$. In the expression $8y^{-4}$, the power applies only to y, so the expression $8y^{-4}$ can be read as *eight times the negative fourth power of y*. The negative fourth power of y can be rewritten as the reciprocal of the fourth power of y, $y^{-4} = \frac{1}{y^4}$. Therefore,

$$8y^{-4}$$
$$= 8 * \frac{1}{y^4}$$
$$= \frac{8}{y^4}$$

When $y = {}^-10$, this expression is the quotient of 8 and the fourth power of $^-10$. The fourth power of a negative number is positive, and the quotient of 8 and a positive number is positive. Therefore, $8y^{-4}$ is positive when $y = {}^-10$.

b. $$8y^{-4}$$
$$= 8 * ({}^-10)^{-4} \qquad \text{Remember when raising a negative number to a power, the number } must \text{ be written in parentheses.}$$
$$= 0.0008$$

The result is positive, which is what we expected from part a.

In the activities, you may have discovered or rediscovered several rules or properties that apply to integer exponents. We will now list these properties. As you apply the properties, it is important to understand what each one says and how it is different from the others. Reading the verbal description of each property, as well as the symbolic description, should make it easier to decide which property is appropriate for a given situation.

You might be able to simplify the expressions in this section by just using their meaning. However, it is important now to concentrate on learning the properties and to practice using them to simplify expressions. This will be helpful in later sections when we extend the concepts of exponents to include fractional powers.

Properties of Integer Exponents

Let x and y be nonzero real numbers. Let n and m be integers. Then the following are true.

1. $x^m x^n = x^{m+n}$ When *multiplying powers of* x, add the exponents.

2. $\dfrac{x^m}{x^n} = x^{m-n}$ When *dividing powers of* x, subtract the exponents.

3. $(xy)^m = x^m y^m$ The *power of a product* can be simplified by applying the power to each factor.

4. $\left(\dfrac{x}{y}\right)^m = \dfrac{x^m}{x^n}$ The *power of a quotient* can be simplified by applying the power to each factor in the numerator and each factor in the denominator.

5. $(x^m)^n = x^{m*n}$ When simplifying a *power of a power of* x, multiply the exponents.

We can select from five different properties as we simplify expressions with exponents. Carefully reading the statements on the right should help in deciding which property to apply. Some people create their own *incorrect* rule that says if parentheses are present in an expression, then the exponents should be multiplied. Consider the following four expressions.

$$B^3 * B^4$$
$$B^3 B^4$$
$$B^3 \cdot B^4$$
$$B^3(B^4)$$

All four expressions represent the *product* of B^3 and B^4. When we multiply powers of B we need to add the exponents. Therefore, all four of the expressions simplify to $B^{3+4} = B^7$. Notice that the last expression is written with parentheses, yet we still add the exponents.

We only multiply the exponents when we are simplifying a *power of a power*. A power of a power, like $(B^3)^4$, always contains parentheses and simplifies to B^{3*4}. However, other expressions that are not powers of powers may also contain parentheses.

We use these properties to simplify expressions containing exponents. When we say **simplify** an expression, we mean to perform all of the indicated numerical and algebraic operations. Perform the operations within each term as it is read, and then add or subtract like terms. When we simplify expressions that include numerical values, we may obtain approximate results; in these cases we continue to use the correct notation for approximately equal in our simplification process.

Example 3

Simplify each of the following expressions. Express your answers using positive exponents only. Assume all variables represent nonzero numbers. Numerically verify your results.

a. $w^6 * w^{-10}$

c. $2x^3(8x^{-5})$

e. $(4B^3)^{-2}$

b. $y^3(y^{-4})(y)$

d. $\dfrac{m^4}{m^{-8}}$

f. $\dfrac{3}{x^{-2}} - (3x)^2$

Solution a.

$$w^6 * w^{-10}$$

$$= w^{6+{-10}} \qquad \text{When multiplying powers of } w, \text{ add the exponents.}$$

$$= w^{-4}$$

$$= \frac{1}{w^4} \qquad \text{Rewrite } w^{-4} \text{ as the reciprocal of the fourth power of } w.$$

VERIFY: Let $w = 5$.

$$w^6 * w^{-10} \overset{?}{=} \frac{1}{w^4} \qquad \text{Is the original expression equal to our simplified expression?}$$

$$5^6 * 5^{-10} \overset{?}{=} \frac{1}{5^4} \qquad \text{Substitute } w = 5.$$

$$0.0016 = 0.0016 \qquad ✔ \quad \text{Evaluate each side.}$$

We conclude that $w^6 * w^{-10}$ simplifies to $\frac{1}{w^4}$.

b.

$$y^3(y^{-4})(y)$$

$$= y^{3+{-4}+1} \qquad \text{When multiplying powers of } y, \text{ add the exponents. If no exponent is given, then}$$
$$\qquad\qquad\quad \text{it is 1, that is, } y = y^1.$$

$$= y^0$$

$$= 1$$

VERIFY   Let $y = 8$.

$$y^3(y^{-4})(y) \overset{?}{=} 1$$
$$8^3(8^{-4})(8) \overset{?}{=} 1$$
$$1 = 1 \quad \checkmark$$

We conclude that $y^3(y^{-4})(y)$ simplifies to 1.

c. $2x^3(8x^{-5})$

$= 16 * x^{3 + -5}$ Multiply the numerical coefficients, and add exponents.

$= 16x^{-2}$

$= \dfrac{16}{x^2}$ Rewrite with a positive exponent. Note that the exponent applies only to x. Therefore, 16 remains in the numerator.

VERIFY   Let $x = 6$.

$$2x^3(8x^{-5}) \overset{?}{=} \frac{16}{x^2}$$

$$2 * 6^3(8 * 6^{-5}) \overset{?}{=} \frac{16}{6^2}$$

$$0.4444 = 0.4444 \quad \checkmark$$

We conclude that $2x^3(8x^{-5})$ simplifies to $\frac{16}{x^2}$.

d. $\dfrac{m^4}{m^{-8}}$

$= m^{4 - -8}$ When dividing powers of m, subtract the exponents.

$= m^{12}$

VERIFY   Let $m = 3$.

$$\frac{m^4}{m^{-8}} \overset{?}{=} m^{12}$$

$$\frac{3^4}{3^{-8}} \overset{?}{=} 3^{12}$$

$$531{,}441 = 531{,}441 \quad \checkmark$$

We conclude that $\frac{m^4}{m^{-8}}$ simplifies to m^{12}.

e. $(4B^3)^{-2}$

$= 4^{-2}(B^3)^{-2}$ Apply the exponent to each of the factors.

$= 4^{-2}B^{-6}$ Multiply the exponents.

$= \dfrac{1}{4^2 B^6}$ Rewrite with positive exponents.

$= \dfrac{1}{16B^6}$

VERIFY   Let $B = 5$.

$$(4B^3)^{-2} \overset{?}{=} \frac{1}{16B^6}$$

$$(4 * 5^3)^{-2} \overset{?}{=} \frac{1}{16 * 5^6}$$

$$0.000004 = 0.000004 \quad \checkmark$$

We conclude that $(4B^3)^{-2}$ simplifies to $\frac{1}{16B^6}$.

You could also simplify the step $4^{-2}B^{-6}$ to $0.0625B^{-6}$. Then rewriting the negative power, $0.0625B^{-6}$ is written as $\dfrac{0.0625}{B^6}$.

f. $\dfrac{3}{x^{-2}} - (3x)^2$

$= 3x^2 - 3^2x^2$ Rewrite with positive exponents in the first term, and apply the power to each of the factors in the second term.

$= 3x^2 - 9x^2$

$= {}^-6x^2$ Combine like terms.

VERIFY   Let $x = 4$.

$$\dfrac{3}{x^{-2}} - (3x)^2 \overset{?}{=} {}^-6x^2$$

$$\dfrac{3}{4^{-2}} - (3*4)^2 \overset{?}{=} {}^-6*4^2$$

$${}^-96 = {}^-96 \quad \checkmark$$

We conclude that $\dfrac{3}{x^{-2}} - (3x)^2$ simplifies to ${}^-6x^2$.

It is important to note that when we numerically verify a result, we are showing that for a specific value, the original expression and the result are equivalent. This process does not show if the expression is simplified as far as possible. For example, in part f of Example 3, if we had not combined the like terms it would still verify that $\dfrac{3}{x^{-2}} - (3x)^2 = 3x^2 - 9x^2$.

As we apply the properties of exponents to simplify expressions with integer exponents, it is important to know exactly which property to apply, and it is equally important to know when the properties do not apply. The next example will give us some practice with making these decisions.

Example 4

First, try to decide whether each of the following equations is true for all values of the variable. Then substitute numerical values into the equation to verify your conjecture. If the statement is false, rewrite the right-hand side of the equation to make it true. Assume that all variables are nonzero.

a. $2m^{-3} = \dfrac{1}{2m^3}$ b. $\dfrac{x^5}{x^{-5}} = x^{10}$ c. $(a + b)^{-2} = \dfrac{1}{a^2} + \dfrac{1}{b^2}$

Solution   a. Many simplified results "look correct." In this first example the expression $2m^{-3}$ is rewritten as the reciprocal because of the negative exponent, so we might assume that this is a true statement. However, it is always a good idea to verify this to determine if our guess is correct.

Let $m = 5$.

$$2m^{-3} \overset{?}{=} \dfrac{1}{2m^3}$$

$$2*5^{-3} \overset{?}{=} \dfrac{1}{2*5^3}$$

$$0.016 \neq 0.004 \quad \times$$

When we substitute $m = 5$ into the statement, the statement is *false*. So our guess was not correct! Let's see if we can determine why this statement is false. A negative exponent means the same as the reciprocal of the base to the positive exponent. In the expression $2m^{-3}$, the base is m, not $2m$. We should read the expression $2m^{-3}$ as *two times the negative third power of* m. Then the correct statement is

$$2m^{-3} = \frac{2}{m^3}$$

When we rewrite the right-hand side, we can also write it as $2m^{-3} = 2 * \frac{1}{m^3}$. The expressions $2 * \frac{1}{m^3}$ and $\frac{2}{m^3}$ are equivalent.

b. In this example, the numerator and denominator might look as though they should cancel to 1. We might conclude that this statement is false. Again, we should verify this numerically.

Let $x = 4$.

$$\frac{x^5}{x^{-5}} \overset{?}{=} x^{10}$$

$$\frac{4^5}{4^{-5}} \overset{?}{=} 4^{10}$$

$$1{,}048{,}576 = 1{,}048{,}576 \quad ✔$$

This statement is *true*. Let's apply the property of exponents for division to see why this statement is true. When dividing powers of x, the property states that we subtract the exponents, therefore $x^{5--5} = x^{10}$. So, the statement is true for all values of x except 0.

c. In this example, it appears that the $^-2$ has been applied to each of the terms in the parentheses. Then the terms are rewritten with positive exponents. This statement *looks* true. However, looking through the properties of exponents, there is no rule for the power of a sum, so we should definitely check this guess numerically.

Let $a = 4$ and $b = 5$.

$$(a + b)^{-2} \overset{?}{=} \frac{1}{a^2} + \frac{1}{b^2}$$

$$(4 + 5)^{-2} \overset{?}{=} \frac{1}{4^2} + \frac{1}{5^2}$$

$$0.01235 \neq 0.1025 \quad ✗$$

The statement is *false*. In the properties, the power of a product can be simplified, but there is no property for simplifying a power of a sum or difference. We can rewrite this statement using the definitions of exponents.

$$(a + b)^{-2}$$

$$= \frac{1}{(a + b)^2} \qquad \text{Rewrite with positive exponent.}$$

$$= \frac{1}{(a + b)(a + b)}$$

We can also multiply out the denominator. So, we rewrite the right-hand side as either

$$\frac{1}{(a + b)(a + b)} \qquad \text{or} \qquad \frac{1}{a^2 + 2ab + b^2}$$

In the next example, we will see how the properties of exponents can help us determine if we are entering complex expressions correctly in our calculators.

Example 5   In evaluating the expression

$$\frac{2\pi * 1.7}{(0.000112)(214)^3}$$

a team of three students got three different results. Estimate the expression to help decide which student's answer is correct.

Student 1 $\dfrac{2\pi * 1.7}{(0.000112)(214)^3} \approx 9.35 * 10^{11}$

Student 2 $\dfrac{2\pi * 1.7}{(0.000112)(214)^3} \approx 776{,}000$

Student 3 $\dfrac{2\pi * 1.7}{(0.000112)(214)^3} \approx 0.00973$

Solution   First we will round each of the numbers to one significant digit so that we can do the arithmetic mentally.

$$\frac{2\pi * 1.7}{(0.000112)(214)^3}$$

$$\approx \frac{2 * 3 * 2}{0.0001 * 200^3} \qquad \text{Because } \pi \approx 3.14, \text{ we round } \pi \text{ to } 3.$$

Because some of the numbers are either very small (0.0001) or very large (200^3), we rewrite them in scientific notation.

$$= \frac{2 * 3 * 2}{1 * 10^{-4} * (2 * 10^2)^3}$$

$$= \frac{12}{1 * 10^{-4} * 8 * 10^6} \qquad \text{Simplify } 2 * 3 * 2 \text{ and } (2 * 10^2)^3.$$

$$= \frac{12}{8 * 10^2} \qquad \begin{array}{l}\text{Simplify the denominator by multiplying 1 and 8 and using the rules} \\ \text{of exponents to multiply } 10^{-4} \text{ and } 10^6.\end{array}$$

$$= \frac{12}{8} * \frac{1}{10^2} \qquad \text{Separate the power of 10 factor.}$$

$$= 1.5 * 10^{-2} \qquad \text{Evaluate } \frac{12}{8}, \text{ and rewrite } \frac{1}{10^2} \text{ as a power of 10.}$$

$$= 0.015$$

From this estimate, we can see that student 3's answer is the closest

$$0.015 \approx 0.00973$$

and we can conclude that student 3 has the correct answer. Try entering the original expression in your calculator to see if your result agrees.

In this section, we reviewed the definitions for zero and negative exponents. Using these definitions, we saw how to evaluate expressions. A clear understanding of what the notation means is needed to evaluate these expressions correctly. We reviewed the properties of **integer exponents.** We saw how to use the properties to **simplify** an expression and ways to verify that our simplification is correct. In this section it is important to practice simplifying expressions using the properties because, in later sections, writing out their meanings will not help in simplifying expressions.

Problem Set 3.1

1. Translate each of the following English phrases into correct symbolic mathematics. Do not simplify.

   a. the fifth power of negative six

   b. the square of negative four

   c. the opposite of the square of four

   d. the square of the reciprocal of x

   e. the negative second power of negative five

   f. the opposite of the negative third power of five

   g. two times the square of n

   h. the negative second power of twice n

2. Explain the difference between $(-3)^4$ and -3^4. How is each expression read?

3. Without using a calculator, evaluate the following numerical expressions.

   a. 5^3 b. 5^{-3} c. 5^0

4. Without using a calculator, evaluate the following numerical expressions.

   a. -2^3 b. -2^{-3} c. -2^2 d. -2^{-2} e. -2^0

5. Without using a calculator, evaluate the following numerical expressions.

   a. $(-3)^3$ b. $(-3)^{-3}$ c. $(-3)^4$ d. $(-3)^{-4}$ e. $(-3)^0$

6. Without using a calculator, evaluate the following numerical expressions.

   a. -8^2 d. $(-6)^{-2}$ g. $(-2)^{-3}$

   b. -10^{-4} e. 7^{-2} h. $\left(\frac{1}{4}\right)^{-2}$

   c. $(-5)^3$ f. $(-1)^{-8}$

7. Evaluate each of the expressions in parts a–e using your calculator. Write the results as fractions, then answer the questions for each part.

   a. $\left(\frac{7}{9}\right)^{-2}$ is equivalent to what fraction squared?

   b. $\left(\frac{3}{5}\right)^{-2}$ is equivalent to what fraction squared?

   c. $\left(\frac{21}{53}\right)^{-1}$ is equivalent to what fraction?

   d. $\left(\frac{2}{3}\right)^{-3}$ is equivalent to what fraction cubed?

   e. $\left(\frac{2}{5}\right)^{-4}$ is equivalent to what fraction raised to the fourth power?

   f. $\left(\frac{x}{y}\right)^{-4}$ is equivalent to what fraction raised to the fourth power?

   g. $\left(\frac{a}{b}\right)^{-n}$ is equivalent to what fraction raised to the nth power?

8. Without using a calculator, evaluate the following numerical expressions.

   a. $\left(\frac{2}{3}\right)^4$ b. $\left(\frac{2}{3}\right)^{-4}$ c. $\left(\frac{2}{3}\right)^0$

9. Without using a calculator, evaluate the following numerical expressions.

   a. $\left(\frac{1}{6}\right)^{-2}$ b. $(-5)^{-2}$ c. $\left(\frac{-2}{3}\right)^{-3}$

10. a. Complete the following table.

x	x^2	x^3	x^4	x^5
$^-3$	$(^-3)^2 = 9$			
$^-2$				
$^-1$				
0				
1				
2				
3				

b. Graph the following functions by plotting points from the preceding table. Verify your graphs using a graphing calculator.

$$y = x \qquad y = x^2 \qquad y = x^3 \qquad y = x^4 \qquad y = x^5$$

c. What do you expect the graph of $y = x^6$ to look like? $y = x^7$? Be as specific as possible.

d. If n is even, what does the graph of $y = x^n$ look like? If n is odd, what does the graph of $y = x^n$ look like?

11. Rewrite each numerical expression in the form x^n, where n is either a negative or positive integer and x is an integer.

a. $\left(\frac{1}{2}\right)^5$ **b.** $\left(\frac{1}{4}\right)^3$ **c.** $\frac{1}{5^3}$ **d.** $\frac{1}{10^{-3}}$

12. a. Substitute $x = 9$ into the expression x^{-2}.

b. Substitute $B = {}^-3$ into the expression B^{-4}.

c. Substitute $w = {}^-2$ into the expression w^{-3}.

d. Substitute $y = {}^-10$ into the expression $(^-y)^4$.

e. Substitute $t = 4$ into the expression $^-t^2$.

f. Substitute $R = {}^-1$ into the expression $^-R^{-2}$.

g. Substitute $M = \frac{-1}{10}$ into the expression M^{-6}.

13. Evaluate the expressions you wrote in the previous problem.

14. a. Without calculating the value of the expression, determine whether $^-5m^{-2}$ is positive or negative when $m = {}^-8$. Explain.

b. Evaluate the expression $^-5m^{-2}$ when $m = {}^-8$. Show your substitution.

15. a. Without calculating the value of the expression, determine whether x^2y^3 is positive or negative when $x = {}^-5$ and $y = {}^-10$. Explain.

b. Evaluate the expression x^2y^3 when $x = {}^-5$ and $y = {}^-10$. Show your substitution.

16. Evaluate the expression $x^2 - y^3$ when $x = {}^-4$ and $y = 5$. Show your substitution.

17. Evaluate the expression $\sqrt{b^2 - 4ac}$ when $b = {}^-2$, $a = 1$, and $c = {}^-15$. Show your substitution.

18. a. If m and t are both negative numbers, is the following expression

$$\frac{8m^4t^3}{16m}$$

positive or negative? Explain how you decided.

b. Evaluate the following expression

$$\frac{8m^4t^3}{16m}$$

when $m = {}^-3$ and $t = {}^-5$. Show your substitution.

19. Simplify each of the following expressions. Assume that all variables are nonzero real numbers. Express the results using only positive exponents. Verify that your result is correct by numerical substitution.

a. $x^2 * x^6 - 5(x^2)^4$

b. $p^3(p^5) + 3(p^4)^2$

c. $20R^2(4R - 3T^3)$

d. $A^7 * A^{-5}$

e. $(z^{-5})^2$

f. $\dfrac{m^7}{m^{-7}}$

g. $(4w^{-4})^2$

20. Simplify each expression below. Assume that all variables are nonzero real numbers. Express the results using only positive exponents. Verify that your result is correct by numerical substitution.

a. $3k(10 - k) + (5k)^2$

b. $\dfrac{10a^6}{(-5a^3)^2}$

c. $(x + 3)(x + 1)$

d. $\dfrac{24p}{8p^{-5}}$

e. $10w^{-5} \cdot 4w^7$

f. $(4x)^2 + x^5 * x^{-3} - \dfrac{x^2}{x^4}$

g. $3x^{-2}(5x^5 + 2) - x^3$

21. Rewrite each expression so that the exponent is positive.

a. x^{-5} **b.** $^-x^{-1}$ **c.** $\dfrac{4}{x^{-3}}$ **d.** $6x^{-5}$ **e.** $\left(\dfrac{4}{x}\right)^{-2}$ **f.** $(5x)^{-3}$

22. Rewrite each expression in the form ax^n, where a is the numerical coefficient and n is either a negative or positive integer. For example, we can rewrite $\dfrac{2}{x^4}$ as $2x^{-4}$.

a. $\dfrac{3}{x^5}$ **b.** $\dfrac{2}{x^{-5}}$ **c.** $\dfrac{32}{(2x)^3}$ **d.** $\left(\dfrac{1}{x}\right)^4$

23. Always True?   First, try to decide whether each of the following equations is true for all values of the variable. Then substitute numerical values into the equation to verify your conjecture. If the statement is false, rewrite the right-hand side of the equation to make it true. Assume that all variables are nonzero.

a. $(AB)^2 = A^2B^2$

b. $3AB^2 = (3AB)^2$

c. $(A + B)^3 = A^3 + B^3$

d. $\left(\dfrac{A}{B}\right)^3 = \dfrac{A^3}{B^3}$

e. $(A + B)^{-2} = \dfrac{1}{A^2} + \dfrac{1}{B^2}$

f. $(3A^2B)^3 = 9A^6B^3$

g. $A^{2m} * A^m = A^{3m}$

24. Always True?   First, try to decide whether each of the following equations is true for all values of the variable. Then substitute numerical values into the equation to verify your conjecture. If the statement is false, rewrite the right-hand side of the equation to make it true. Assume that all variables are nonzero.

a. $3 * 3^2 = 9^2$ **b.** $4A^{-1} = \dfrac{1}{4A}$ **c.** $2mk^3 = (2mk)^3$ **d.** $\left(\dfrac{x}{y}\right)^{-3} = \left(\dfrac{y^3}{x}\right)$

25. Always True?   First, try to decide whether each of the following equations is true for all values of the variable. Then substitute numerical values into the equation to verify your conjecture. Assume that all variables are nonzero.

a. $m^5 + m^{-4} = m^{-4}(m^9 + 1)$ **c.** $k^n + k^n = 2k^n$

b. $A^n + A^n = A^{2n}$ **d.** $x^5 + x^{-4} = x$

26. Simplify each expression. Assume that all variables are nonzero real numbers. Express the results using only positive exponents. Verify that your result is correct by numerical substitution.

a. $m^{-6} * m^{-4}$ **d.** $\dfrac{1}{5x^{-3}}$ **g.** $\left(\dfrac{A}{B}\right)^{-2}$

b. $w^4(w^{-5})(w)$ **e.** $3x^0$ **h.** $\dfrac{(7Q^{-4})^2}{7Q^6}$

c. $5x^{-3}$ **f.** $(-5x)^0$ **i.** $\left(\dfrac{7Q^{-4}}{7Q^6}\right)^2$

27. Simplify each expression. Assume that all variables are nonzero real numbers. Express the results using only positive exponents. Verify that your result is correct by numerical substitution.

a. $10y^2(4y^3)$ **e.** $x^m(x^4)$

b. $(8x^3)^{-2}(3x^{-3})$ **f.** $(Km)^4$

c. $\dfrac{5w^7}{w^2}$ **g.** $(-4w^3)^2 + \dfrac{2w^4}{w^{-2}} - 3w^{-2}(w^8)$

d. $\dfrac{R^{x+3}}{R^x}$ **h.** $12x(xk^2 - 4x) + (10xk)^2$

28. Rewrite each expression as a sum or difference. Verify your results numerically.

a. $(2x + 5y)^2$ **d.** $\dfrac{10x^2y - 5xy^2}{5xy}$

b. $(A + 6B)(A - 6B)$ **e.** $\dfrac{2GH^4 + (4GH^4)^2}{5GH^3}$

c. $\dfrac{y^3 - 5y^2 + y}{y}$

29. Evaluate the following expressions for the given values of the variables. Show your substitutions. Express approximate results to the hundredths place.

a. $-x^2y^3$ for $x = -2$ and $y = -3$

b. $\dfrac{7(r^3t)^2}{14t}$ for $r = 2.4$ and $t = -5.1$

c. $a^{-2} + b^{-3}$ for $a = -0.25$ and $b = -0.5$

d. $\dfrac{1}{a^2 + b^3}$ for $a = -0.25$ and $b = -0.5$

30. Estimate the value of the following expressions. Next, evaluate the original expression using your calculator, round results to two significant digits, and compare this result to your estimate.

a. $(5.98 * 10^{24})(6{,}370{,}000)^2$ **b.** $\dfrac{3.37 * 10^{-4}}{(0.550)(0.0032)}$ **c.** $\dfrac{(6.9 * 10^{-11})(1.87 * 10^9)^2}{3}$

31. For each figure, write the expression for the shaded area. Simplify your expressions.

a.

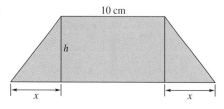

b.

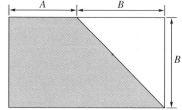

3.2 Probability and Counting Techniques

1. In this activity, you will be playing a game and deciding whether the game is fair. You will need a partner to play this game. In the game, two coins are flipped. Player A gets 1 point if the two coins match. Because there are two ways to get a match, player B will get 2 points if the two coins do not match.

   a. Do you think that this is a fair game? Why or why not? If you are not sure, guess.

   b. Next you will play the game to help you determine if your decision was correct or to help you make your decision. You and your partner must first decide which player will be player A and which will be B. Flip two coins. Record your results in the following table. Continue playing the game for about 3 minutes.

	Tally	Total Points
Player A: Match		
Player B: No match		

   c. Based on your results from playing the game, do you think the game is fair? Why or why not?

   d. The **experimental probability** of an event is the ratio of the number of times the event occurs during the experiment and the total number of trials. That is,

   $$\text{experimental probability of an event} = \frac{\text{number of occurrences of the event}}{\text{total number of trials}}$$

   Determine the experimental probability of getting a match and the experimental probability of getting no match. Do these probabilities support your answer from part c? Explain.

2. Consider the numbers 1, 2, 3, 4, 5, and 6.

   a. List all of the two-digit numbers that can be made using the whole numbers from 1 to 6. It is helpful to list the numbers in an organized manner. One way to do this is to start with all of the numbers whose first digit is a 1, then proceed to the numbers whose first digit is a 2, and so forth.

11	21
12	22
13	.
.	.
.	.
.	

   b. How many two-digit numbers can be made using the whole numbers from 1 to 6?

   c. List all of the three-digit numbers that can be made using the whole numbers from 1 to 6. It is helpful to build on the list that you created from part a.

   d. How many three-digit numbers can be made using the whole numbers from 1 to 6?

   e. Based on your results from parts a–d, how many four-digit numbers can be made using the whole numbers from 1 to 6?

3. Repeat Activity 2 with the added restriction that a digit can only be used once in a given number.

Discussion 3.2

In this section, we will be looking at a couple of counting techniques to see how they can be used to determine probabilities. In Activity 1, we were asked to decide whether a game was fair. To help us make this decision we **simulated** the game. A **simulation** is a procedure in which an experiment that closely resembles the situation is conducted repeatedly. Through a simulation we may collect data to help us make informed decisions.

We were asked to determine whether the game in Activity 1 was fair. One way to decide whether a game is fair is to determine the probability of each event in the game. There are two types of probabilities: experimental and theoretical.

> ### Definition _____
> The **experimental probability** of an event is the ratio of the number of observed occurrences of the event and the total number of trials.
>
> $$\text{experimental probability of an event} = \frac{\text{number of occurrences of the event}}{\text{total number of trails}}$$

> ### Definition _____
> The **theoretical probability** of an event is the ratio of the number of ways to obtain the event and the total number of possible outcomes, assuming all outcomes are equally likely to occur.
>
> $$\text{theoretical probability of an event} = \frac{\text{number of ways to obtain the event}}{\text{total number of possible outcomes}}$$

For example, if a coin is tossed 100 times and 60 heads occur, then the experimental probability of obtaining a head is $\frac{60}{100} = \frac{3}{5}$. However, exactly two outcomes are possible when a coin is flipped, and there is only one way to obtain a head. This means that the theoretical probability is $\frac{1}{2}$. So, you can think of the theoretical probability as the probability of an ideal experiment. In this scenario, ideally we would obtain 50 heads after tossing the coin 100 times.

Example 1

A six-sided die is tossed 20 times, and the following results are recorded.

Number	1	2	3	4	5	6
Tally	|||	卌	||	|||	|||	||||

a. Determine the experimental and theoretical probability of obtaining a 6.

b. Determine the experimental and theoretical probability of obtaining a number less than 5.

Solution   a. In the table we observe 4 occurrences of the number 6 out of a total of 20 tosses. Therefore, the experimental probability of obtaining a 6 is $\frac{4}{20}$, or $\frac{1}{5}$.

Probabilities are often expressed as decimals or percents. As a decimal $\frac{1}{5}$ is 0.2; as a percent it is 20%.

Because there is one way to obtain a 6 out of six possible outcomes, the theoretical probability of obtaining a 6 is $\frac{1}{6}$. As a decimal this is approximately 0.167; as a percent this is $\approx 16.7\%$.

b. In the table we observe 13 occurrences of a number less than 5 out of a total of 20 tosses. Therefore, the experimental probability of obtaining a number less than 5 is $\frac{13}{20}$. As decimal this is 0.65; as a percent it is 65%.

Because there are four ways to obtain a number less than 5 (1, 2, 3, or 4) out of six possible outcomes, the theoretical probability of obtaining a number less than 5 is $\frac{4}{6}$, or $\frac{2}{3}$. As a decimal this is approximately 0.667; as a decimal this is $\approx$ 66.7%.

Whether we choose to use the fraction, decimal, or percent depends on the situation. The fraction is the most natural form to obtain from the data because that is how the probabilities are defined. However, when we compare two or more probabilities, the decimal or percent forms are easier to use. In part b of Example 1, if we were to compare the experimental and theoretical probabilities we would be comparing $\frac{13}{20}$ and $\frac{4}{6}$. As fractions these are difficult to compare. As decimals, we can see that the experimental probability of 0.65 is slightly smaller than the theoretical probability of 0.667.

In Activity 1, you were asked to determine the experimental probability of obtaining a match and of getting no match. You probably found that both probabilities were close to one-half. Now, let's determine the theoretical probabilities for each of these events.

NOTE: From now on the word *probability* means *theoretical probability* unless otherwise stated.

Sample Spaces and Tree Diagrams

To determine the indicated probabilities we must determine all of the possible outcomes of tossing two coins. This is referred to as the sample space.

> Definition _____
> The set of all possible outcomes of an experiment is called the **sample space.**

To help us determine the sample space, imagine tossing two coins—one is a penny and one is a dime. The outcome for which the penny lands heads and the dime lands tails is different from the outcome in which the penny lands tails and the dime lands heads.

This means that the sample space for tossing two coins is

    HH   HT   TH   TT

where H stands for heads and T stands for tails.

We can visualize this in the following tree diagram. A **tree diagram** is a visual way to list the sample space and determine probabilities.

Two coins are used in this experiment, and we will be making two observations. First we observe the outcome of the first coin, then we observe the outcome of the second coin. We say that this experiment has two **stages.** In the first stage we want to display all of the possible outcomes for the first coin. Two outcomes are possible: heads or tails. So, we draw the first stage of the tree diagram with two **branches.**

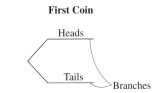

First Coin

Heads

Tails

Branches

First Stage of Tossing Two Coins Experiment

If the first coin lands heads up, the second coin could be a head or a tail. Therefore, for the second stage we need to draw two branches off of the first heads' branch. Similarly, if the first coin lands tails up, the second coin could be a head or a tail. So the first stage tails' branch has two branches off of it.

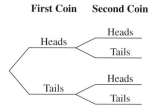

First Coin   Second Coin

Heads
Tails
Heads
Tails

Heads

Tails

First and Second Stages of Tossing Two Coins Experiment

To read the outcomes from a tree diagram, we begin at the point on the left and trace along each path. In the following figure, one path is marked, and the outcome is recorded to the right of the path. Note that HT means you get heads on the first coin and tails on the second.

First Coin   Second Coin   Outcome

Heads

Heads
Tails → HT

Tails

Heads
Tails

Finally, we complete drawing our tree diagram by recording the outcome of each of the four branches.

First Coin   Second Coin   Outcome

Heads

Heads → HH
Tails → HT

Tails

Heads → TH
Tails → TT

Next, we assign probabilities to each of the four outcomes. The chance of heads or tails is equally likely for each toss of the coins. Therefore, the probability of each outcome is equal. Because four outcomes are possible, the probability for each of the outcomes is $\frac{1}{4}$.

Using this information, there are two ways to obtain a match out of four possible outcomes and two ways to obtain no match out of four possible outcomes. Therefore, the probability of a match is $\frac{2}{4}$, or $\frac{1}{2}$, and the probability of no match is $\frac{2}{4}$, or $\frac{1}{2}$. Symbolically we can write these probabilities as

$$P(\text{match}) = \frac{1}{2} \quad \text{and} \quad P(\text{no match}) = \frac{1}{2}$$

Many people believe that the probability of a match when tossing two coins is $\frac{2}{3}$. This common misconception comes from thinking that there are two ways to get a match, HH or TT, but only one way to get a no match, with one H and one T. This is the reason why the game in Activity 3 awarded one point for a match and two points for a no match. We can see from both our experimental and theoretical probabilities that the probability of obtaining a match or a no match are equal. Therefore, the game in Activity 3 is not fair. How can the game be made fair?

Example 2

Determine the probability of obtaining HHH when three coins are tossed.

Solution   We know that there is only one way to obtain HHH. Therefore, we only need to determine the total number of possible outcomes to determine the desired probability. We can build a tree diagram similar to the one we built earlier.

First Coin	Second Coin	Third Coin	Outcome
Heads	Heads	Heads	HHH
		Tails	HHT
	Tails	Heads	HTH
		Tails	HTT
Tails	Heads	Heads	THH
		Tails	THT
	Tails	Heads	TTH
		Tails	TTT

Two outcomes are possible for the toss of the first coin. For each of these two outcomes are possible, giving $2 * 2$, or 4, outcomes for the results of the first two coins. Then, for each of these four results two outcomes are possible, giving $4 * 2$, or $(2 * 2) * 2 = 8$, total possible outcomes when tossing three coins.

Because there is one way to obtain HHH out of a total of 8 possibilities, the probability of obtaining HHH when tossing three coins is $\frac{1}{8}$.

Counting Techniques

In Activity 2, we determined the number of two-digit, three-digit, and four-digit numbers that could be formed using the whole numbers from 1 to 6. In the first two situations, we wrote out all the possibilities to answer the question. This is not always a feasible approach. Let's use a tree diagram to try to generalize the pattern.

The first of the two trees shown in the following figures indicates that six numbers are possible for the first digit. Then, for each of these, six numbers are possible for the second digit, resulting in a total of $6 * 6 = 36$ possible two-digit numbers using the whole numbers from 1 to 6. Similarly, the partial tree diagram adjacent to that tree shows that six numbers are possible for the first digit. For each of the first six, six possibilities exist for the second digit, for a subtotal of $6 * 6$, and for each of these 36, six numbers are possible for the third digit. Therefore, a total of $6 * 6 * 6 = 216$ three-digit numbers can be formed using the whole numbers from 1 to 6.

	First Digit	Second Digit	Result

First Digit: 1
- 1 → 11
- 2 → 12
- 3 → 13
- 4 → 14
- 5 → 15
- 6 → 16

First Digit: 2
- 1
- 2 → etc.
- 3
- 4
- 5
- 6

First Digit: 3
- 1
- 2
- 3
- 4
- 5
- 6

First Digit: 4
- 1
- 2
- 3
- 4
- 5
- 6

First Digit: 5
- 1
- 2
- 3
- 4
- 5
- 6

First Digit: 6
- 1
- 2
- 3
- 4
- 5
- 6

	First Digit	Second Digit	Third Digit	Result

First Digit: 1
- Second 1
  - 1 → 111
  - 2 → 112
  - 3 → 113
  - 4 → 114
  - 5 → 115
  - 6 → 116
- Second 2
  - 1 → 121
  - 2 → 122
  - 3 → etc.
  - 4
  - 5
  - 6
- Second 3
  - 1
  - 2
  - 3
  - 4
  - 5
  - 6
- Second 4
- Second 5   etc.
- Second 6

First Digit: 2
- Second 1
  - 1
  - 2
  - 3
  - 4
  - 5
  - 6
- Second 2
  - 1
  - 2
  - 3
  - 4
  - 5
  - 6
- Second 3
  - 1
  - 2
  - 3
  - 4
  - 5
  - 6
- Second 4
- Second 5
- Second 6

First Digit: 3
First Digit: 4   etc.
First Digit: 5
First Digit: 6

Example 3

How many seven-digit numbers can be formed using the whole numbers from 1 to 9?

Solution   As you might imagine the tree diagram rapidly becomes difficult to draw. However, we can still use the reasoning involved in the tree diagram.

Nine possible numbers can be used for the first digit. For each of these nine, nine numbers are possible for the second digit. For each of these, nine possibilities exist for the third digit, and so forth. This results in a total of

$$\boxed{9} * \boxed{9} * \boxed{9} * \boxed{9} * \boxed{9} * \boxed{9} * \boxed{9} = 9^7 = 4{,}782{,}969$$

1st   2nd   3rd   4th   5th   6th   7th

Using the digits from 1 to 9, we see that 4,782,969 seven-digit numbers can be formed.

Example 4

How many seven-digit numbers can be formed using the whole numbers from 1 to 9 without repeating any digits?

Solution   We can analyze this problem as we did in Example 3. Nine possible numbers can be used for the first digit. For each of these nine, eight numbers are possible for the second digit because we cannot repeat the first digit. For each of these, seven possibilities exist for the third digit because we cannot repeat the first or second digit, and so on. This results in a total of

$$\boxed{9} * \boxed{8} * \boxed{7} * \boxed{6} * \boxed{5} * \boxed{4} * \boxed{3} = 181{,}440$$

1st   2nd   3rd   4th   5th   6th   7th

There are 181,440 seven-digit numbers that can be formed using the digits from 1 to 9 without repeating any digits.

In Examples 3 and 4 we use a principle called the multiplication principle for counting. Informally, the **multiplication principle** allows us to calculate the number of ways an outcome that takes place in multiple stages can occur without sketching a tree diagram. The total number of outcomes can be found by multiplying the number of outcomes in each individual stage.

This section examined two fundamental ideas. One involves probabilities, both experimental and theoretical. The **experimental probability** of an event is the ratio of the number of times that an event occurs during an experiment and the total number of trials. The **theoretical probability** of an event is the ratio of the number of ways to obtain the event and the total number of possible outcomes, assuming all outcomes are equally likely to occur. You can think of the theoretical probability as the probability associated with an ideal experiment. In everyday language, the word *probability* refers to theoretical probability. We found that it is necessary to count the total number of outcomes of a situation to determine certain probabilities. This brought us to the second idea, which is the multiplication principle for counting. Informally stated, the **multiplication principle** states that when something takes place in multiple stages, the total number of outcomes can be found by multiplying the number of outcomes in each individual stage.

Problem Set 3.2

1. **The Keys.**   Keys of different shapes are designed by choosing different patterns for different parts. The keys for General Motors cars have six parts.
   a. Originally two patterns were used for each part. How many key designs were possible?
   b. Later three patterns were used for each part. How many key designs were possible using three patterns?
   c. If four patterns were used for each part, how many key designs are possible?
   d. General Motors cars come with two different types of keys: one for the door and one for the ignition. How many different sets of two keys are available if three patterns are used for each part of each key?

2. **Travel Wardrobe.**   A catalog advertises mix-and-match wardrobe separates for travel. You can select from a striped or plain T-shirt; a pair of pants, jumper, or long skirt; and a sweater or a jacket. Assume that an outfit consists of three items, one from each group. How many different outfit combinations can be put together for a trip if you purchase all of these items?

3. **Phone Numbers in Liberal.**   Phone numbers in the United States consist of ten digits with three parts: an area code (the first three digits), an office code (the next three digits), and a line number (the last four digits).

   a. Liberal, Missouri, only has one office code. How many phone numbers can Liberal have before they need to add another office code?

   b. If Liberal were to add two office codes, for a total of three, how many phone numbers are possible?

   c. The state of Missouri has three area codes. The office code can start with any digit from 2 through 9. How many phone numbers are possible in the state of Missouri if no additional area codes are added?

4. **The Combination Lock.**   A typical dial combination lock has a lock combination consisting of three numbers. Determine the number of combinations if the lock consists of numbers from 0 to 39? 0 to 49?

5. **Eye Color.**   A father and mother both carry genes for brown and blue eyes. When a child is conceived, it receives one gene for eye color from each of its parents.

   a. List all of the eye color gene combinations for a child from these parents.

   b. Assume that if at least one of the genes is brown, the child's eyes will be brown because brown eye color is dominant over blue. Determine the probability that a child from these parents will have brown eyes. (Actually determining eye color is more complicated than this.)

6. **A Family of Four**

   a. What is the probability that a family with four children has three girls and one boy, with the youngest being the boy?

   b. What is the probability that a family with four children has four girls?

   c. If a family has three children, all of which are girls, what is the probability that the fourth child is a boy?

7. **Pick 4.**   To play Pick 4, a state lottery game, you must choose four numbers from 0 to 9. Numbers can be repeated. Four numbers are randomly drawn to determine the winners. To win, your numbers must match the state's numbers in the order in which they are drawn. How many different outcomes are possible for Pick 4? What are your chances of winning?

8. **TLC.**   Ashley, who lives in Oregon, recently received a new license plate for her car. The license plate number is TLC 692. About a year later, Ashley was surprised to see another Oregon car that had a license plate number that also began with TLC. Most license plate numbers in the state of Oregon begin with three letters and end with a three-digit number.

   a. How many Oregon license plate numbers are possible that begin with the letters TLC? What assumptions are you making?

   b. What proportion of the total number of license plates in Oregon begin with the letters TLC? What assumptions are you making?

   c. Certain license plate numbers cannot be used because some letter combinations are not allowed. If this restriction is taken into account, is the actual proportion of license plates beginning with TLC larger or smaller than the proportion you found in part b? Explain your response.

9. Spinners.   A probability experiment consists of spinning each of the spinners shown and recording the numbers in order.

1st Spinner   2nd Spinner   3rd Spinner

   a. List the sample space for this experiment.

   b. What is the probability that the result is 123, that is, a 1 on the first spinner, 2 on the second, and 3 on the third?

   c. What is the probability that the numbers on all three spinners match?

   d. What is the probability that the sum of the numbers from the three spinners is 7?

10. A Conference Trip.   A college's honor society has five officers. The three male officers are named Alex, Bob, and Chet. The two female officers are named Dona and Elisa. The club is trying to decide on two officers to be sent to a national conference.

   a. If they decide that one male and one female should be selected from the officers to go to the conference, list all of the possible combinations who could attend. What is the probability that Dona is selected to go?

   b. If they decide that any two officers should be able to go to the conference, list all of the possible combinations who could attend. What is the probability that Dona is selected to go?

11. Sum of Two Dice

   a. Read the following experiment, and guess which column will contain the most X's.

   b. Perform the experiment with a partner.

   c. Determine the experimental probability of each sum.

   d. Determine all of the possible outcomes when the red and white dice are thrown.

   e. Determine the theoretical probability of each sum.

The Experiment

Let one person roll two dice (one white and one red), and announce the *sum* of the roll to their partner. The partner makes an X in the appropriate box on the following chart, starting at the bottom of each column. Stop when the X's reach the top in any one column.

Sum	2	3	4	5	6	7	8	9	10	11	12
Total number of times the sum occurred											

12. Product of Two Dice

    a. Read the following experiment, and guess which column will contain the most X's.

    b. Perform the experiment with a partner.

    c. Determine the experimental probability of each product.

    d. Determine all of the possible outcomes when the red and white dice are thrown.

    e. Determine the theoretical probability of each product.

The Experiment

Let one person roll two dice (one white and one red), and announce the *product* of the roll to their partner. The partner makes an X in the appropriate box on the following chart, starting at the bottom of each column. Stop when the X's reach the top in any one column.

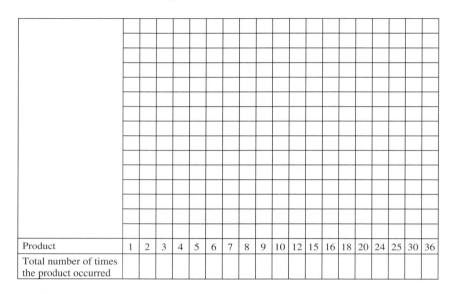

Product	1	2	3	4	5	6	7	8	9	10	12	15	16	18	20	24	25	30	36
Total number of times the product occurred																			

13. Craps

    a. Consider a pair of dice, one of which is white and the other is red. Determine all of the possible outcomes when the two dice are thrown.

    b. In the game of craps, two dice are thrown. The player throwing the dice wins if the sum of the dice is 7 or 11 and loses if the sum is 2, 3, or 12. What is the probability of winning on a given throw? What is the probability of losing on a given throw?

Chapter Three Summary

We know that a *positive exponent* is used as a shorthand notation for *repeated multiplication*. For example, x^4 means $x * x * x * x$. Because a negative number is the opposite of a positive number and division is the inverse of multiplication, a *negative exponent* is used as a shorthand notation for *repeated division*. That is, x^{-4} represents

$$\frac{1}{x^4} \quad \text{or, equivalently,} \quad \frac{1}{x * x * x * x}$$

Through applications and the properties of exponents we deduced the following.

> If x is a nonzero real number,
>
> $$x^{-n} = \left(\frac{1}{x}\right)^n = \frac{1}{x^n}$$

The following are key ideas for working with integer exponents.

- Remember that exponents apply to the item that is to the exponent's immediate left. For example, in the expression $8 * 3^2$, the exponent applies only to the 3. In the expression -5^4, the exponent applies only to the 5. Therefore, -5^4 is read as *the opposite of the fourth power of five* and simplifies to -625. In the expression $(-5)^4$, the exponent applies to -5 and simplifies to 625.

- When mentally evaluating numerical expressions that include negative exponents, rewrite the expression so that the exponents are positive using the preceding definition.

- When substituting a negative number for a variable that is raised to a power, include parentheses around the negative number, and place the exponent just to the right of the parentheses.

- When simplifying algebraic expressions involving negative exponents, it is *not* necessary to rewrite the exponents so that they are positive. Instead, apply the properties of exponents using the exponents given. If the directions ask you to write your results using only positive exponents, do this after you simplify the expression using the properties of exponents. Then, verify your result numerically.

- The two properties of exponents that are often confused are $x^m x^n$ and $(x^m)^n$. Using the language that accompanies these properties can help us keep them straight.

  $x^m x^n$ When *multiplying powers of x, add* the exponents.

  $(x^m)^n$ When simplifying a *power of a power of x, multiply* the exponents.

- Remember that *no* properties apply an exponent to a sum or difference. The only way to rewrite a power of a sum or difference is to write out the meaning of the exponent. Then perform the resulting multiplication. For example,

$$(x + 5)^2$$
$$= (x + 5)(x + 5)$$
$$= x^2 + 10x + 25$$

In Section 3.2 we studied probability and counting techniques. The probability of an event gives us an indication of how likely the event is to occur. We can determine the probability of an event by performing the experiment many times and recording the results. This gives us the experimental probability of an event. The **experimental probability** of an event is the ratio of the number of times the event occurs during the experiment and the total number of trials.

If we know that each outcome of an experiment is equally likely, we can compute the theoretical probability of an event in the experiment. The **theoretical probability** of an event is the ratio of the number of ways to obtain the event and the total number of possible outcomes, assuming all outcomes are equally likely.

To calculate the theoretical probability of an event, you need to know all the different outcomes possible. A **sample space** is a list of all of the possible outcomes. We can determine the sample space for an experiment by making an organized list or a **tree diagram**. In general, theoretical probability can predict the results of a probability experiment. For an experiment that has a large number of trials, the experimental probability is close to the theoretical probability.

In some situations we can determine the size of a sample space by using the **multiplication principle** rather than drawing a tree diagram. In these situations the total number of outcomes can be found by multiplying the number of outcomes in each individual stage.

Chapter Review

1–3

1. Suppose you are asked to solve each of the following equations. Without actually solving them, rank them in order from easiest to hardest. (Label the easiest #1, second easiest #2, and so on.)

   a. $2(x - 5) + 4 = 8x$ _____

   b. $4w - 18 = 32$ _____

   c. $\dfrac{k}{3} = 7$ _____

   d. $38.50 = 2.50 - 3.25p$ _____

   e. $\dfrac{R}{3} - 5R = 10$ _____

   f. $\dfrac{R}{3} = \dfrac{10 + 5R}{2}$ _____

2. Solve the following equations. Check your solutions.

   a. $\dfrac{w + 10}{2} = \dfrac{w - 10}{6}$

   b. $\dfrac{m}{2} + \dfrac{m + 5}{3} = 10$

   c. $\dfrac{4}{5x - 2} = \dfrac{4}{x}$

   d. $\dfrac{7}{5x} + 4 = \dfrac{24}{5} - \dfrac{6}{10x}$

3. Suppose you are asked to solve each of the following equations for the indicated variable. Without actually solving them, rank them in order from easiest to hardest. (Label the easiest #1, second easiest #2, and so on.)

   a. $PK = R - PT$ for T _____

   b. $PK = R - PT$ for K _____

   c. $PK = R - PT$ for R _____

   d. $PK = R - PT$ for P _____

   e. $4A = \dfrac{3A + 2}{T}$ for A _____

   f. $4A = \dfrac{3A + 2}{T}$ for T _____

4. Solve the following equations for the indicated variable.

   a. $2k - 7 = 5k(4 - p)$ for p

   b. $2k - 7 = 5k(4 - p)$ for k

   c. $\dfrac{x + 3y}{x} = \dfrac{x}{5}$ for y

   d. $m + \dfrac{p}{m} = 5p$ for p

5. In each of the following situations, a relationship is described by defining the independent and the dependent variables. Decide if the relationship described is a function.

   a. The number of minutes played in a basketball game is the independent variable, and the number of points scored in the game is the dependent variable.

   b. The number of college credits that Brook is taking is the independent variable and Brook's tuition is the dependent variable.

6. Determine which of the following graphs represent functions.

a.

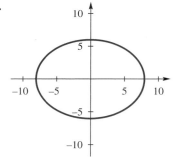

b.

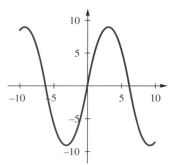

c.

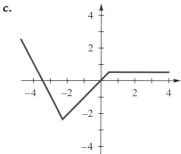

d.

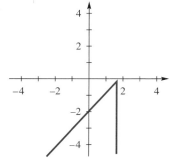

7. Determine which of the following tables represent functions.

a.

Input	Output
0	0
2	10
4	25
−2	10
−4	25

b.

Input	Output
0	0
10	2
25	4
10	−2
25	−4

c.

Input	Output
3	5
15	−6
−4	0
11	4
5	5

8. Determine which of the following equations represent functions.

a. $x + 10 = y$

b. $\sqrt{x + 10} = y$

c. $x - 16 = y^2$

9. The Anniversary Party.   Colleen and Bruce are planning an anniversary party and have received quotes from three caterers.

> Caterer A charges a set price per person for its services.
>
> Caterer B charges a basic amount for setting up and an additional price per person.
>
> Caterer C charges a set price for any party up to 200 guests.

Use the following graph to answer the following questions.

a. Match caterers A, B, and C with the graphs of their lines L, M, and N. Explain how you decided.

b. What is the slope of line N? What does the slope mean in this problem situation?

c. What is the vertical intercept of line N? What does it mean in this problem?

d. Colleen and Bruce have invited about 75 people to their party. Based on cost alone, which caterer should they select?

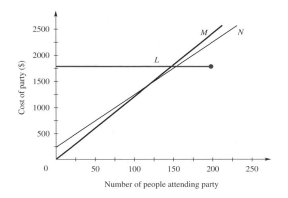

10. Which of the following equations graph as a line?

a. $3x + y = 0$

b. $\dfrac{x}{5} + \dfrac{y}{8} = 2$

c. $y = \dfrac{5}{x + 2}$

d. $y = \sqrt{x} + 10$

e. $y = \sqrt{7}\,x$

f. $\dfrac{12x + 13y}{25} = 80$

11. Without graphing, determine which of the following *could* be the graph of the linear function $y = 23.35 - 0.025x$. Explain how you made your decision.

a.

b.

c.

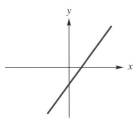

d.

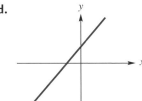

e.

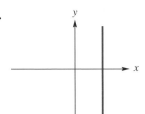

f.

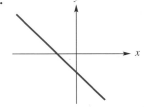

12. Write an expression in terms of the variable that generalizes the pattern.

a.

x	y
1	$3 * 4$
2	$4 * 5$
3	$5 * 6$
4	$6 * 7$
.	
.	
.	
x	

b.

K	P
0	6
1	$6 + 1$
2	$6 + 2^3$
3	$6 + 3^3$
4	$6 + 4^3$
.	
.	
.	
.	
K	

c.

c	d
2	$2^2 + 3 * 1$
4	$4^2 + 3 * 2$
6	$6^2 + 3 * 3$
8	$8^2 + 3 * 4$
.	
.	
.	
c	

d.

R	T
4	6
8	$6 * 6$
12	$6 * 6 * 6$
16	$6 * 6 * 6 * 6$
.	
.	
.	
R	

13. Climbing Mt. Hood.   A group of climbers is planning to climb Mt. Hood in Oregon. They plan to climb the southern route, which starts at Timberline Lodge located approximately 6000 feet above sea level. The altitude at the summit of Mt. Hood is approximately 11,240 feet. Air temperature drops approximately 0.21°F for every 100-foot rise in altitude.

a. If the temperature at Timberline Lodge is 34°F, what is the temperature on the mountain 500 feet above Timberline? 1500 feet above Timberline?

b. If the temperature at Timberline Lodge is 34°F, write an equation for the temperature in terms of the number of feet above the lodge.

c. Draw a graph for this problem. Consider what values are reasonable to include for each of the axes.

d. If the temperature at Timberline Lodge is 34°F, is there a point on the mountain where the temperature drops to freezing? If yes, at what height above sea level does this occur?

e. If the temperature at Timberline Lodge is 34°F, what is the temperature at the summit?

14. The following geometric figure is a rectangle with a quarter circle attached.

a. Write an expression for the perimeter of the figure.

b. Write an expression for the area of the figure.

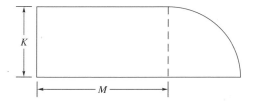

15. Suppose you are asked to write the equation for each line. Without actually writing the equation, rank them in order from easiest to hardest. (Label the easiest #1, second easiest #2, and so on.)

a.

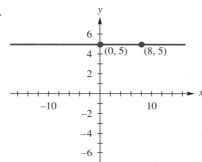

b.

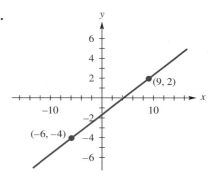

c.

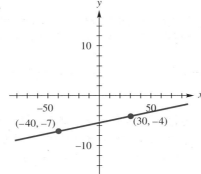

d.

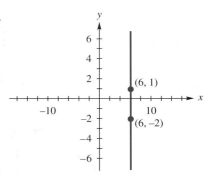

16. For each graph, write the equation and check your equation.

a.

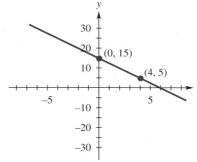

b.

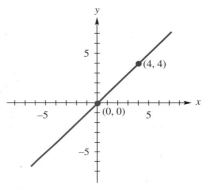

c.

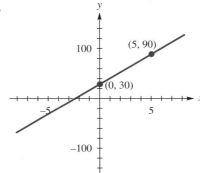

d.

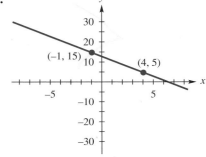

17. The Biology Books.   The cost to produce 400 lab books for biology is $4959. If they add a section, 450 lab books will cost $5571.50 to produce. Assume that the relationship between the cost of the books and the number produced is linear.

   a. Write an equation for the cost to produce the lab books in terms of the number produced.

   b. If the biology department has budgeted $6000 for the lab books, how many can they produce?

18. For each part, sketch the graph and write the equation of the line satisfying the given conditions. Label two points on the line.

   a. With a slope of $^-3$ and going through the point $(2, {}^-3)$

   b. Passing through the points $(5, 36)$ and $(13, 28)$

   c. Passing through the point $({}^-5, 3)$ and parallel to $x + y = 6$

   d. Passing through the point $(4, {}^-2)$ and perpendicular to $y = {}^-4x + 5$

   e. A vertical line passing through the point $(3, 4)$

19. Write the equations of three different lines that are parallel to $2x + 3y = {}^-6$.

20. Write the equations of three different lines that are perpendicular to $y = \dfrac{3x + 5}{4}$.

21. Given the linear equation $y = mx + b$, where $m > 0$ and $b < 0$, sketch a graph that can be the graph of this line.

22. Rock.   Doris has had the soil on one side of her house excavated so she can put in a dog kennel. She plans to put down about 2 inches of $\frac{3}{4}$ minus rock ($\frac{3}{4}$ minus is the size of the rock). Then another 1.5 inches of pea gravel will be put on top of the $\frac{3}{4}$ minus rock. The following figure shows the area where the rock is to go. How many yards of each type of rock does Doris need to buy? (*Note:* Rock is ordered by the yard, where 1 yard is really 1 cubic yard.)

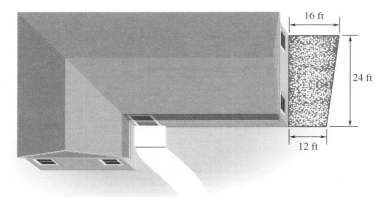

23. Without using a calculator, evaluate the following numerical expressions.

   a. $^-7^2$

   b. $^-2^{-3}$

   c. $(^-5)^2$

   d. $(^-10)^{-4}$

   e. 6^{-2}

   f. $\left(\dfrac{1}{2}\right)^{-3}$

   g. $\left(\dfrac{5}{3}\right)^{-2}$

   h. $(^-1)^{-8}$

24. Without using a calculator, evaluate the following numerical expressions.

   a. Substitute $x = 4$ into the expression x^{-2}.

   b. Substitute $B = {}^-5$ into the expression B^{-2}.

   c. Substitute $w = {}^-2$ into the expression ${}^-w^{-4}$.

   d. Substitute $y = {}^-3$ into the expression $({}^-y)^4$.

   e. Substitute $t = 4$ into the expression ${}^-t^2$.

   f. Substitute $R = {}^-10$ into the expression ${}^-R^{-3}$.

   g. Substitute $M = \frac{-1}{10}$ into the expression M^{-5}.

25. Simplify each expression. Assume that all variables are nonzero real numbers. Express the results using only positive exponents. Verify your results.

   a. $x^3 * x^7 - (4x^5)^2$ **f.** $w^7(w^{-5})(w)$

   b. $10b^2(4b - 3b^3)$ **g.** $(x + 3)(x + 1)$

   c. $8k^{-5} * k$ **h.** $(z^{-5})^2$

   d. $5x^{-2}$ **i.** $\dfrac{w^{-4}}{w^4}$

   e. $\dfrac{1}{5x^{-2}}$ **j.** $3k(10 - k) + 5k^2$

26. Chuck's Boat. Chuck's boat is rated at 325 horsepower. However, when the temperature of the engine is 100°F and at an altitude of about 1000 feet above sea level, his boat's power is only 290 horsepower. Chuck noticed on a recent vacation to a high desert lake that the power seemed to drop significantly on his boat. With a little research he discovered that at 100°F and at an altitude of about 4500 feet above sea level, the boat's power was only 250 horsepower. (*Note:* Altitude above sea level and temperature both affect the horsepower of a boat. An engine temperature of 100°F is normal for a nice summer day, but it decreases the engine's horsepower. For the remainder of this problem, we assume a constant engine temperature of 100°F.)

   a. Assuming that the relationship between the altitude and horsepower for a constant temperature is linear, write an equation for the power of Chuck's boat in terms of altitude above sea level. You need at least three significant digits for the slope and the vertical intercept in your equation.

   b. Graph the relationship between horsepower for Chuck's boat and altitude.

   c. What is the vertical intercept? What does it mean in the problem situation?

   d. Explain what the ordered pair (3000, 267) means in this problem situation.

   e. What is the slope of your line? Interpret the meaning of the slope in this problem.

27. Solve and check the following systems of equations.

   a. $\begin{cases} 2x - y = 10 \\ {}^-x + 3y = {}^-10 \end{cases}$ **d.** $\begin{cases} 4.1x - 3.2y = {}^-15.6 \\ {}^-2.6x + 1.7y = 6.6 \end{cases}$

   b. $\begin{cases} x = 3y + 1 \\ y = 13 - 5x \end{cases}$ **e.** $\begin{cases} y = 2.3x - 52.4 \\ 5.6x + 2.2y = 23.3 \end{cases}$

   c. $\begin{cases} \dfrac{x + y}{5} = 4.2 \\ 2x - y = 24 \end{cases}$ **f.** $\begin{cases} 3x - 2y = 44 \\ 2x + 5y = 4 \end{cases}$

28. A student graphed the system

$$\begin{cases} y = 0.346x + 4 \\ y = 0.341x + 7 \end{cases}$$

in the standard graphing window and observed that the lines were parallel. The student concluded that this system had no solution.

a. Explain how you know the student is wrong and what the student did wrong.

b. Find the solution to the system.

29. Alison's Purchases.   A receipt shows that Alison spent a total of $74.50 on five Whatzits and nine Thingamagigs. Another receipt shows that she spent a total of $63.60 on 14 Whatzits and two Thingamagigs. What is the cost of a Whatzit, and what is the cost of a Thingamagig?

30. Solve and check the following systems of equations.

a. $\begin{cases} 2x - y + 6z = 19 \\ {}^-x + 4y - 5z = {}^-10 \\ 3x + y - z = 11 \end{cases}$
 b. $\begin{cases} 5x - 2y + z = {}^-19 \\ {}^-2x - y - 3z = {}^-4 \\ x + 2y - z = 7 \end{cases}$

31. Cornstalks.   Georgette planted corn seed at the beginning of the summer. About 3 weeks after she planted them they had germinated and were about one inch tall. Five weeks later she measured them again and they were about 28 inches tall. Assuming this relationship is linear, write an equation for the relationship between the height of the cornstalks and the number of weeks since they were planted. Use your equation to determine when the cornstalks will reach 6 ft.

32. Glamour Garments makes two styles of outfits. The sporty suit uses two zippers and five designer buttons. The evening gown uses one zipper and eight designer buttons. They currently have in stock 171 zippers and 576 buttons. How many of each style outfit can they make without ordering more zippers or buttons?

33. Graph the following equations.

a. $y = \dfrac{20 - 3x}{5}$

b. $y = \sqrt{x + 1}$

34. Find the area and perimeter of the shaded figure. The cutout shape on the base is a semicircle.

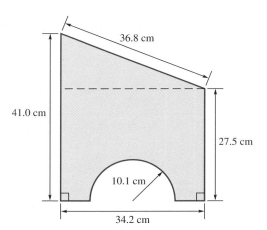

35. State Representatives.   The following table includes the population, rounded to the nearest thousand, and the number of congressional representatives for some states.

   a. Draw a graph with population as the independent variable and number of representatives as the dependent variable. Draw an eyeball fit line through the data points.

   b. Write an equation for your line.

   c. What is the slope of your line? What does it mean in this problem situation?

   d. Wisconsin's population is approximately 5,224,000. Use your equation to predict the number of representatives for Wisconsin.

   e. Look up the number of representatives for Wisconsin. Did your equation predict the correct number?

State	Population	Number of Representatives
Alabama	4,352,000	7
Alaska	614,000	1
Arizona	4,669,000	5
Arkansas	2,538,000	4
California	32,667,000	52
Colorado	3,971,000	5
Connecticut	3,274,000	6
Delaware	744,000	1
District of Columbia	523,000	1
Florida	14,916,000	22
Georgia	7,642,000	11
Hawaii	1,193,000	2
Idaho	1,229,000	2
Illinois	12,045,000	18
Indiana	5,899,000	10

36. a. If a and b are both negative numbers, is the following expression positive or negative? Explain how you decided.

$$\frac{5(ab^2)}{-2a}$$

   b. Evaluate the following expression when $a = -5$ and $b = -3$. Show your substitution.

$$\frac{5(ab^2)}{-2a}$$

37. Evaluate the following expressions for the given values of the variables. Show your substitutions. Express approximate results to the thousandths place.

   a. $4x^{-2} + y^3$ for $x = -3.1$ and $y = 4.23$

   b. $\dfrac{5(ab^2)^3}{-2a^4}$ for $a = 5$ and $b = -3.25$

   c. $\dfrac{-8m^4}{m^2 + 3k}$ for $m = 0.26$ and $k = 1.04$

38. Old Faithful.   The Old Faithful Geyser is not as predictable as many people believe. The time between eruptions can vary between 40 and 95 minutes. One possible way to predict the time between eruptions is to use the duration of an eruption to predict the time until the next one.

Draw an eyeball fit line through the data points. Write an equation for your line. Use your equation to predict the time between eruptions after an eruption that lasts 5.0 minutes.

Duration (min.)	Time Between Eruptions (min.)
4.4	78
3.9	74
4.0	68
4.0	76
3.5	80
4.1	84
2.3	50
4.7	93
1.7	55
4.9	76
1.7	58
4.6	74
3.4	75
4.3	80
1.7	56
3.9	80
3.7	69
3.1	57
4.0	90
1.8	42

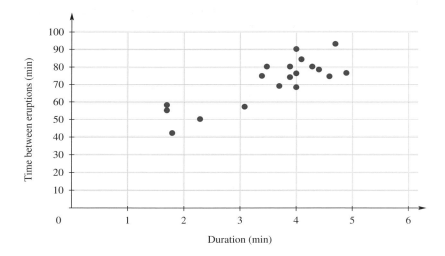

39. Double Diego.   Diego has invested $6500 in an account that pays an annual interest rate of 6.49%. If the account compounds interest annually and Diego leaves all of his money, including interest, in the account, how long will it take for the money to double?

40. Simplify each of the following expressions. Assume that all variables are nonzero real numbers. Express the results using only positive exponents. Verify your results.

a. $\dfrac{36p}{9p^{-5}}$

b. $10w^{-5} * 4w^7$

c. $\dfrac{8x^2y - xy^2}{xy}$

d. $3w^4 * 8w^{-5} * w$

e. $12x^0$

f. $3x^{-3}(4x^5 - 1) + x^2$

g. $(a - 4b)(a - 6b)$

h. $(8x^3)^{-2}(3x^{-3})$

i. $(-2w^3)^4 + \dfrac{2w^4}{w^{-3}} + 6w^{-2}(w^{14})$

j. $12x(xk^2 - 4x) + (10xk)^2$

41. Tetrahedron Die.   A tetrahedron is a four-sided die (in the shape of a pyramid).

a. What is the probability of rolling a 1 with a tetrahedron?

b. If one red and one green tetrahedron are rolled, how many different outcomes are possible for the experiment?

c. If two tetrahedron are rolled, what is the probability that they both land on 1's?

d. If two tetrahedron are rolled, what is the probability that they are a pair?

e. If three tetrahedron are rolled, what is the probability that all three land on 1's?

42. Two Different Dice.   A four-sided die and a traditional six-sided die are both tossed.

a. What is the probability that they both land on 1's?

b. What is the probability that they both land on the same number?

43. Pigs' Ears.   Pigs' ears are often notched to identify each individual pig as well as its litter. Each ear is divided into four regions plus the tip. Each of the four regions can have zero, one, or two notches. The tip of the ear can have zero or one notch.

a. The right ear is usually used to identify the litter to which a pig belongs. How many different litters can be identified with notches?

b. How many different notch patterns are possible using both ears?

c. If we assume that a pig must have at least one notch in one of its ears, how does that change your answer to part b?

Chapter Four

Linear Functions Extended

In this chapter, the concept of linear functions will be extended in a variety of ways. We will look at solving linear inequalities both graphically and algebraically. The absolute value function will be introduced, and equations involving absolute values will be solved.

Descriptive statistics can be used to make sense of the large data sets we encounter in many applications. We will explore several statistical measures and their uses.

4.1 Linear Inequalities

Activity Set 4.1

1. **a.** Is $^-3 < 8$ true or false?
   b. Is $4 > 10$ true or false?
   c. Is $^-3 > ^-5$ true or false?
   d. Graph $x < 7$ on a number line.
   e. Graph $w \geq 3$ on a number line.

2. **a.** Is $x = 4$ a solution to $x + 1 < 5$?
   b. Is $x = 3$ a solution to $x + 1 < 5$?
   c. Is $x = 2$ a solution to $x + 1 < 5$?
   d. Is $x = 5$ a solution to $x + 1 < 5$?
   e. Describe all the values of x that are solutions to $x + 1 < 5$.

3. Follow the directions. Does the inequality remain true?
   a. Add 7 to both sides of the inequality; $5 < 8$
   b. Add 7 to both sides of the inequality; $5 > ^-8$
   c. Add $^-7$ to both sides of the inequality; $5 < 8$
   d. Add $^-7$ to both sides of the inequality; $5 > ^-8$
   e. If you add or subtract the same amount from both sides of an inequality, does the inequality remain true?

4. Follow the directions. Does the inequality remain true?
   a. Multiply both sides of the inequality by 7; $49 < 84$
   b. Multiply both sides of the inequality by $\frac{1}{7}$; $49 > ^-84$
   c. Multiply both sides of the inequality by $^-7$; $49 < 84$
   d. Multiply both sides of the inequality by $-\frac{1}{7}$; $49 > ^-84$
   e. If you multiply or divide both sides of an inequality by a given number, does the inequality always remain true? If not, describe when it remains true and when it does not.

5. *Use the following graph* to answer all of the questions. Do *not* graph these on your own calculator.

   a. Identify three solutions to the equation $y = {}^-2.5x + 9.5$.

   b. Identify three solutions to the equation $y = 1.2x + 5.8$.

   c. Identify the solution to the equation ${}^-2.5x + 9.5 = 1.2x + 5.8$.

      What is the variable in your solution?

   d. Approximate the value of the expression ${}^-2.5x + 9.5$ when $x = {}^-2$.

      Approximate the value of the expression $1.2x + 5.8$ when $x = {}^-2$.

      When $x = {}^-2$, which of the two expressions has a larger value?

   e. Approximate the value of the expression ${}^-2.5x + 9.5$ when $x = 4$.

      Approximate the value of the expression $1.2x + 5.8$ when $x = 4$.

      When $x = 4$, which of the two expressions has a larger value?

   f. Describe all the values of x that are solutions to

$$^-2.5x + 9.5 > 1.2x + 5.8$$

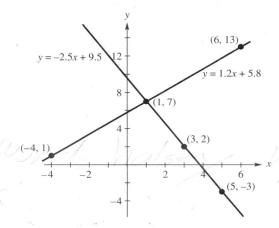

Discussion 4.1

In the activities, we found that when we add or subtract the same amount from both sides of an inequality, the inequality remains true. However, it is not as simple when we multiply or divide. In this case we need to know the sign of the number we are multiplying or dividing by. If we multiply or divide both sides of an inequality by a positive number, then the inequality remains true. If we multiply or divide by a negative number, then the inequality is not true until we reverse its direction.

We will use these ideas to solve linear inequalities algebraically. The properties are summarized in the following box.

> ### Properties of Inequalities
> 1. *Addition and Subtraction.* If we add (subtract) the same real number to (from) both sides of an inequality, the direction of the inequality does not change.
> 2. *Multiplication (Division) by a Positive Number.* If we multiply or divide both sides of an inequality by a positive number, the direction of the inequality does not change.
> 3. *Multiplication (Division) by a Negative Number.* If we multiply or divide both sides of an inequality by a negative number, the direction of the inequality must be reversed.

Notice that the we can perform the exact same operations on an inequality that we can on an equation, except when we multiply or divide by a negative number we must remember to reverse the direction of the inequality.

Example 1

Solve each linear inequality algebraically. Report your solution using inequality notation, and graph your solution on a number line. Check your solution.

  a. $4x + 10 \leq 25$ b. $12 - 3w < 2w + 40$

Solution a. $4x + 10 \leq 25$

  $4x \leq 15$ Subtract 10 from both sides. The direction of the inequality remains the same.

  $\dfrac{4x}{4} \leq \dfrac{15}{4}$ Divide both sides by 4. Because 4 is a positive number, the direction of the inequality remains the same.

  $x \leq 3.75$

From the algebra it looks like our solution is all values of x that are less than or equal to 3.75. Using inequality notation, we can write the solution as $x \leq 3.75$. We can graph this on a number line. Notice that the circle is closed (that is, a solid dot) at 3.75. This indicates that 3.75 is included in the solution set.

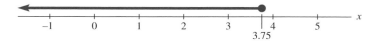

However, to be sure we must check our solution. Checking a solution to an inequality is somewhat different from checking the solution to an equation. Typically, the solution to an inequality consists of a range of values. It is not possible to check every value, so we check values in different regions.

In this example, we call 3.75 the **border point** because it borders the shaded region and the unshaded region. To check a solution to an inequality, we check the border point, a point that is in the solution region (shaded), and a point that is not in the solution region (unshaded).

When we check the border point, both sides of the inequality should be equal. When we check a point in the solution region, the inequality should be true, and when we check a point that is not in the solution region the inequality should be false.

CHECK

Border point: $x = 3.75$ Both sides of the inequality should be *equal*.

$$4 * 3.75 + 10 \stackrel{?}{=} 25$$
$$25 = 25 \qquad ✔$$

Point in region: $x = 0$ The inequality should be *true*.

$$4 * 0 + 10 \stackrel{?}{\leq} 25$$
$$10 \leq 25 \qquad ✔$$

Point not in region: $x = 5$ The inequality should be *false*.

$$4 * 5 + 10 \stackrel{?}{\leq} 25$$
$$30 \nleq 25 \qquad ✔ \qquad \text{This is a false statement, which is what we want.}$$

Therefore, our solution $x \leq 3.75$ checks.

b. $12 - 3w < 2w + 40$

$12 - 5w < 40$ Subtract $2w$ from both sides. The direction of the inequality remains the same.

$ {}^-5w < 28$ Subtract 12 from both sides. The direction of the inequality remains the same.

$\dfrac{^-5w}{^-5} > \dfrac{28}{^-5}$ Divide both sides by $^-5$. Because we are dividing by a negative number, we reverse the direction of the inequality.

$w > {}^-5.6$ This represents the solution written as an inequality.

From the algebra it looks like our solution is all values of w greater than $^-5.6$. We can graph this on a number line. Notice that the circle is open at $^-5.6$. This indicates that $^-5.6$ is not part of the solution.

CHECK   Even though $^-5.6$ is not part of the solution set, we must check to see that it is the correct border point. When a border point is substituted into an inequality, both sides of the inequality should produce the same value, regardless of the inequality statement. Therefore, when we check a border point, we are checking to see that both sides of the inequality are *equal*.

Border point: $w = {}^-5.6$ Both sides of the inequality should be *equal*.

$$12 - 3 * {}^-5.6 \stackrel{?}{=} 2 * {}^-5.6 + 40$$
$$28.8 = 28.8 \qquad ✔$$

Point in region: $w = {}^-2$ The inequality should be *true*.

$$12 - 3 * {}^-2 \stackrel{?}{<} 2 * {}^-2 + 40$$
$$18 < 36 \qquad ✔$$

Point not in region: $w = {}^-6$ The inequality should be *false*.

$$12 - 3 * {}^-6 \stackrel{?}{<} 2 * {}^-6 + 40$$
$$30 \nless 28 \qquad ✔ \qquad \text{This is a false statement, which is what we want.}$$

Therefore, our solution $w > {}^-5.6$ checks.

In Example 1, we solved two different linear inequalities algebraically. However, just as with linear equations and 2×2 linear systems, we can also solve these inequalities graphically. Let's consider the graph from Activity 5.

Example 2

Use the graph to answer the following questions.

a. Identify the solution to the system

$$\begin{cases} y = {}^-2.5x + 9.5 \\ y = 1.2x + 5.8 \end{cases}$$

b. Identify the solution to the equation $^-2.5x + 9.5 = 1.2x + 5.8$.

c. Identify the solution to the inequality $^-2.5x + 9.5 > 1.2x + 5.8$.

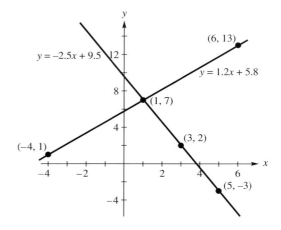

Solution a. We solved systems of linear equations in Section 2.4. We know the solution to the system is the point of intersection. In this case, that is the point $(1, 7)$. We could also write our solution as $x = 1$ and $y = 7$.

b. The solution to the equation $^-2.5x + 9.5 = 1.2x + 5.8$ is similar to the solution to the system. The difference is that a 2×2 system consists of two variables, whereas the equation we are solving here consists of only one variable. Therefore, the solution to $^-2.5x + 9.5 = 1.2x + 5.8$ is just $x = 1$.

c. In solving inequalities graphically, it is helpful to color code the lines and equations. To make things easier to visualize, we color the equation $y = {}^-2.5x + 9.5$ and its corresponding line blue.

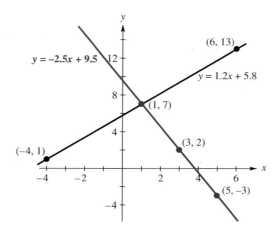

We know from part b that the two sides of the inequality, $^-2.5x + 9.5 > 1.2x + 5.8$, are equal when $x = 1$. Based on this, we know that when x is any number other than 1, the two sides are not equal. Therefore, $x = 1$ is our border point, and our job is to determine if the inequality we are solving is true when x is less than 1 or when x is greater than 1. To do this we try to visualize the situation using our color-coded graph.

To visualize this we start at the border point $x = 1$ on the horizontal axis. We can put an open circle at $x = 1$ because we want $-2.5x + 9.5$ to be greater than but not equal to $1.2x + 5.8$. Then, we can imagine a person standing on the x-axis somewhere to the right of 1. This person looks straight up or down and determines if the original inequality is true.

For instance, let's imagine that the person is standing on the x-axis at 5. From the graph, the person sees that the expression $1.2x + 5.8$ (which is black) produces a positive value of about 12, and the expression $-2.5x + 9.5$ (which is blue) produces a value of -3 when $x = 5$. Therefore, the expression $1.2x + 5.8$ is larger than the expression $-2.5x + 9.5$ when $x = 5$. That is, at $x = 5$, the black line is above the blue line. However, in the original inequality, $-2.5x + 9.5 > 1.2x + 5.8$, we want $-2.5x + 9.5$ (the blue line) to be larger. We want the blue line to be above the black line. The inequality is false when $x = 5$. Notice, that as long as the person is standing to the right of $x = 1$, the black line is above the blue line. This means that the expression $1.2x + 5.8$ is larger than the expression $-2.5x + 9.5$. So, all the values to the right of 1 are not solutions.

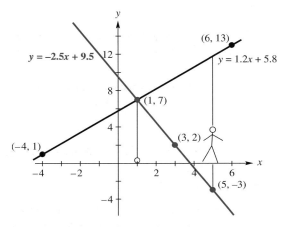

Let's try having the person stand to the left of $x = 1$, say at $x = -2$. At $x = -2$, the person sees that the expression $-2.5x + 9.5$ (blue) produces a value of about 14, whereas the expression $1.2x + 5.8$ (black) produces a value of about 3. So, at $x = -2$, $-2.5x + 9.5$ is larger than $1.2x + 5.8$, which makes the inequality true. As long as the person is anywhere to the left of $x = 1$, the blue line is above the black line; therefore, $-2.5x + 9.5$ is larger than $1.2x + 5.8$, which is what we want.

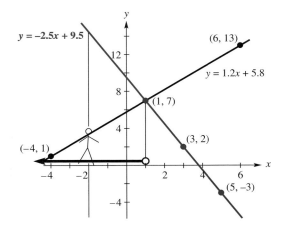

The solution to $-2.5x + 9.5 > 1.2x + 5.8$ is all values of x less than 1. Using inequality notation, this is reported as $x < 1$. Following is a graph of the solution on a number line.

NOTE: It is important to note that the two-dimensional graph (*x*- and *y*-axis) was used to solve the inequality in part c. However, it is not the solution. The original inequality only included the variable *x*. The number line is a graphical representation of the solution, and $x < 1$ is an algebraic representation of the solution. Therefore, drawing the two-dimensional graph is showing the work we did to solve the inequality, and drawing the number line graph is a representation of the solution. The process of solving an inequality by reading a two-dimensional graph is not trivial. At this point, you may be wondering why are we bothering. It seems easier to solve these inequalities algebraically. You are right; it is often easier to solve linear inequalities algebraically. However, many nonlinear inequalities cannot be solved algebraically and therefore, we must learn to read the solution from their corresponding two-dimensional graphs. We encourage you to practice reading the solution from the graphs now, when you have an alternative. This allows you to gain the confidence to solve inequalities graphically when an algebraic solution is not an option.

Example 3

Solve the inequality $4 - 3(x - 10) \le 16$ graphically. Report your solution using inequality notation. Also graph your solution on a number line. Check your solution.

Solution   To solve the inequality graphically, we must graph the functions $y = 4 - 3(x - 10)$ and $y = 16$. The first graphs as a line with a negative slope and a vertical intercept of (0, 34). Can you see why? The second function graphs as a horizontal line. If we graph these functions from $^-10$ to 10 on the horizontal axis and $^-10$ to 40 on the vertical axis, we get a reasonable view. Again we can use color to help us. Let's color the line $y = 16$ blue and the other line black.

The first thing we need to do after we obtain the graphs is find the intersection point, which in this case is (6, 16). This means that $x = 6$ is the border point. Because we want $4 - 3(x - 10)$ (black) to be less than *or equal to* 16 (blue), the border point is part of the solution to the inequality. Therefore, we put a closed circle at $x = 6$. Next we need to decide if the solution to the inequality is the values of *x* to the left or the right of the border point.

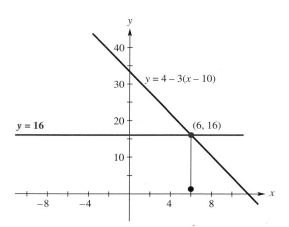

Suppose a person is standing to the left of the border point at $x = 2$. Then, the person sees that $4 - 3(x - 10)$ produces a value of about 28, which is larger than 16. If the person is standing to the right of the border point at $x = 10$, he or she sees that $4 - 3(x - 10)$ produces a value of about 5, which is less than 16.

Because the original inequality is $4 - 3(x - 10) \leq 16$, we want the values of $4 - 3(x - 10)$ (black) to be less than 16 (blue). Therefore, values to the left of the border point are not solutions, but those to the right are.

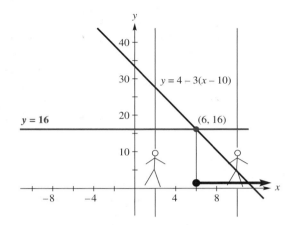

From the graph, the solution appears to be $x \geq 6$. Following is the number line graph of the solution.

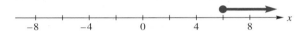

Finally, we need to check our solution. As we saw in Example 1, we need to check three points. In Example 1, we substituted each value into the inequality. This can also be done by making a table.

CHECK

	x	$4 - 3(x - 10)$	$\leq$	16	Conclusion
Border Point (Expressions should be equal.)	6	16	=	16	✔
Solution Region (Inequality should be true.)	8	10	$\leq$	16	✔
Nonsolution Region (Inequality should be false.)	0	34	$\nleq$	16	✔

We can see that our solution, $x \geq 6$, checks.

Because you already know their solutions, you might try solving the inequalities in Example 1 graphically. This will provide you with additional practice reading solutions from two-dimensional graphs.

Example 4

Sails per Week. The Windless Surfing Company has determined its cost and revenue from making custom sails to be

$$\text{cost} = 400 + 18.10 * S \qquad \text{and}$$
$$\text{revenue} = 43.33 * S$$

where S is the number of sails produced per week.

For what values of S is revenue > cost?

Solution We can solve this problem either graphically or algebraically. Here we do it both ways so that we can compare the two methods. To solve the problem graphically, we need to graph both the cost and the revenue function. Both functions graph as lines. The cost function has a vertical intercept of (0, 400) and a slope of 18.10. The revenue function has a vertical intercept of (0, 0) and a slope of 43.33. These two functions are shown on the following graph, with the revenue function shown in blue and the cost function in black.

Using a calculator, we can see that the border point is $S \approx 15.85$. So, to solve the inequality we need to determine whether S is less than or greater than 15.85. If we are to the left of 15.85, say at $S = 10$, the cost function (black) produces a value of about 575, which is greater than the revenue (blue) function's value of about 400. If we are to the right of 15.85 at $S = 25$, then the blue revenue function lies above the black cost function and, therefore, produces a larger value. Because we want to determine the values of S for which the revenue is larger than the cost, then $S > 15.85$. However, in the context of the problem we assume that the company produces and sells a whole number of sails per week. Because our solution is all values greater than 15.85, it makes sense to round 15.85 up to 16. Therefore, the revenue is greater than the cost when the number of sails produced each week is at least 16.

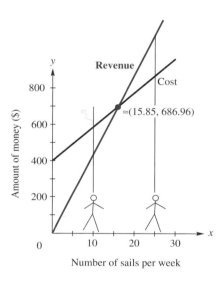

Let's solve this same problem algebraically and compare our solutions. We want revenue > cost, so we substitute the corresponding expressions into this inequality and solve for S

$$\text{revenue} > \text{cost}$$
$$43.33S > 400 + 18.10S$$
$$25.23S > 400$$
$$\frac{25.23S}{25.23} > \frac{400}{25.23}$$
$$S > 15.85$$

Again, we round 15.85 up to 16 and conclude that S must be greater than or equal to 16. Because we solved this problem in two different ways and arrived at the same solution, we omit the numerical check.

In this section we looked at solving **linear inequalities** both algebraically and graphically. We found that solving linear inequalities algebraically is similar to solving linear equations. The only difference is that if we multiply or divide both sides of the inequality by a negative number, we must reverse the direction of the inequality in that step.

The graphical technique for solving linear inequalities translates the linear inequality in one variable to two linear functions in two variables. We then use a two-dimensional graph to solve the inequality and express the solution in terms of our original variable. It is important to master this skill of solving problems from the graph now because we will use it again in other contexts.

Problem Set 4.1

1. Rewrite each expression as a sum or difference. Verify your results.

   a. $(3A - 5B)^2$ b. $\dfrac{3m^4 - 2m^2 + m}{m}$ c. $\dfrac{3x^2y - 12x}{xy}$ d. $\dfrac{5AB^3 + (3B^2)^3}{4B^3}$

2. Simplify each expression. Assume that all variables are nonzero real numbers. Express the results using only positive exponents. Verify your results.

   a. $b^5 * b^{-3}$

   b. $\dfrac{x^3 * x^4}{x^{-3}}$

   c. $\dfrac{x^3 + x^4}{x^{-3}}$

   d. $\dfrac{x^4 - x^{-3}}{x^{-3}}$

   e. $2\pi(5m)^2 + 4(3m^2) - (3m)(5m)$

   f. $\dfrac{5w}{3w^{-3}}$

   g. $(8m^2 * n^3)^{-2}$

   h. $\left(\dfrac{a}{b^2}\right)^{-2}$

   i. $\dfrac{(25x^{-3}y)^2}{xy^3}$

3. a. Solve the inequality $6 + 5x > 21$.

   b. Complete a table similar to the one shown here to demonstrate your check.

	x	6 + 5x	>	21	Conclusion
Border Point					
Solution Region					
Nonsolution Region					

4. a. Solve the inequality $^-2(x - 3) \le 3x - 24$.

   b. Complete a table similar to the one shown here to demonstrate your check.

	x	−2(x − 3)	≤	3x − 24	Conclusion
Border Point					
Solution Region					
Nonsolution Region					

5. Solve each inequality algebraically. Graph your solution on a number line. Check your solution. (Be sure to check the border point, a point that lies in the solution region, and one that does not lie in the solution region.)

   a. $3x + 7 \geq 16$

   b. $^-x - 9 < x + 10$

   c. $\frac{8}{3} \leq \frac{5}{3}R$

   d. $\frac{2}{3}x - 13 > ^-25$

   e. $23 \leq ^-4(3w + 7)$

   f. $2.1x - 17 < 4.7x - 10$

   g. $\dfrac{x - 3}{2} - \dfrac{x}{4} > 5$

   h. $60 - 25C \geq 20C - 48$

6. In Problem 5a, you solved $3x + 7 \geq 16$ algebraically. Solve this same inequality graphically using a two-dimensional graph. Describe how you used the graph to determine the solution.

7. In Problem 5b, you solved $^-x - 9 < x + 10$ algebraically. Solve this same inequality graphically using a two-dimensional graph. Describe how you used the graph to determine the solution.

8. Solve each inequality graphically. Include a sketch of your two-dimensional graph. Report your solution using inequality notation. Graph your solution on a number line. Check your solution.

   a. $3(x - 3) < ^-6$

   b. $12 + x \geq 8(x - 4)$

   c. $1 - x > \dfrac{^-2}{5}x + \dfrac{3}{5}$

   d. $2x - 16 < \dfrac{3}{4}(x - 12) - 5$

9. Solve the following inequalities using any method. Report your solution using inequality notation, graph the solution on a number line, and check your solution.

   a. $7 - 3x \leq ^-4$

   b. $\dfrac{w + 4}{5} < \dfrac{^-w}{3}$

   c. $15.6 + 0.22m \leq 0.37m$

   d. $\dfrac{3x}{2} - \dfrac{2(x + 1)}{3} \geq 3 - \dfrac{x}{2}$

   e. $105.5 < 30.4 - 6.2(5.5 - x)$

   f. $4150 + 18A \geq ^-3520 + 77A$

   g. $29.7P - 2.8(1.04 + P) > 31.2P$

   h. $206m - 0.0043 \geq 0.0836$

10. a. *Use the given graph to solve the following equation.*

$$\frac{x}{2} - \frac{10x - 25}{10} = 3(x + 3) - (x + 14)$$

b. *Use the given graph to solve the following system of equations.*

$$\begin{cases} y = 3(x + 3) - (x + 14) \\ y = \dfrac{x}{2} - \dfrac{10x - 25}{10} \end{cases}$$

c. *Use the given graph to solve the inequality.*

$$\frac{x}{2} - \frac{10x - 25}{10} > 3(x + 3) - (x + 14)$$

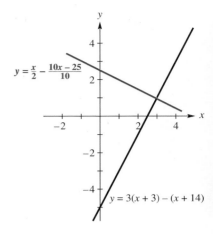

11. a. *Use the given graph to solve the following equation.*

$$3(x + 3) - (x + 14) = 0.6x - 2.5 - x$$

b. *Use the given graph to solve the following system of equations.*

$$\begin{cases} y = 3(x + 3) - (x + 14) \\ y = 0.6x - 2.5 - x \end{cases}$$

c. *Use the given graph to solve the following inequality.*

$$3(x + 3) - (x + 14) \leq 0.6x - 2.5 - x$$

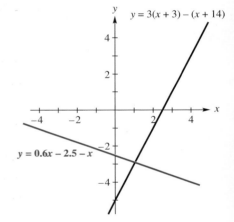

12. a. *Use the given graph to solve the following equation.*

$$1.5x - 1 = \frac{8 + 3x}{2}$$

b. *Use the given graph to solve the following system of equations.*

$$\begin{cases} y = 1.5x - 1 \\ y = \dfrac{8 + 3x}{2} \end{cases}$$

c. *Use the given graph to solve the following inequality.*

$$1.5x - 1 < \frac{8 + 3x}{2}$$

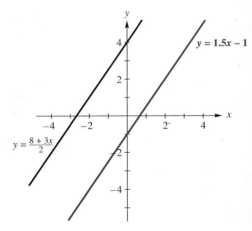

13. Write an expression for the shaded area in the figure. Simplify your expression. Assume the figures are nested rectangles.

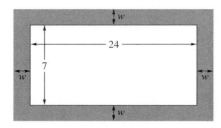

14. The graph represents the profit (in hundreds of dollars) in terms of the number of blocks sold (in thousands) for two toy companies.

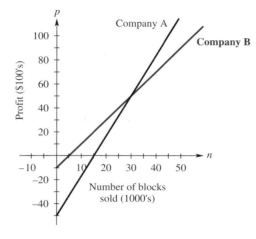

a. *Use the graph* to approximate the profit for each company if 20,000 blocks are sold. Which company's profit is larger?

b. *Use the graph* to approximate the profit for each company if 45,000 blocks are sold. Which company's profit is larger?

c. *Use the graph* to determine how many blocks need to be sold for the profit of Company A to be larger than the profit of Company B.

d. Write the equations for the profit for Company A and Company B. Let n represent the number of blocks sold in 1000's and p the profit in $100's.

e. Using your results from part d, write an inequality in terms of n that represents the statement "*The profit of company A is greater than the profit of company B.*"

f. Solve the inequality you wrote in part e. How does your solution compare with your estimate? What are the units on your answer?

15. The Shark Problem.   You can swim at the rate of 20 meters per minute. A shark can swim at the rate of 65 meters per minute. You are swimming offshore and observe a shark in the water near you. You immediately start swimming toward shore. The shark starts swimming toward the same point on the shore, a distance of 150 meters for the shark.

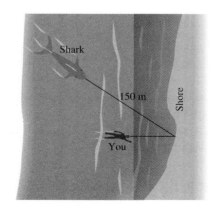

a. If your distance to the shore is 45 meters, how long will it take you to reach the shore? How long will it take the shark? Do you survive?

b. Write a linear inequality to determine the "safe" distance from shore for the swimmer. Solve your inequality for your "safe" distance from shore, and graph the solution on a number line.

16. The Bike Ride.   Two friends agree to meet on Saturday and go for a bike ride on a 40-mile loop. Beth arrived on time and, after waiting 15 minutes, started out alone. Erik arrived 2 hours late, 1 hour and 45 minutes after Beth started biking. Erik knows he can ride 20.0 miles per hour and Beth usually rides 7.5 miles per hour.

- **a.** For each bike rider, write an equation for the distance traveled in terms of the elapsed time. Graph both of these on a single coordinate system.

- **b.** What is the slope of the equation that models Beth's ride? What does the slope represent in this problem situation?

- **c.** What is the vertical intercept of the equation that models Beth's ride? What does it represent in this problem situation?

- **d.** What is the vertical intercept of the equation that models Erik's ride? What does it represent in this problem situation?

- **e.** One hour after Erik begins biking, has he caught up to Beth? Does Erik catch Beth before the end of the loop? Write an inequality to find the interval of time when Beth is ahead of Erik. Solve this inequality, and graph the solution on a number line.

- **f.** Assuming Beth is still talking to Erik when he catches up to her, we can assume that he slows down and rides the rest of the bike ride at her pace. Redraw the first graph to model this possible scenario.

17. All the World's a Stage.   Northeastern Community College (NCC) is selling tickets to a drama production. The student tickets sell for $2.50, and the general admission tickets sell for $3.75. They need to make at least $7000 from the sale of tickets to pay for their expenses. The theater at NCC has a maximum seating capacity of 2500.

- **a.** Write an equation for the number of tickets (student tickets and general admission tickets) that must be sold to just pay for expenses from ticket sales.

- **b.** Write an equation for the number of tickets (student tickets and general admission tickets) that can be sold to completely fill the theater at NCC.

- **c.** Label a graph with the number of student tickets on the vertical axis and the number of general admission tickets on the horizontal axis. Graph your equations from part a and b on this coordinate system.

- **d.** Is it possible to not sell out the theater and still make a profit? If possible, list at least one combination of ticket sales in which they do not sell out but do make a profit.

- **e.** Is it possible to sell out the theater and not make a profit? If possible, list at least one combination of ticket sales in which they do sell out but do not make a profit.

18. Designer Windsocks.   Windless Surfing Company's marketing manager observes that if the price of designer windsocks is $45, then on average 75 windsocks are sold per month. Experience has shown that if the price is raised from $45, then for every $2 increase in price, three fewer windsocks are sold each month.

- **a.** Write an equation for the number of windsocks sold per month in terms of the price.

- **b.** What price results in zero windsocks per month being sold?

- **c.** If the manager wants to sell at least 50 windsocks each month, what range of possible prices for the windsocks must she set?

19. The following table shows the data from two linear functions. Determine the values of the input variable such that

output from **function 1** ≤ output from **function 2**

Input	Output	
	Function 1	*Function 2*
10	6	−2
15	8	4
20	10	10
25	12	16
30	14	22
35	16	28

20. The following tables show the data from two linear functions. Determine the approximate values of the input variable such that

output from **function 1** > output from **function 2**

Function 1		Function 2	
Input	*Output*	*Input*	*Output*
16	87	16	97.33
16.5	88.5	16.5	97.50
17	90	17	97.67
17.5	91.5	17.5	97.83
18	93	18	98.00
18.5	94.5	18.5	98.17
19	96	19	98.33
19.5	97.5	19.5	98.50
20	99	20	98.67
20.5	100.5	20.5	98.83
21	102	21	99.00
21.5	103.5	21.5	99.17
22	105	22	99.33
22.5	106.5	22.5	99.50
23	108	23	99.67
23.5	109.5	23.5	99.83
24	111	24	100.00
24.5	112.5	24.5	100.17

4.2 Compound Inequalities

Discussion 4.2

In English, a compound sentence is formed by connecting two sentences with one of the words *and, but, or,* or *nor.* In mathematics, some applications require us to connect more than one inequality. The resulting inequality is called a **compound inequality.** These inequalities are connected with the words *and* or *or.* However, when the word *or* is used in a mathematical context, it has a slightly different meaning than when it is used in English. In English, we might say "you may choose to go to a movie tonight or go to a baseball game." In this context, *or* is used to indicate that you may choose either the movie or the game but not both. In mathematics, *or* is used to mean that either condition may be true or *both* may be true. The mathematical definitions of *and* and *or* follow. We will see how these definitions are applied in several examples.

Definition _____

In mathematics, the word **and** is used when *both* conditions must be true.

Definition _____

In mathematics, the word **or** is used when *either condition or both conditions* are true.

Example 1

Graph the solution represented by each compound inequality on a number line.

  a. $x < 10$ and $x \geq 4$
  b. $w < {}^-1$ or $w > 5$
  c. $0 \leq t$ and $t \geq 15$

Solution

For each part, we can begin by graphing each portion of the compound inequality on a number line. Then we determine how the two inequalities are combined depending on whether they are connected with the word *and* or with the word *or.*

  a. $x < 10$ and $x \geq 4$

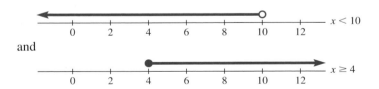

and

Because the two inequalities are connected with the word *and,* we need to determine all values of x that are both less than 10 and greater than or equal to 4. If we pick a point to the left of 4, say $x = 2$, then only the first inequality is true, If we pick a point to the right of 10, say $x = 12$, then only the second inequality is true. To make both inequality statements true, we must choose values between 4 and 10. The point $x = 4$ is also included because it satisfies both statements as well. Therefore, the graph of the compound inequality includes 4 and all numbers between 4 and 10. A graphical representation of $x < 10$ and $x \geq 4$ is shown on the following number line.

It is important to realize that the first two number lines were used to determine the solution that the compound inequality represents. However, the graph of the compound inequality is the third number line, which combines the two.

b. $w < {}^-1$ or $w > 5$

or

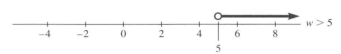

Because an *or* statement is true when any portion of the statement is true, the solution represented by $w < {}^-1$ or $w > 5$ can be illustrated by shading those numbers less than $^-1$ and also shading those numbers greater than 5, as shown on the following number line.

c. $0 \le t$ and $t \ge 15$

The first half of this inequality says that t is greater than or equal to 0, and the second half says that t is greater than or equal to 15. Each of the statements is graphed on a separate number line below.

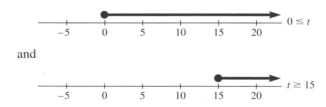

Because the two inequalities are connected by *and,* the only numbers that satisfy the compound inequality are those that make both inequalities true. If $t = 5$, then the first inequality is true, but the second is not. If $t = 15$, the first inequality is true as well as the second. Similarly, if $t = 20$, both inequalities are true. Therefore, the graph of $0 \le t$ and $t \ge 15$ is represented by shading all numbers greater than or equal to 15 as shown on the following number line.

In Example 1a, the graph of $x < 10$ and $x \geq 4$ was represented by shading all of the numbers between 4 and 10 including 4. The graph of $x < 10$ and $x \geq 4$ is shown on the following number line.

$x < 10$ and $x \geq 4$
also written as
$4 \leq x < 10$

In this type of situation, in which the variable falls between two numbers, the compound inequality can be written in a shorthand notation. For instance, $x < 10$ and $x \geq 4$ can be written as $4 \leq x < 10$.

It is important to note that this shorthand notation can only be used when the variable lies *between* two numbers on the number line. This means that the shorthand notation is equivalent to a compound inequality that represents an *and* statement. It can never be used to represent an *or* statement.

Example 2

Normal Temperature.   Normal internal body temperature for a person is considered to be 98.6° but, in practice, a person's temperature is considered normal if it is at or above 97.5° and at or below 99.7°. Draw a number line graph of the interval. Write the interval for normal temperature using inequality notation.

Solution   Another way to state the condition is to say that a person's temperature is considered normal if it falls *between* 97.5° and 99.7°, including 97.5° and 99.7°. This is graphed on the following number line.

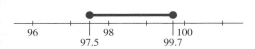

Because this graph is between 97.5 and 99.7, we can either write the statement using the shorthand notation for between two numbers, or we can write the statement as an inequality connected with *and*.

If T represents the normal temperature for a person, then the shorthand notation can be written as

$97.5 \leq T \leq 99.7$

Written as a compound inequality, this is

$T \geq 97.5$ and $T \leq 99.7$

Example 3

Sturgeon.   In Oregon, a sturgeon must be tossed back into the river if it is shorter than 42 inches or longer than 5 feet 6 inches. On a number line, graph the sizes of sturgeon that must be tossed back. Write the compound inequality represented by your graph.

Solution   Because 5 feet 6 inches is equivalent to 66 inches, sturgeon must be tossed back of they are shorter than 42 inches or longer than 66 inches. This is graphed on the following number line.

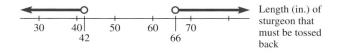

This graph can be written algebraically using a compound inequality connected with *or*. If *L* represents the length of a sturgeon in inches, then the sturgeon is tossed back if

$$L < 42 \qquad \text{or} \qquad L > 66$$

Example 4

Graph the solution represented by each compound inequality on a number line.

  a. $-8 < x < 0$
  b. $5 > k \qquad \text{or} \qquad k > 15$
  c. $x > 3 \qquad \text{or} \qquad x \geq -1$

Solution   a. $-8 < x < 0$

This is the shorthand notation for a graph that is shaded between two numbers. This statement says that *x* is between -8 and 0, not including either endpoint. This is graphed on the following number line.

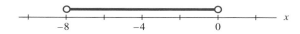

b. $5 > k \qquad \text{or} \qquad k > 15$

For this compound inequality, we start by graphing each part of the inequality. The first part, $5 > k$, says that *k* is less than 5. The second part says that *k* is greater than 15.

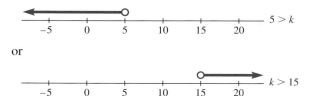

Because the two inequalities are connected with *or*, we want all of the values that satisfy either of the statements or both of them. Therefore, the graph of $5 > k$ or $k > 15$ includes the values of the two graphs. The graph of $5 > k$ or $k > 15$ is shown in the following number line.

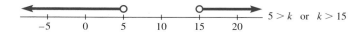

c. $x > 3$ or $x \geq {}^-1$

Each part of the compound inequality is graphed on one of the following number lines.

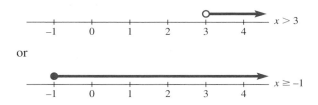

or

For a statement that contains the word *or*, we need to make sure at least one of the inequalities is true. If we pick a number greater than 3, say $x = 4$, then both statements are true and, thus, the original compound statement is true. If we pick a number between $^-1$ and 3, say $x = 2$, then the second inequality is true, which also satisfies the compound statement. If $x = 3$, then the second statement is true and, again, the compound statement is true. Therefore, the graph of $x > 3$ or $x \geq {}^-1$ is represented by shading all the values to the right of $^-1$, including $^-1$. This graph is shown on the following number line.

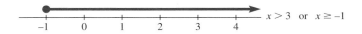

In this section, we looked at compound inequalities. A **compound inequality** consists of two or more inequalities connected with *and* or *or*. It is important to remember that the mathematical meaning of these words is slightly different from their meaning in an English sentence. Mathematics uses compound inequalities to write solutions to problems. Therefore, we need to become proficient in our ability to use and understand compound inequality statements.

Problem
Set
4.2

1. Simplify the following expressions. Write your results using only positive exponents. Verify your results. Assume that all variables are nonzero.

   a. $n^0(n^{-3})(n^5)$

   b. $\dfrac{A^6}{A^{-4}}$

   c. $(4r^{-2})^3 + 3r(5r^{-7})$

   d. $7R^3(2R^{-2} + 5R^{-1}) - 10R^2$

   e. $(T^2)^{-3} - T^{-7}(7T)$

   f. $\dfrac{10xy^2}{x^{-2}y}$

2. For each part, graph the solution to the inequality on a number line.

   a. $x \geq 25$

   b. $0 > k$

   c. $^-4 \leq p$

   d. $p < {}^-10$

3. For each part, graph the solution to the compound inequality on a number line.

   a. $x \leq 2$ and $x \geq {}^-5$

   b. $m < 2$ or $m > 10$

   c. $x > 0$ and $x > {}^-14$

   d. $4 < x < 10$

   e. $^-2.5 \leq R \leq 2.5$

4. For each compound inequality, graph the solution on a number line.

a. $8 < x$	and	$x > 0$		**d.** $^-10 \le x \le 25$	and	$x > 8$
b. $m \le {^-4}$	or	$m \ge {^-1}$		**e.** $3 - x > {^-5}$	and	$x > 0$
c. $k < 4$	or	$k \ge 3$		**f.** $5 + 2x \le {^-3}$	or	$5 + 2x \ge 3$

5. Write an inequality or compound inequality to describe each of the following graphs.

a.

b.

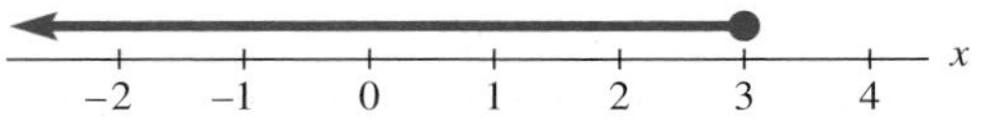

c.

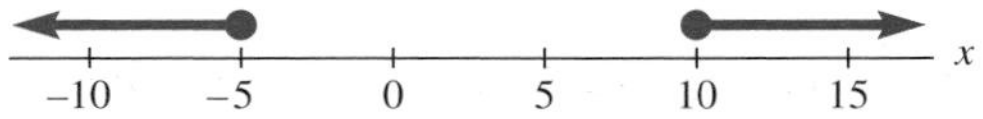

d.

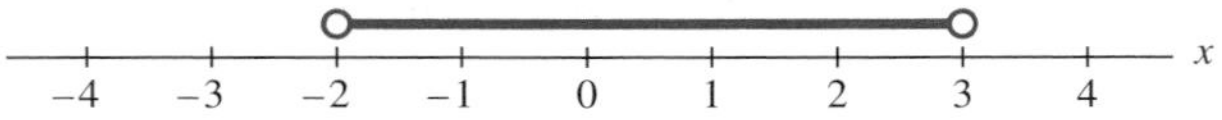

6. Write a compound inequality for each of the following graphs.

a.

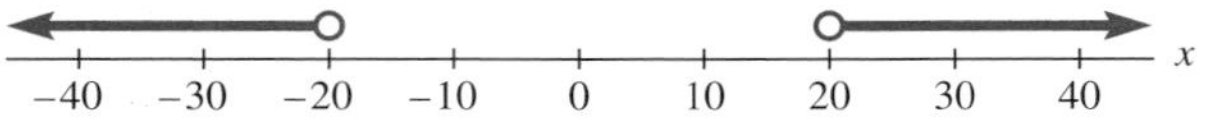

b.

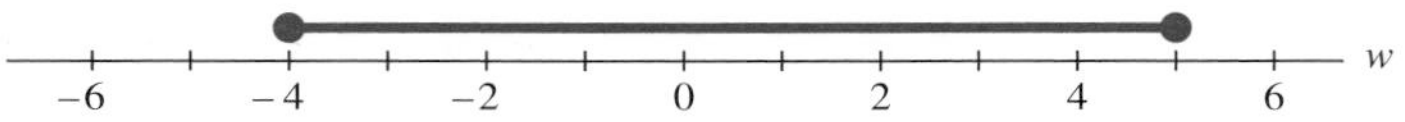

c.

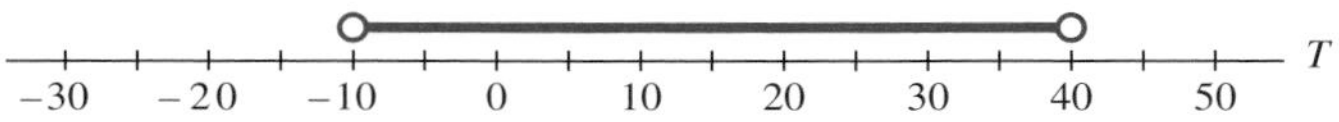

d.

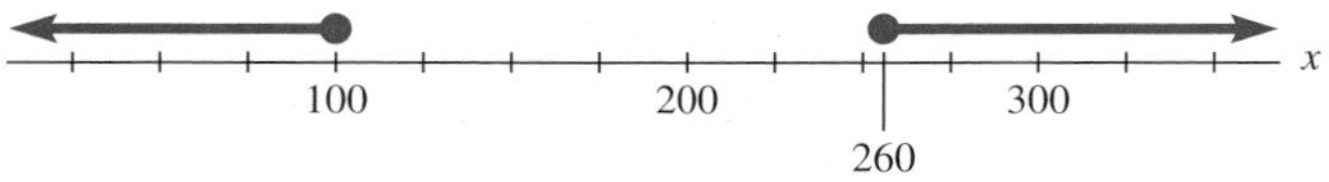

7. Write a compound inequality for each of the following graphs.

a.

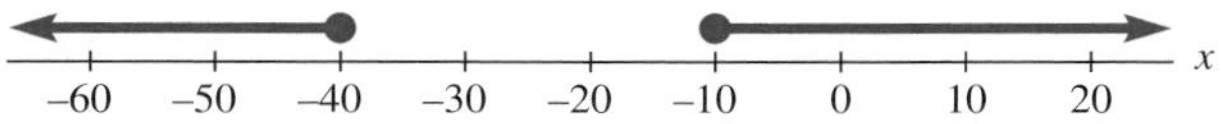

b.

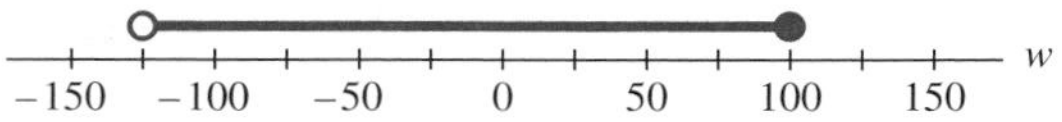

c.

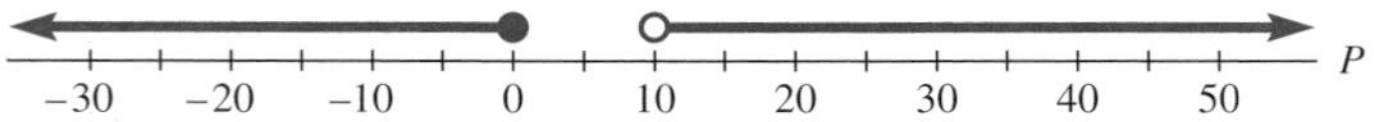

d.

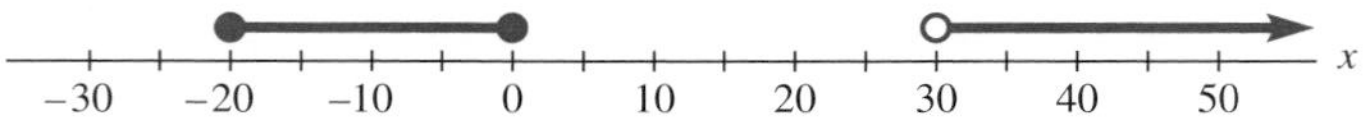

8. Where Is the Third Planet?   Three destinations on a game board, planet A, planet B, and planet C, do not lie in a straight line. As a player we do not know where each destination planet is located, but we do know that A is 10 units from B and B is 15 units from C. Determine how far A can be from C. (*Hint:* The sum of the lengths of any two sides of a triangle must be greater than the length of the third side.)

9. Shooting for a "B."   Suppose your scores on four lab exams were 86%, 82%, 73%, and 78%. According to the syllabus, the lowest possible percent for a "B" is 80%. All lab exams are equally weighted.

   a. What score do you need on the fifth exam to guarantee a "B" for a lab grade?

   b. Let x represent the score on your fifth exam. Write an inequality to determine possible values of x to guarantee the "B."

   c. Solve your inequality from part b, and graph your solution on a number line. Be sure to consider the maximum possible score when writing your solution.

10. *Use the graph* to solve the following inequality.

$$2.5(2x - 4) + 160 \leq \frac{-10(30 - 3x)}{3} + 300$$

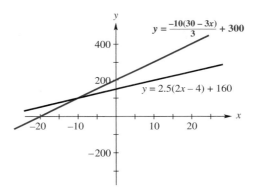

11. The "W."   The following graph defines a function. Use the graph to approximate the answers to the following questions.

   a. Approximate the value of this function when $x = 2$, when $x = 0$, and when $x = {}^-7$.

   b. Approximate the values of x that make the function equal to zero.

   c. Approximate the values of x that make the function greater than zero.

   d. Approximate the values of x that make the function less than zero.

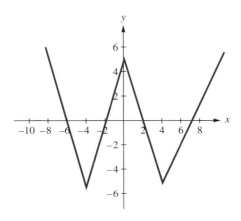

12. a. *Use the graph* (do not graph your own) to solve the following equation.

$$-0.5(x - 2)^2 + 8 = 0$$

b. *Use the graph* (do not graph your own) to solve the following inequality.

$$-0.5(x - 2)^2 + 8 < 0$$

c. *Use the graph* (do not graph your own) to solve the following inequality.

$$-0.5(x - 2)^2 + 8 > 2$$

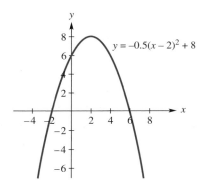

4.3 Absolute Value Functions—Graphically

Activity

Set

4.3

1. a. Complete the table. Remember that the distance
    between any two points is always positive.

   b. Plot the points from the table.

Input	Distance Between the Input and 3
⁻4	7
1	
2	
3	
4	
5	
10	

2. a. Complete the table, and plot the points.

   b. How does your graph compare with your graph in the previous
      activity? What does this tell you about the meaning of $|x - 3|$?

x	$|x - 3|$
⁻4	7
1	
2	
3	
4	
5	
10	

3. a. Graph the function $y = x - 3$. How does this graph compare with the graph of $y = |x - 3|$?

   b. Use your calculator to graph the two functions $y = x + 6$ and $y = |x + 6|$. How do these
      graphs compare?

   c. Use your calculator to graph the two functions $y = 2x - 10$ and $y = |2x - 10|$. How do
      these graphs compare?

   d. Use your calculator to graph the two functions $y = 0.25x - 5$ and $y = |0.25x - 5|$. How
      do these graphs compare?

   e. Use your calculator to graph the two functions $y = 12 - 3x$ and $y = |12 - 3x|$. How do
      these graphs compare?

4. Based on your results from Activities 1–3, what do you expect the graph of $y = |x - 7|$ to
    look like? Be as specific as possible. Graph this function to verify your conjecture.

5. For each of the following functions, predict what you expect the graph to look like and then
    verify your conjecture.

   a. $y = |x - 10|$ **c.** $y = |x + 12|$ **e.** $y = |3x + 42|$

   b. $y = |x + 3|$ **d.** $y = |x|$ **f.** $y = |0.5x - 13|$

6. Solve each equation or inequality graphically. Graph your solution on a number line. Write
    your solution using equality or inequality notation.

   a. $|x - 2| = 5$ **c.** $|x - 10| > 4$ **e.** $|x + 5| = -1$

   b. $|x + 3| \leq 5$ **d.** $|2x - 8| > 16$ **f.** $|x + 5| \geq -1$

Discussion 4.3

From your previous experience you may know that the absolute value function "makes things positive." For example, $|4| = 4$ and $|{}^{-}4| = 4$. Therefore, if we put a positive number into $|x|$, then we get out exactly what we put in. If we put a negative number into $|x|$, then we do *not* get out what we put in. Instead, we get out the opposite of what we put in, because absolute value makes the negative number positive. This is stated more formally in the following algebraic definition of the absolute value function.

> ### Definition
> **Algebraic Definition of the Absolute Value Function**
>
> $$|x| = \begin{cases} x & \text{if } x \geq 0 \\ {}^{-}x & \text{if } x < 0 \end{cases}$$

A second definition of the absolute value function relates to what you learned in the activities. In the activities you may have found that $|x - 3|$ represents the distance between x and 3 on a number line. Similarly, $|x - 6|$ represents the distance between x and 6. Because $|x|$ is the same as $|x - 0|$, $|x|$ represents the distance between x and the origin. This is stated in the following geometric definition of the absolute value function.

> ### Definition
> **Geometric Definition of the Absolute Value Function.**   The absolute value of x, $|x|$, represents the distance, on a number line, between the number x and zero.

At this point, we now have two definitions of the absolute value function. The geometric definition provides a description of the distance represented by an absolute value, and the algebraic definition provides a mathematical way to rewrite an absolute value expression. Both definitions play a role in our understanding of this new function.

Now let's look at the graphs of some of these functions. Consider $y = |x - 6|$. From the geometric definition, we know that $|x - 6|$ represents the distance between x and 6 on a number line. Therefore, if $x = 6$, then the distance, or output, is 0. This means that the point (6, 0) is on the graph. Next if x is 4 or 8, then the distance is 2, which means the points (4, 2) and (8, 2) are on the graph. If we continue this process we obtain the following graph, which is in the shape of a V.

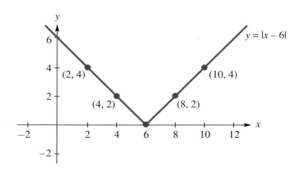

For another way to view this graph, consider the graph of $y = x - 6$. Notice this is a linear function, not an absolute value function. The graph of $y = x - 6$ is a line with a vertical intercept at $(0, {}^-6)$ and a slope of 1 as seen on the following graph.

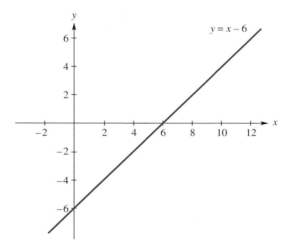

Now, let's consider how the two functions $y = x - 6$ and $y = |x - 6|$ are related. Suppose $x = 8$. Then, the linear function $y = x - 6$ produces a value of $8 - 6 = 2$, and the absolute value function $y = |x - 6|$ produces a value of $|8 - 6| = 2$ as well. However, if $x = 4$, then the linear function $y = x - 6$ produces a value of $4 - 6 = {}^-2$, whereas the absolute value function $y = |x - 6|$ produces a value of $|4 - 6| = 2$. Therefore, the point $(4, {}^-2)$ is on the graph of the linear function, whereas, the point $(4, 2)$ is on the graph of the absolute value function. Similarly, $(1, {}^-5)$ is on the graph of $y = x - 6$, and $(1, 5)$ is on the graph of $y = |x - 6|$. Graphically, we can see that when each point on the line $y = x - 6$, which lies below the horizontal axis, is reflected about the x-axis, it lies on the graph of $y = |x - 6|$.

To summarize, because the absolute value function only outputs positive values, the graph of $y = |x - 6|$ can be obtained by graphing the corresponding linear function $y = x - 6$ and reflecting (about the x-axis) the portion of the line that is below the x-axis. This process is illustrated on the following graph.

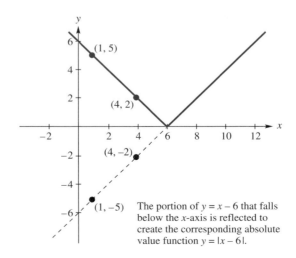

The portion of $y = x - 6$ that falls below the x-axis is reflected to create the corresponding absolute value function $y = |x - 6|$.

Example 1

Graph the function $y = |2x + 20|$.

Solution We know that the function $y = |2x + 20|$ graphs as a **V**. To determine the location of the **V**, we start by graphing the corresponding linear function $y = 2x + 20$. The linear function $y = 2x + 20$ graphs as a line with a vertical intercept at (0, 20) and a slope of 2. Without considering the scale, we can sketch the basic shape.

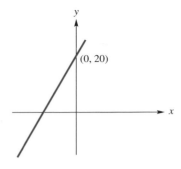

Because we need to reflect the portion of the graph that lies below the *x*-axis, we must determine the horizontal intercept. The *y*-coordinate of any horizontal intercept is 0. Therefore, we can find the horizontal intercept by setting $y = 0$ and solving for *x*.

$$0 = 2x + 20$$
$$^-20 = 2x$$
$$^-10 = x$$

The horizontal intercept of $y = 2x + 20$ is at ($^-10$, 0). With this information we can complete the graph of $y = 2x + 20$ and then reflect the portion of the graph of $y = 2x + 20$ that lies below the *x*-axis to graph the absolute value function $y = |2x + 20|$. The following graph shows both the linear function $y = 2x + 20$ and the absolute value function $y = |2x + 20|$. The dashed line indicates the portion of the line that is reflected.

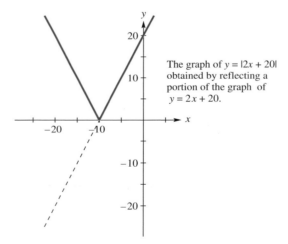

The graph of $y = |2x + 20|$ obtained by reflecting a portion of the graph of $y = 2x + 20$.

Therefore, the graph of $y = |2x + 20|$ is the following **V**.

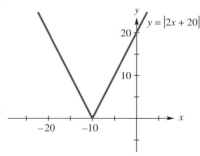

Example 2

Solve each equation or inequality graphically. Graph your solution on a number line. Write your solution using equality or inequality notation.

a. $|x + 6| = 8$ b. $|x + 6| < 8$ c. $\left|\frac{1}{3}x - 7\right| \geq 4$

Solution   a. To solve the equation $|x + 6| = 8$ graphically, we need to graph the two functions $y = |x + 6|$ and $y = 8$. The absolute value function $y = |x + 6|$ graphs as a V, and $y = 8$ graphs as a horizontal line. To determine the location of the V, we first consider the corresponding linear function $y = x + 6$. The line $y = x + 6$ has a vertical intercept at $(0, 6)$ and a slope of 1. If we set $y = 0$, we can find the horizontal intercept, which is at $(^-6, 0)$. From this we can make a quick sketch of the line to determine the graph of the absolute value function.

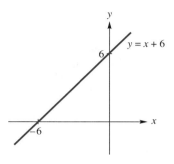

With this graph, we know that the absolute value function $y = |x + 6|$ is a V, where the vertex of the V is at $(^-6, 0)$.

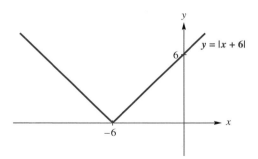

To graph both $y = |x + 6|$ and $y = 8$ we need to extend our window.

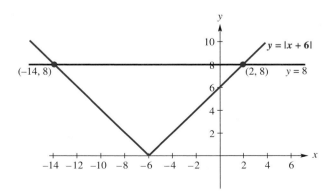

Now that we have the graphs, we can solve the original equation, $|x + 6| = 8$. The intersection points are $(^-14, 8)$ and $(2, 8)$. Therefore, the solution to the equation is

$$x = {}^-14 \quad \text{or} \quad x = 2$$

To check our solution, we can substitute each of the values into the original equation.

CHECK $x = {}^-14$ **CHECK** $x = 2$

$|{}^-14 + 6| = 8$ $|2 + 6| = 8$

$\qquad 8 = 8$ ✔ $\qquad 8 = 8$ ✔

Our solution checks. We can write the solution as $x = {}^-14$ or $x = 2$. It is graphed on the following number line.

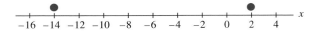

b. To solve the inequality $|x + 6| < 8$, we can use our graph from part a. However, for this part, we are solving an inequality rather than an equation. We know that the intersection points give us our border points. Therefore, $x = {}^-14$ and $x = 2$ are our border points. We can put open circles at $x = {}^-14$ and $x = 2$ because we want the absolute value to be less than but not equal to 8. These two border points divide the horizontal axis into three regions. We need to check each of the regions and determine the region(s) where the inequality is true.

To solve the inequality $|x + 6| < 8$ we need to determine the values of x that make the absolute value less than 8. If x is less than $^-14$, say at $^-17$, then the absolute value function produces values larger than 8. If x is between $^-14$ and 2, say at $^-8$, then the absolute value produces values less than 8. Because we want the absolute value to be less than 8, the values between $^-14$ and 2 are solutions. If x is greater than 2, say at $x = 5$, then the absolute value produces values larger than 8.

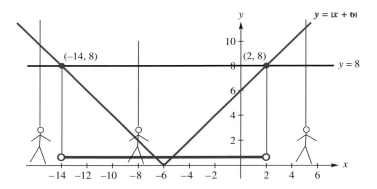

From the graph, the solution is all values between $^-14$ and 2. We can write this using inequality notation as

$$^-14 < x < 2$$

or alternatively as

$$x > {}^-14 \quad \text{and} \quad x < 2.$$

This solution is graphed on the following number line.

c. To solve $\left|\frac{1}{3}x - 7\right| \geq 4$ graphically, we need to graph $y = \left|\frac{1}{3}x - 7\right|$ and $y = 4$. The absolute value function $y = \left|\frac{1}{3}x - 7\right|$ graphs as a V, and the linear function $y = 4$ graphs as a horizontal line. To determine the location of the V, we consider the corresponding linear function $y = \frac{1}{3}x - 7$, which is a line with a vertical intercept at $(0, {}^-7)$ and a slope of $\frac{1}{3}$. To find the horizontal intercept of the line, we set $y = 0$ and solve for x.

$$0 = \frac{1}{3}x - 7$$
$$7 = \frac{1}{3}x$$
$$21 = x$$

The horizontal intercept is at $(21, 0)$. This allows us to sketch the graph of the line $y = \frac{1}{3}x - 7$. We can then use this to graph the absolute value function $y = \left|\frac{1}{3}x - 7\right|$.

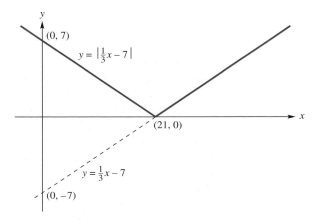

Now that we know what the absolute value function looks like, we can create a window that shows both and $y = \left|\frac{1}{3}x - 7\right|$ and $y = 4$.

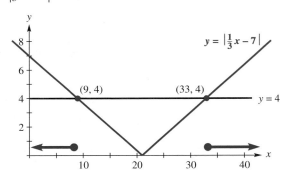

From the graph we can determine the solution to the inequality $\left|\frac{1}{3}x - 7\right| \geq 4$. First, the intersection points are $(9, 4)$ and $(33, 4)$. This tells us that our border points are at $x = 9$ and $x = 33$. We include the border points in our solution because the inequality is greater than or *equal to*. Next, we check each of the three regions that are created by the border points. We find that all values less than or equal to 9 are solutions and all values greater than or equal to 33 are solutions. This can be written using inequality notation as

$$x \leq 9 \qquad \text{or} \qquad x \geq 33$$

The solution is graphed on the following number line.

In Example 2, we solved both absolute value equations and inequalities graphically. However, we can also solve the equations algebraically with a little common sense. Let's consider a simple absolute value equation like $|x| = 5$. Because the absolute value function makes the output positive, the input could have been 5 or $^-5$. Therefore, the solution to $|x| = 5$ is $x = 5$ or $x = ^-5$. We will use this idea in the next example.

Example 3

Solve $|2.5x + 18| = 7$ algebraically. Check your solution.

Solution   It is important to note that we cannot simply drop or ignore the absolute value symbols. However, if an absolute value is equal to 7, then what's inside that absolute value could be 7 or $^-7$. Therefore, we can write

$$2.5x + 18 = 7 \qquad \text{or} \qquad 2.5x + 18 = ^-7$$

This yields two equations, and we must solve each of these to determine the solution to the absolute value equation.

$$
\begin{aligned}
2.5x + 18 &= 7 & \text{or} & & 2.5x + 18 &= ^-7 \\
2.5x &= ^-11 & \text{or} & & 2.5x &= ^-25 \\
x &= ^-4.4 & \text{or} & & x &= ^-10
\end{aligned}
$$

Next we check both of these values.

CHECK $x = ^-4.4$ **CHECK** $x = ^-10$

$|2.5 * ^-4.4 + 18| = 7$ $|2.5 * ^-10 + 18| = 7$

$\qquad\qquad 7 = 7$ ✔ $\qquad\qquad 7 = 7$ ✔

The solution to $|2.5x + 18| = 7$ is $x = ^-4.4$ or $x = ^-10$.

NOTE: We *cannot* use the algebraic method in Example 3 to solve absolute value inequalities. For now, we solve the inequalities graphically only.

In this section we have explored **absolute value functions** of the form $y = |mx + b|$. We know that these functions graph as V's. We can determine the location of the V by graphing the corresponding linear function $y = mx + b$ and then reflecting the portion of this line that falls below the horizontal axis. We used this information to solve absolute value equations and inequalities graphically. And finally, using a basic understanding of the absolute value function we were able to solve absolute value equations algebraically.

Problem
Set
4.3

1. Translate each distance statement into an absolute value equation or inequality.

   a. The distance between x and 0 is 10 units.

   b. The distance between w and $^-4$ is 5 units.

   c. The distance between x and 0 is greater than or equal to 5 units.

   d. The distance between x and 7 is less than 4 units.

2. Without using your calculator, determine whether each function graphs as a line or a V.

   a. $y = |4x + 15|$ **d.** $y = x$

   b. $y = 0.35x - 14$ **e.** $y = |100 - 5x|$

   c. $y = 23$ **f.** $y = |x|$

3. Predict what the graph of each function looks like. Then, graph the function on your calculator to verify your prediction.

   a. $y = |x + 9|$ **c.** $y = |x|$

   b. $y = |5x - 32|$ **d.** $y = |100 - 5x|$

4. Solve each absolute value equation algebraically. Numerically check your solution.

   a. $|x - 3| = 8$ **e.** $\left|\frac{5}{3}t + 4\right| = {}^-1.2$

   b. $|w + 6.2| = 3$ **f.** $|3.2x - 45.5| = 75.4$

   c. $|12 - 2x| = 3$ **g.** $\left|z - 10\frac{1}{2}\right| = 2\frac{1}{2}$

   d. $|0.25q + 1| = 2$ **h.** $|x + 1| = {}^-3$

5. Solve each absolute value equation or inequality graphically (include a sketch of the two-coordinate graph). Write your solution using equality or inequality notation. Graph your solution on a number line.

   a. $|x - 7| < 3$ **c.** $|x| \leq 7.5$

   b. $|x + 3| > 15$ **d.** $|8.5 - x| = 15$

6. Solve each absolute value equation or inequality graphically (include a sketch of the two-coordinate graph). Write your solution using equality or inequality notation. Graph your solution on a number line.

   a. $|M + 12| \geq 6$ **d.** $|0.25q + 1| > 2$

   b. $\left|\frac{x}{2} + 5\right| \geq 2\frac{1}{2}$ **e.** $|W - 5| \geq {}^-1.2$

   c. $|45 - 3x| = 18$ **f.** $|x + 1| < {}^-3$

7. Weight of M&M's.   The true weight of a 16 ounce bag of plain M&M's can vary by small amounts. About 95% of all bags have weights given by the absolute value inequality

   $$|w - 16.0| \leq 0.98$$

   where w represents the actual weight of the bag. Determine the range of weights for 16-ounce bags of M&M's.

8. Women's Heights.   The height of 95% of the women in a certain population satisfies the absolute value inequality

$$|h - 64.5| \leq 5.20$$

where h represents height measured in inches. Determine the range of the heights.

9. Use the following graph to approximate the solution to each absolute value equation or inequality. Write your solution using equality or inequality notation. Graph your solution on a number line.

a. $|77x + 153| = 115.5$ **b.** $|77x + 153| > 115.5$

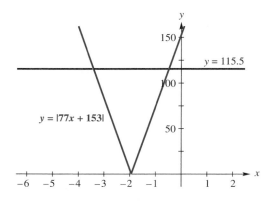

10. Solve each of the following absolute value inequalities graphically. Graph the solution on a number line. Include a sketch of the two-coordinate graph that you use to solve the inequality.

a. $x + 15 \geq |48 - 4x|$ **b.** $|0.4x + 15| > 6 - 0.2x$ **c.** $|x + 12| \leq 5 + x$

11. *Use the following graph* to solve each absolute value inequality. Write your solution using equality or inequality notation. Graph your solution on a number line.

a. $|mx + b| = c$ **b.** $|mx + b| > c$ **c.** $|mx + b| \leq c$ **d.** $|mx + b| < {}^-10$

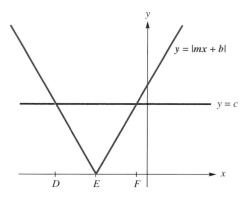

4.4 Absolute Value Functions—Analytically

Discussion 4.4

In the previous section we saw how to solve absolute value equations and inequalities graphically. In this section we will look at solving these equations and inequalities analytically using a distance concept. You may recall from the activities in the previous section that the table of values and the graphs for the function $y = |x - 3|$ and the function described as "the distance between the input and 3" were the same. We can solve absolute value statements by translating them from absolute value notation into distance statements.

Absolute Value as Distance

$|x - c|$ represents the distance between x and c on a number line.

$|x + c|$ represents the distance between x and *the opposite* of c on a number line.

Example 1

Solve each absolute value inequality analytically using a distance statement. Report your solution using inequality notation. Graph your solution on a number line.

a. $|x - 3| \leq 2$ b. $|w + 4| > 6$

Solution

a. First we rewrite the inequality $|x - 3| \leq 2$ as a distance statement.

The distance between x and 3 is less than or equal to 2 units.

In other words, x is *at most* 2 units away from 3 on a number line. Two units to the left of 3 is the point 1 and two units to the right of 3 is the point 5. What other points satisfy the inequality? One unit from 3 are the points 2 and 4, because this distance is less than 2, the points 2 and 4 are in the solution. So, our solution appears to be all of the points between 1 and 5, including 1 and 5. The solution is shown on the following graph.

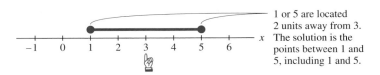

1 or 5 are located 2 units away from 3. The solution is the points between 1 and 5, including 1 and 5.

Using inequality notation, we write the solution as $1 \leq x \leq 5$ or as $x \geq 1$ and $x \leq 5$.

To check our solution, we need to check the border points 1 and 5. Notice that two border points divide the number line into three regions. We should check a point in each of the regions as well as the border points. We expect our solution to be in the region between 1 and 5. We do not expect points to the left of 1 or to the right of 5 to solve the inequality. A table helps us organize this information.

CHECK

	x	$|x - 3|$	$\leq$	2	Conclusion
Border Point (should be equal)	1	2	=	2	✔
Border Point (should be equal)	5	2	=	2	✔
Solution Region (should be true between 1 and 5)	4	1	$\leq$	2	✔
Nonsolution Region (should be false left of 1)	0	3	$\nleq$	2	✔
Nonsolution Region (should be false right of 5)	7	4	$\nleq$	2	✔

b. $|w + 4| > 6$ can be interpreted as

The distance between w and the opposite of 4 (or ⁻4) is greater than 6 units.

In other words, w is more than 6 units away from ⁻4 on a number line. We can then use this distance statement to graph the solution. Starting at the point ⁻4 on a number line, we find that 6 units to the left is at ⁻10 and 6 units to the right is at 2. These are the border points for our solution. However, because w is more than 6 units away from ⁻4, these values are not a part of the solution. To be farther away from ⁻4, w needs to be to the left of ⁻10 or to the right of 2. This can be seen on the following number line.

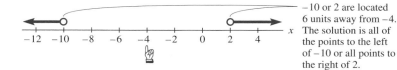

⁻10 or 2 are located 6 units away from ⁻4. The solution is all of the points to the left of ⁻10 or all points to the right of 2.

Using inequality notation, we write the solution as

$w < {}^-10$ or $w > 2$

CHECK

	w	$|w + 4|$	$>$	6	Conclusion
Border Point (should be equal)	⁻10	6	=	6	✔
Border Point (should be equal)	2	6	=	6	✔
Solution Region (should be true left of ⁻10)	⁻12	8	>	6	✔
Solution Region (should be true right of 2)	5	9	>	6	✔
Nonsolution Region (should be false between ⁻10 and 2)	0	4	$\ngtr$	6	✔

We conclude that our solution is $w < {}^-10$ or $w > 2$.

Suppose we want to solve the inequality $|2x - 1| \leq 3$ analytically. Using the distance statement, we can translate this inequality as *The distance between 2x and 1 is less than or equal to 3 units.* In this situation, we can graph the possible values for *2x* and then solve for *x*. Because *2x* can be at most 3 units from 1, *2x* can be three units to the right of 1, at 4, three units to the left of 1, at ⁻2, or anywhere in between. We obtain the following graph for the values of *2x*.

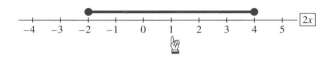

Therefore, $2x \geq {}^-2$ and $2x \leq 4$. Next, solve each of these inequalities for *x*.

$$2x \geq {}^-2 \qquad \text{and} \qquad 2x \leq 4$$
$$x \geq {}^-1 \qquad \text{and} \qquad x \leq 2$$

The solution is $x \geq {}^-1$ and $x \leq 2$, which can also be written as $^-1 \leq x \leq 2$. It is important to note that the preceding number line is not the graph of the solution. It is an intermediate step to help us solve for *x*. The final solution is graphed on the following number line.

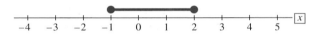

CHECK

	x	$\lvert 2x - 1 \rvert$	$\leq$	3	Conclusion
Border Point *(should be equal)*	⁻1	3	=	3	✔
Border Point *(should be equal)*	2	3	=	3	✔
Solution Region *(should be true between ⁻1 and 2)*	1	1	$\leq$	3	✔
Nonsolution Region *(should be false left of ⁻1)*	⁻3	7	$\nleq$	3	✔
Nonsolution Region *(should be false right of 2)*	5	9	$\nleq$	3	✔

We conclude that our solution is $^-1 \leq x \leq 2$.

Example 2

Solve the inequality $\left|\frac{1}{2}w + 3\right| > 5$.

Solution

The inequality $\left|\frac{1}{2}w + 3\right| > 5$ can be rewritten as the distance statement

The distance between $\frac{1}{2}w$ and $^-3$ is greater than 5 units.

The location of $\frac{1}{2}w$ must be more than 5 units from $^-3$. Five units to the right of $^-3$ is at 2; five units to the left of $^-3$ is at $^-8$. Because $\frac{1}{2}w$ must be more than 5 units away from $^-3$, $\frac{1}{2}w$ must be greater than 2, or $\frac{1}{2}w$ must be less than $^-8$. With this we obtain the following graph for $\frac{1}{2}w$.

From the graph, we can see that

$$\frac{1}{2}w < ^-8 \qquad \text{or} \qquad \frac{1}{2}w > 2.$$
$$w < ^-16 \qquad \text{or} \qquad w > 4 \qquad \text{Solve each inequality for } w.$$

CHECK

	w	$\left\|\frac{1}{2}w + 3\right\|$	$>$	5	Conclusion
Border Point (should be equal)	$^-16$	5	$=$	5	✔
Border Point (should be equal)	4	5	$=$	5	✔
Solution Region (should be true left of $^-16$)	$^-20$	7	$>$	5	✔
Nonsolution Region (should be false between $^-16$ and 4)	0	3	$\not>$	5	✔
Solution Region (should be true right of 4)	10	8	$>$	5	✔

We conclude that $w < ^-16$ or $w > 4$ is the solution to $\left|\frac{1}{2}w + 3\right| > 5$. Following is the graph of the solution.

Example 3

Coke.   When a bottling company fills a 2-liter bottle of Coke, they are allowed a tolerance of 15 milliliters. This means that a 2-liter bottle can have 15 milliliters more or less than 2 liters and still meet industry standards. Graph the possible amount of Coke in a 2-liter bottle on a number line. Write the amount algebraically as a compound inequality, and write the amount as an absolute value inequality.

Solution   First we need to convert the measurements to a single unit. We convert to milliliters, although it is just as easy to convert all of the measurements to liters.

$$2 \text{ L} = 2000 \text{ mL}$$

Then the amount of Coke in a 2-liter bottle is between

$$2000 \text{ mL} - 15 \text{ mL} = 1985 \text{ mL} \qquad \text{and} \qquad 2000 \text{ mL} + 15 \text{ mL} = 2015 \text{ mL}$$

Next, we graph this information on a number line.

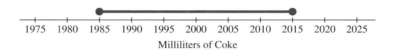

Milliliters of Coke

Let v = volume in a 2-liter bottle of Coke in milliliters; then the solution can be written as the compound inequality

$$1985 \le v \le 2015$$

To write the solution as an absolute value inequality, we need to know where the center of the graph is and the distance from the center to the endpoints. From the problem situation, we know that the graph is centered at 2 L or 2000 mL and that the endpoints are 15 mL to the left and right of 2000 mL

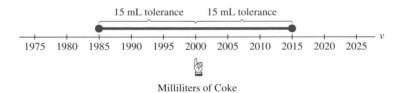

Milliliters of Coke

It might be helpful to state this as the distance statement

   The distance between v and 2000 is less than or equal to 15.

Symbolically, this is written as

$$|v - 2000| \le 15$$

Notice that in Example 3, the compound inequality statement tells us the extreme values that are acceptable for the volume of a 2-liter bottle. The absolute value inequality tells us the goal, or center value, and how far off the volume could be from this amount.

Problem
Set
4.4

1. Translate each distance statement into an absolute value equation or inequality.

   a. The distance between x and 0 is 5 units.

   b. The distance between w and 15 is 10 units.

   c. The distance between y and $^-9$ is greater than or equal to 2 units.

   d. The distance between $3x$ and $^-7$ is less than 4 units.

2. Solve each absolute value equation or inequality analytically using a distance statement. Report your solution using equality or inequality notation and graph your solution on a number line.

   a. $|x - 10| = 5$

   b. $|T - 6| < 5$

   c. $|y + 17| \leq 20$

   d. $|2m + 7| < 8$

3. Solve each absolute value equation or inequality analytically using a distance statement. Report your solution using equality or inequality notation and graph your solution on a number line.

   a. $|w - 10| > 0.5$

   b. $|x + 2.3| \geq 1.7$

   c. $|x - 1005| \leq 10$

   d. $|R - 5| > ^-2$

   e. $|0.25w - 1.75| \geq 1$

   f. $|x + 3| < ^-2$

4. **The Life of a Hard Disk.**   The lifetime L of a hard disk in a computer, for 95% of those produced by a manufacturer, satisfies the absolute value inequality

   $$|L - 27{,}350| \leq 5880$$

   where L is measured in hours of use. Determine the range of hours for the lifetime of a hard disk produced by this manufacturer.

5. **A Few Good Men and Women.**   A high school counselor was interested in how many students are planning to volunteer for military service after graduation. She randomly selected and questioned 50 students, and 15 answered that they planned to enlist after graduating. Using statistical techniques, she found that she could predict the percentage P of students that would enlist in military service after graduation. Her computations resulted in the inequality

   $$|P - 30| \leq 12.3$$

   Determine the range for the percent of students who plan to enlist in the military services after graduation. If 580 seniors graduate from this school, what is the range for the number of graduates planning to enlist?

6. **Airplane Specifications.**   An airplane manufacturer is producing a ball bearing whose diameter is to measure 1.7500 inches. The manufacturing process is required to be extremely accurate. The tolerance allowed on the ball bearing is one ten-thousandth of an inch.

   a. On a number line, graph the range of acceptable diameters for the ball bearing.

   b. Write a compound inequality for the range of acceptable diameters for the ball bearing.

   c. Write an absolute value inequality for the range of acceptable diameters for the ball bearing.

   d. Write an absolute value inequality for the range of diameters of ball bearings that are rejected by the manufacturer.

7. Fishing Line.   Fishing line is rated by pounds of strain at which it is expected to break. If a fishing line is rated at 60 pounds, the line is expected to break if a fish of about 60 pounds pulls at it. In reality, we do not expect all 60-pound lines to break at exactly 60 pounds. One manufacturer sets the specifications for their 60-pound line to have a tolerance of 5 pounds.

    a. On a number line, graph the interval for the weights at which the 60-pound line is expected to break.

    b. Write a compound inequality representing the interval of the weights at which the 60-pound line is expected to break.

    c. Write an absolute value inequality representing the interval of the weights at which the 60-pound line is expected to break.

8. The following are graphs of the solutions to absolute value equations and inequalities. For each graph, write the solution using equality or inequality notation, and determine an *absolute value* equation or inequality that has the given solution.

    a.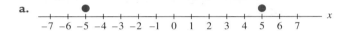

    b.

    c.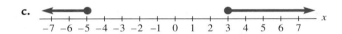

    d.

    e.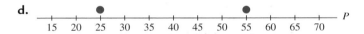

    f.

Activity
Set
4.5

4.5 Statistics

1. Use the following set of data to answer the questions.

   10, 13, 19, 15, 17, 11, and 13

   a. What is the mean, the median, and the mode for this data set?

   b. What percent of the data falls above the mean? Below the mean?

   c. The data set is represented on the following number line. Locate the mean and the median on this number line. What observations can you make about their locations?

   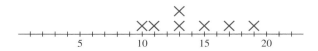

   d. Compute the difference between each data entry and the mean to fill in the following table. What is the sum of the differences between each value and the mean?

Data Entry	Data Entry − Mean
10	10 − 14 = ⁻4
11	
13	
13	
15	
17	
19	
Sum of the differences =	

2. Use the following set of data to answer the questions.

   10, 13, 33, 15, 17, 11, and 13

   a. What is the mean, the median, and the mode for this data set?

   b. What percent of the data falls above the mean? Below the mean?

   c. What percent of the data falls above the median? Below the median?

   d. Represent the data set on the following number line, similar to Activity 1. Locate the mean and the median on this number line.

   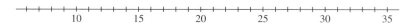

   e. Compute the difference between each data entry and the mean to fill in the following table. What is the sum of the differences between each value and the mean?

   f. What did you discover about the sum of the differences between the data entries and the mean?

   g. Discuss which of the measures of central tendency is affected by data in which a few of the values are extremely higher or lower than the others.

Data Entry	Data Entry − Mean
10	10 − 16 = ⁻6
11	
13	
13	
15	
17	
33	
Sum of the differences =	

3. a. Write three different data sets with at least seven data entries whose mean is 25 and whose median is 25.

   b. Write three different data sets with at least seven data entries whose mean is 25 and whose median is 20.

4. The numbers of items produced per shift by two different factories are listed in the following table. Which factory has more variability in the production levels? Explain how you arrived at your conclusion.

| *Production Level for Factory 1* | 85 | 89 | 90 | 34 | 79 | 86 |
| *Production Level for Factory 2* | 46 | 59 | 35 | 72 | 91 | 67 |

Discussion 4.5

Today we receive more and more information in the form of statistical data. We gather data in the workplace, in marketing and advertising, in politics and issues that affect public policy, and even in our recreation and sports. One of the goals of a statistician is to organize the vast amount of data received and summarize that data. In this section we will look at some of the ways statistics can be used to summarize data.

The first step in summarizing a list of data is usually to list the data in ascending or descending order. It is then easier for a reader to determine the minimum and maximum values. This is known as **sorting** the data.

Sorting the data gives us, at a glance, more information than reading through long lists of unorganized numerical data. However, we are still looking at *all* of the data in the list. Consider how large the list would be of all two-bedroom apartments available for rent in Los Angeles. This would be too much information to easily read from a list. One method of summarizing data is to report a single value for the measure of central location. Three numbers report the central location of a set of data: the mean (or arithmetic mean), the median, and the mode. When summarizing a set of data, we will use one of the three different measures of central location.

The mean, commonly referred to as the average, is easily computed and takes into consideration all of the data values. However, the mean is affected by extremely high or low values.

Definition

The **mean** of a set of data is the average found by adding all of the data entries and then dividing by the number of data entries.

$$\text{mean} = \frac{\text{sum of the data entries}}{\text{number of data entries}}$$

or symbolically as

$$\bar{x} = \frac{x_1 + x_2 + x_3 + \cdots + x_n}{n}.$$

where $x_1, x_2, x_3, \ldots, x_n$ represent the data entries and n represents the number of data entries.

Definition _____

The **median** is the value that falls in the center of a data set sorted in ascending or descending order. The median splits the data into approximately the upper 50% and the lower 50%.

How to Find the Median

To find the median, first sort the data. Then,

- If the number of data entries is odd, the value that falls in the middle is the median.
- If the number of data entries is even, the average of the two middle values is the median.

The mode is important when looking at the data value that occurs the most often. If no value occurs more than once, there is no mode for the data set. Some data sets will have more than one value for the mode.

Definition _____

The **mode** is the value that occurs most often in a set of data.

Example 1   List A: 3, 5, 8, 6, 4

List B: 23, 79, 84, 109, 96, 79

    a. Determine the mean and median of List A.

    b. Determine the mean and median of List B.

    c. Compare the mean and the median of List A. Compare the mean and the median of List B.

Solution   a. The mean of List A is found using the formula

$$\bar{x} = \frac{3 + 5 + 8 + 6 + 4}{5}$$

$$\bar{x} = 5.2$$

The mean of List A is 5.2.

To find the median of List A, we must first sort the data. Then we locate the value that falls in the middle.

List A

    3

    4

    5 . . .   The median is 5.

    6

    8

b. The mean of List B is

$$\bar{x} = \frac{23 + 79 + 84 + 109 + 96 + 79}{6}$$

$$\bar{x} \approx 78.3$$

The mean of List B is approximately 78.3.

To find the median of List B, we first sort the data. Because the number of data entries in List B is even, no data value falls directly in the middle. To determine the median we locate the middle two values and compute the average of these two values.

List B

  23

  79

  79 ⎫
      ⎬ ··· The median is $\frac{79 + 84}{2} = 81.5.$
  84 ⎭

  96

  109

c. The mean and median of List A are about equal. They both could be used as a measure of central tendency for the data list.

The mean and median of List B differ more than the results for List A. This difference may not be significant because the numbers in the second list are larger. To help us decide, let's look at the results in a different way. In List A, three data values fall below the mean of 5.2 and two values are above the mean. This mean falls close to the middle of the data. In List B, one data value falls below the mean of 78.3 and five values are above. The mean does not serve as a good measure of central location for this data list. The number 23 in List B appears to be an outlier, that is, a value far removed from the rest of the data in the list. The mean of a data set is affected by such extreme values in the set.

Different choices of measures of central tendency are more useful in certain applications. A fisheries expert wanting to test the weights of salmon migrating upstream to spawn is most likely to use the mean because the data is not likely to include many extreme values, and this number is easily computed. A city real estate agent summarizing the costs of houses sold during the month of June may want to use the median because costs of housing can often include outliers, homes that sell for much more (or much less) than the rest of the market. A clerk in a dress shop records the sizes of dresses sold to help predict future buying. The clerk is most interested in the mode of the sizes. It is our job to determine the best choice to use as the measure of central location to summarize a set of data.

In comparing or summarizing data sets, we are also interested in determining a number to communicate the amount of variability within the data set. A measure of variability should tell us something about how spread out and how predictable the data are. The range and standard deviation are two ways to measure variability. The range is the difference between the maximum and the minimum data values. It is easily computed, but only the two extreme values are considered when computing the range.

Definition _____

The **range** of a set of data is the difference between the maximum and the minimum data values.

  range = maximum data entry − minimum data entry

The standard deviation of a data set measures variability associated with each of the data entries. For this reason it is more difficult to compute. However, the standard deviation gives us more information about the overall variability in a data set.

Definition _____

The **variance** of a set of data is the sum of the squares of the differences between each data entry and the mean, divided by the number of data entries.

$$\text{variance} = \frac{\text{sum of } (x_i - \overline{x})^2}{\text{number in data set}}$$

Definition _____

The **standard deviation** of a set of data is equal to the square root of the variance.

$$\text{standard deviation} = \sqrt{\text{variance}} = \sqrt{\frac{\text{sum of } (x_i - \overline{x})^2}{\text{number in data set}}}$$

Example 2

Comparing Heights of Players.   The following are frequency graphs for the heights of the girls on two different tournament volleyball teams. Find the mean and the standard deviation for each team. Use these results to compare the data sets.

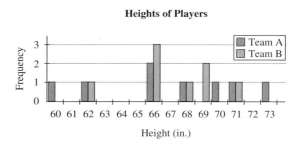

Heights of Players

Solution   The mean height of the players on Team A is

$$\overline{x} = \frac{60 + 62 + 2 * 66 + 68 + 70 + 71 + 73}{8}$$

$$\overline{x} = 67 \text{ inches}$$

Compute the mean of Team B using the formula or the statistical function on your calculator. The mean of Team B is 67.125 inches, or $67\frac{1}{8}$ inches.

Our next task is to compute the standard deviation for each team. We look at Team A first. The first step in computing the standard deviation is to compute the differences between each data entry and the mean of 67 inches. This work is shown in the following table.

Heights (in.) x_i	Differences (in.) $x_i - \overline{x}$
60	$60 - 67 = {}^-7$
62	$^-5$
66	$^-1$
66	$^-1$
68	1
70	3
71	4
73	6

We observed in the activities that the sum of the right column in the preceding table is always zero. This is because the mean acts as a balance point for the data set.

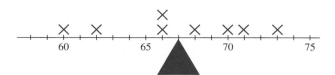

The fulcrum for the balance is at the mean of 67.

Because the sum of the differences always equals zero, we square the column of differences and then sum that column.

Heights (in.) x_i	Differences (in.) $x_i - \bar{x}$	Square of Differences (in.^{2}) $(x_i - \bar{x})^2$
60	$60 - 67 = {}^-7$	$({}^-7)^2 = 49$
62	${}^-5$	25
66	${}^-1$	1
66	${}^-1$	1
68	1	1
70	3	9
71	4	16
73	6	36
Sum of squares of differences = 138 in.^{2}		

The variance is the average of the squares of the differences for the data set. To calculate this we divide the sum of the squares of the differences by the number of data entries.

$$\text{variance} = \frac{138 \text{ in.}^2}{8} = 17.25 \text{ in.}^2$$

Because we are comparing heights of volleyball players, it seems strange to be working with numbers with square inches as their units. It is reasonable to use standard deviation and not variance as a measure of variability for this data set. The **standard deviation** is the square root of the variance.

$$\text{standard deviation} = \sqrt{17.25 \text{ in.}^2} \approx 4.2 \text{ in.}$$

We can repeat this process for team B.

Heights (in.) x_i	Differences (in.) $x_i - \bar{x}$	Square of Differences (in.^{2}) $(x_i - \bar{x})^2$
62	$62 - 67.125 = {}^-5.125$	$({}^-5.125)^2 \approx 26.27$
66	$^-1.125$	1.27
66	$^-1.125$	1.27
66	$^-1.125$	1.27
68	0.875	0.77
69	1.875	3.52
69	1.875	3.52
71	3.875	15.02
Sum of squares of differences = 52.91 in.^{2}		

The variance is $\dfrac{52.91 \text{ in.}^2}{8} \approx 6.61 \text{ in.}^2$

The standard deviation is $\sqrt{6.61 \text{ in.}^2} \approx 2.6 \text{ in.}$

We found that for Team A, the mean of the heights of the players is 67 inches, and the standard deviation is approximately 4.2 inches. For Team B, the mean of the heights of the players is $67\frac{1}{8}$ inches, and the standard deviation is approximately 2.6 inches. The mean heights are very close. The standard deviation of Team A is much larger than the standard deviation for Team B. There is more variability in the heights on Team A.

Let's review the steps we just followed to compute the standard deviation.

Strategy

Finding Standard Deviation

1. Compute the mean of the data set.
2. Compute the difference between each data value and the mean of the data set.
3. Square each of the differences.
4. Sum the squares of the differences, and divide by the number of data entries. This number is the variance of the data set.
5. Compute the square root of the variance. This number is the standard deviation.

Many calculators compute the statistical values of mean and standard deviation. The variable $\bar{x}$ usually represents the mean. The variable σ or σ_x is used for the standard deviation. Some also compute the median. These values provide statisticians with numerical values to summarize sets of data.

One method to interpret standard deviation is to use the fact that for any data set at least 75% of the data falls within two standard deviations of the mean. That means that at least 75% of the data falls between

$$mean - 2 * standard\ deviation \qquad and \qquad mean + 2 * standard\ deviation$$

This is a graphical representation of the values that fall within two standard deviations of the mean.

At least 75% of the data

$\bar{x} - 2\sigma \qquad \bar{x} \qquad \bar{x} + 2\sigma$

Example 3

Apartment Rentals.   Ads in a local paper listed the following monthly rents for a two-bedroom apartment.

$1200, $600, $1350, $650, $595, $1100, $795, $750, $995, $700, $725, $558, and $750

a. Compute the mean and the standard deviation.

b. Determine the value that is two standard deviations above the mean and the value that is two standard deviations below the mean.

c. Determine the percentage of the data that actually falls within two standard deviations of the mean.

Solution

a. Using the technology or the step-by-step process, we can determine that the mean monthly rent for a two-bedroom apartment is about $828 and the standard deviation is approximately $242.

b. To determine the values that are two standard deviations above the mean and two standard deviations below the mean, we compute:

$$\text{mean} + 2 * \text{standard deviation} = \$828 + 2 * \$242 = \$1312$$
$$\text{mean} - 2 * \text{standard deviation} = \$828 - 2 * \$242 = \$344$$

c. We expect that at least 75% of the data falls between a low value of $344 and a high value of $1312. Looking at the data, we see that one value, $1350, does not fall between $344 and $1312. Therefore, 12 out of 13 of the values fall within two standard deviations of the mean. As a percent this is $\frac{12}{13} \approx 0.923$, or 92.3%.

In this section, we saw how the numerical values of the **mean,** the **median,** and the **mode** can be used to describe the central location for a set of data. The **range** and the **standard deviation** are measures of variability in a set of data. If we are given the mean and the standard deviation of a data set, we can say that at least 75% of all of the data fall within two standard deviations of the mean. The values for central location and variability are used to provide a summary for a set of data.

Problem Set 4.5

1. **Your Final Grade.**   Your test scores for the term are 23, 79, 84, 100, 96, and 88. Each of these test scores is out of a possible 100 points. Your final grade is determined by the following scale. Do you want your grade to be based on the mean, median, or the mode of these scores? Explain.

Grading Scale	
90–100%	A
80–89%	B
70–79%	C
60–69%	D
Below 60%	F

2. **Computing a Bonus.**   Your sales for the last six days are $120, $555, $240, $120, $265, and $0. You boss has just announced that a bonus will be given to all employees based on one of the measures of central location using the sales for these six days. Would you rather your bonus be based on the mean, the median, or the mode? Explain.

3. **Best Bet.**   The results of a game of chance are displayed in the following table. It is your task to guess the next outcome for this game. In deciding on the "best bet" for the next outcome of the game, are you more interested in the mean, the median, or the mode of the outcomes of the game? Explain.

Outcomes	1	2	3	4	5
Frequency	15	20	5	3	1

4. Light Beer.   The number of calories in "light" beer were recorded for 12-oz samples of different brands. The results are 106, 99, 101, 103, 108, 107, 107, and 106.

   a. Compute the mean and the standard deviation for this data set.

   b. Sketch a number line representation of the data.

   c. Locate the mean on the number line.

   d. Locate the endpoints of the interval within two standard deviations of the mean.

   e. What percent of the data actually falls within two standard deviations of the mean?

5. a. Which do you expect to be higher, the standard deviation of the reaction time for 30 randomly selected adults or the reaction time of 30 adults who have not slept for 48 hours? Explain.

   b. Which do you expect to be higher, the standard deviation of the I.Q.'s for 30 randomly selected adult mathematics students or the I.Q.'s of 30 randomly selected adults? Explain.

6. Games of Chance.   The outcomes for two different games of chance are recorded in the following frequency graphs.

   a. Find the range for each game. Can you use the range to determine which game has more variability?

   b. Find the standard deviation for each game. Can you use the standard deviation to determine which game has more variability?

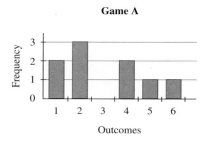
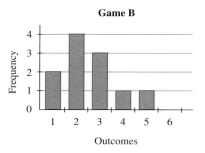

7. Recess Anxiety.   An anxiety level index is invented for third-grade students, and the results for each of 14 randomly selected students are obtained in a classroom setting and a recess setting. A scale from 0 to 10, where 0 represents low anxiety and 10 represents high anxiety, is created to quantify observed behaviors that indicate when students were feeling anxious.

   a. Use the mean to decide whether more anxiety occurs in the classroom or during recess. Explain your results.

   b. Compute the standard deviation to determine whether the variability in the anxiety levels is higher in the classroom or at recess. Explain your results.

Classroom	Recess	Classroom	Recess
8.2	9.7	9.1	8.7
8.8	8.1	4.1	2.5
7.0	6.8	3.8	6.7
5.1	4.2	5.8	4.0
3.9	4.0	4.9	3.4
4.7	3.6	8.7	7.9
8.1	7.1	8.2	7.0

8. SAT Scores.   The mean score for the College Entrance Examination Board Scholarship Aptitude Test is 500, with a standard deviation of 100. What can you say about the spread of the scores?

9. Data Project.   Gather a set of data of interest to you. The set should include at least 25 numerical entries. Decide on the best measures of central location and variability to summarize your data set. Write a brief paragraph to describe the data set. Include the values of the central location and variability in your paragraph, and interpret these results in the context of your data set.

10. A Television Survey.   Ask ten college students to estimate the average number of hours they spend watching television each week, and record this data. Ask ten random adults who are not college students to estimate the average number of hours they spend watching television each week, and record this data. Compare the two sets in terms of the central tendency for each set and the variability for each set.

11. Quality Control.   In the process of manufacturing a new ink jet printer, the number of hours required to complete the process for the first 12 machines were recorded.

Machine Number	1	2	3	4	5	6	7	8	9	10	11	12
Number of Hours in Production Process	59	64	57	62	52	58	65	71	63	71	63	64

a. Graph the data from the table.

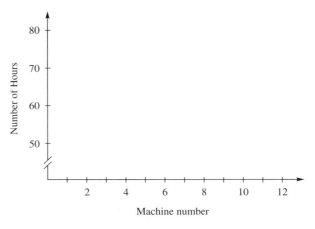

b. Compute the mean of the number of hours in the production process. Locate this value on the vertical axis. Draw a horizontal line through this point on your graph.

c. Compute the standard deviation of the number of hours in the production process. Compute the mean minus two standard deviations and the mean plus two standard deviations. Locate these two points on the vertical axis. Draw a dotted horizontal line through each of these points.

d. Do all of the data points lie between the two dotted lines?

12. Is a College Degree Worth the Paper It's Printed On?   The following data lists the lifetime earnings estimate by educational attainment. This information is based on 1992 data.

Level of Education Finished		Lifetime Earnings Estimate ($)
Don't Finish High School	(10th grade)	609,000
High School Graduate	(12th)	821,000
Attended College	(13th)	993,000
Associate Degree	(14th)	1,062,000
Bachelor's Degree	(16th)	1,421,000
Master's Degree	(18th)	1,619,000
Doctorate	(20th)	2,142,000

a. Plot the lifetime earnings in terms of the grade completed on a two-dimensional coordinate system. Draw an eyeball fit line through the data.

b. Write an equation for the lifetime earnings in terms of the last grade completed.

c. Complete column two of the following table using your equation. Compute the difference between the output of the equation and the actual value from the preceding list. Include this information in the third column.

d. Square the results of the third column, and find the sum of that column. Compare your results with others from your class. The best eyeball fit line has the smallest sum.

Grade Completed	Output from Eyeball Fit Line	Difference Output − Data	Square of the Differences
10			
12			
13			
14			
16			
18			
20			

Chapter Four Summary

We started this chapter by **solving linear inequalities algebraically.** With one exception, linear inequalities can be solved algebraically using the same strategies that we use to solve linear equations. The exception is that when we multiply or divide both sides of an inequality by a negative number, the direction of the inequality must be reversed. The solution to a linear inequality is usually an entire region of numbers. To visualize the solution we usually graph the solution on a number line.

We can also solve linear inequalities graphically. To do this we can proceed using the following strategy.

Strategy

Solving a Linear Inequality Graphically

1. Graph each side of the inequality on a graphing calculator. Consider the effects of the parameters on the graph to determine a reasonable graphing window.
2. Find the point of intersection. The horizontal coordinate is the border point of the solution region.
3. Use the graph to check the region to the left of the border point and to the right of the border point. One of these regions is the solution region.
4. Graph the solution region on a number line.
5. Check your solution. Check the border point where both sides of the inequality should be equal. Check the solution region where the inequality should be true. Check the region that is not part of the solution where the inequality should be false.

We added to our collection of functions an **absolute value function** of the form $y = |mx + b|$. We know that these functions graph as V's. We can graph the function $y = |mx + b|$ by graphing the corresponding linear function $y = mx + b$ and reflecting the portion of $y = mx + b$ that falls below the horizontal axis.

Inequalities that involve absolute value functions and linear functions can be solved graphically. The process is similar to the preceding strategy. The only difference is that usually two intersection points create two border points. This divides the number line into three regions that must be checked. The solution to an absolute value inequality is usually a compound inequality.

A **compound inequality** is the combination of two or more inequalities. We use the word **and** and the word **or** to connect them. In mathematics, we use the word *and* to mean that both conditions are true. We use the word *or* to mean that either condition is true or both conditions are true. If we need to write an inequality for a region between two numbers, we can either use a compound inequality containing the word *and* or we can write the inequality using an in-between statement. For example, if the region is between 1 and 8, not including either endpoint, we use either of the following inequality statements to indicate this.

$$x > 1 \qquad \text{and} \qquad x < 8$$
$$1 < x < 8$$

Absolute value equations can be solved graphically. The solutions are the horizontal coordinates of the intersection points. We can also solve these equations analytically using common sense. If the absolute value of an expression is equal to a positive number, then the expression is either that positive number or its opposite. It is important to remember that we can *never* just drop the absolute value signs from an equation or inequality.

In Section 4.4, we solved absolute value equations and inequalities similar to those in Section 4.3. However, rather than solving them graphically, we solved them using the distance definition associated with the absolute value. To do this, we must first translate the absolute value equation or inequality into a "distance statement." We then use this statement to graph the solution on a number line. Typically there is a "center point," and we move in both directions a certain distance, based on our distance statement, from that center point.

In Section 4.5, we looked at different methods of summarizing data with statistics. Summaries of data typically tell us about the central value and how much variation occurs in the data. Three numbers can be used to describe the center or typical value for a set of data: the mean, the median, and the mode.

- The **mean** is an average of a set of numbers. It is found by adding up all of the numbers and then dividing by the number of values.
- The **median** is the number that falls in the middle of a set of data when the data is arranged from least to greatest.
- The **mode** is the value that occurs most often in a set of numbers.

When reporting the central value of a data set, we must select the appropriate measure to use.

- The **mean** works well if the data set has no excessively high or low values.
- The **median** is a good choice for data sets that contain only a few numbers that are excessively high or low.
- The **mode** is the appropriate measure when we are interested in the number that occurs most often.

The **range** or **standard deviation** can measure the variability in a set of data. If we know the mean and standard deviation for a set of data, we can say that at least 75% of all of the data falls within two standard deviations of the mean.

Chapter Five

Exponents

In Chapter 3, we learned a set of properties that allowed us to simplify expressions with integer exponents. In this chapter, we will continue our study of exponents by investigating the meaning of fractional exponents. We will see that fractional exponents are defined so that the properties of exponents previously studied continue to work. Our experience in careful reading, interpreting, and substituting values into expressions with exponents will be useful as we continue to build on this knowledge in this chapter. By expanding our definition of exponents to include fractional exponents, we will now be able to solve some equations involving powers.

5.1 Reciprocal of Integer Exponents

Activity Set 5.1

1. We have a clear idea of what 9^n means when n is an integer, but we have never been given a definition of what 9^n means when n is a fraction. Now we need to explore what $9^{1/2}$ means.

   As we are exploring this new concept, please do not enter the expression in your calculator. First we need to *understand* what fractional exponents mean; later we will use the calculator to evaluate expressions in which the arithmetic is difficult.

   a. Complete the table. *Use your table* to estimate the value of $9^{1/2}$.

n	9^n
-2	
-1	$\frac{1}{9}$
0	
1	

   b. Following is a graph of $y = 9^n$. *Use the graph* to estimate the value of $9^{1/2}$.

   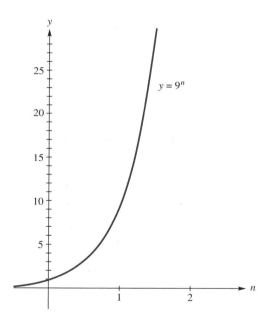

2. We now have a couple of estimates for the value of $9^{1/2}$. You might already guess its exact value, but let's look at $9^{1/2}$ one more way before asking what you think it means. In this activity we use the properties of exponents to help us determine the value of $9^{1/2}$.

   a. Assuming that all of the properties of exponents hold, rewrite $9^{1/2} * 9^{1/2}$ as an expression with a single base. What other number do you know whose product with itself is 9; that is, $? * ? = 9$? What does this tell you about the value of $9^{1/2}$?

   b. Based on your observations from part a, how do you define $9^{1/2}$? How would you define $x^{1/2}$?

3. Now let's look at a different fraction as an exponent and see if what you discovered about the one-half power can be generalized to other fractional exponents.

  a. Assuming that all of the properties of exponents hold, rewrite $8^{1/3} * 8^{1/3} * 8^{1/3}$ as an expression with a single base. What other number do you know that, when used as a factor three times, is 8; that is, $? * ? * ? = 8$? What does this tell you about the value of $8^{1/3}$?

  b. Based on your observations from part a, how would you define $8^{1/3}$? How would you define $x^{1/3}$?

4. How would you define $x^{1/n}$, where n is a positive integer and x a real number?

Discussion

5.1

In Section 3.1, we looked at the meaning of integer exponents. Positive integer exponents are interpreted as repeated multiplication and negative integer exponents as reciprocals of the corresponding positive powers. In this section, we will look at the meaning of exponents that are the reciprocals of integers. That is, we need to define $x^{1/n}$, where n is a nonzero integer.

Whichever way we define $x^{1/n}$, we would like it to satisfy the properties of integer exponents that we discussed in Section 3.1. Using the property that states that when we are multiplying powers of x, we add the exponents, we see that

$$16^{1/2}16^{1/2} = 16^{1/2+1/2} = 16^1 = 16$$

So $16^{1/2}$ is the number whose product with itself is 16. We know that 4 is a number whose product with itself is 16. Another number whose product with itself is 16 is $^-4$. The expression $16^{1/2}$ must only represent one value. Therefore, the expression $16^{1/2}$ is defined to be 4, the positive square root of 16. That is, $16^{1/2} = \sqrt{16} = 4$. The positive square root is often referred to as the principal square root.

Similarly,

$$8^{1/3}8^{1/3}8^{1/3} = 8^{1/3+1/3+1/3} = 8^1 = 8$$

So $8^{1/3}$ is the number that when used as a factor three times is 8. That is, $8^{1/3} = \sqrt[3]{8} = 2$. This agrees with what we observed in the activities and motivates the following definition.

> ## Definition
> If x is a real number and n is a positive integer greater than 1, then $x^{1/n} = \sqrt[n]{x}$, provided $\sqrt[n]{x}$ is a real number.

Notice the proviso: provided $\sqrt[n]{x}$ is a real number. Can you think of cases in which $\sqrt[n]{x}$ is not a real number?

From the definition, we can see that $x^{1/n}$ and $\sqrt[n]{x}$ are two different notations for the same thing, the nth root of x. The expression $x^{1/n}$ is written in **exponential notation (form),** and $\sqrt[n]{x}$ is written in **radical notation (form).** Each notation has its own advantage as you will see. Because we have two different notations, it is important to be able to go back and forth between them.

> **Notations Used with Rational Exponents**
>
> In **exponential notation,** the nth root of x is written as $x^{1/n}$.
>
> In **radical notation,** the nth root of x is written as $\sqrt[n]{x}$.

Many people find it easier to mentally evaluate expressions written in radical form. For example, $64^{1/3}$ can be thought of as $\sqrt[3]{64}$. To evaluate $\sqrt[3]{64}$, we need to find a number that when used as a factor three times produces 64. We know that $4 * 4 * 4 = 64$; therefore $\sqrt[3]{64} = 4$.

Similarly, to evaluate $\sqrt[4]{81}$, we need to find a number that when used a factor 4 times is 81. This is either $3 * 3 * 3 * 3$ or $^-3 * {}^-3 * {}^-3 * {}^-3$. As we discussed earlier with square roots, $\sqrt[4]{81}$ can only have one result. The notation has been defined such that any even root is evaluated as the principal or positive root. Therefore, $\sqrt[4]{81} = 3$.

Example 1

Without using your calculator, evaluate each expression.

    a. $25^{1/2}$ b. $27^{1/3}$ c. $16^{1/4}$ d. $\left(\dfrac{4}{81}\right)^{1/2}$

Solution   a. From the definition, we can rewrite $25^{1/2}$ as $\sqrt{25}$.

$$25^{1/2}$$
$$= \sqrt{25}$$
$$= 5$$

b. The expression $27^{1/3}$ can be rewritten as $\sqrt[3]{27}$.

$$27^{1/3}$$
$$= \sqrt[3]{27}$$
$$= 3$$

c. The expression $16^{1/4}$ can be rewritten as $\sqrt[4]{16}$.

$$16^{1/4}$$
$$= \sqrt[4]{16}$$
$$= 2$$

d.
$$\left(\frac{4}{81}\right)^{1/2}$$
$$= \frac{4^{1/2}}{81^{1/2}} \qquad \text{Apply the power to the numerator and denominator.}$$
$$= \frac{\sqrt{4}}{\sqrt{81}}$$
$$= \frac{2}{9}$$

Notice that when we evaluate expressions with exponents that are reciprocals of integers, it is easier to first translate these expressions into their radical form. Most of us are more comfortable with thinking about what $\sqrt{36}$ means than what $36^{1/2}$ means, even though they mean the same thing.

Example 2

Write out how each expression is read. Then, without using a calculator, evaluate the expression.

a. $-9^{1/2}$ b. $-9^{-1/2}$ c. $(-9)^{-1/2}$

Solution

a. The expression $-9^{1/2}$ can be read as *the opposite of the one-half power of nine*. The one-half power of nine can be rewritten as the square root of nine.

$$-9^{1/2}$$

$$= -\sqrt{9}$$ The one-half power applies only to the 9, so the negative is *not* included under the radical.

$$= -3$$

b. The expression $-9^{-1/2}$ can be read as *the opposite of the negative one-half power of nine*. To start, we can rewrite the negative exponent.

$$-9^{-1/2}$$

$$= -\frac{1}{9^{1/2}}$$ The negative exponent applies only to the 9. The negative sign remains in front of the fraction or in the numerator.

$$= -\frac{1}{\sqrt{9}}$$

$$= -\frac{1}{3}$$

Notice that when we rewrite a negative power, we write it as *the reciprocal of the base* to the positive exponent. We do *not* take the reciprocal of the exponent.

c. The expression $(-9)^{-1/2}$ is read as the negative one-half power of negative nine.

$$(-9)^{-1/2}$$

$$= \frac{1}{(-9)^{1/2}}$$ The exponent applies to -9.

$$= \frac{1}{\sqrt{-9}}$$

Because $\sqrt{-9}$ is not a real number, the expression $(-9)^{-1/2}$ is not a real number.

Example 3

a. If M is a negative number, decide without evaluating the expression whether $7M^{1/3}$ is positive, negative, or not a real number.

b. Show your substitution of $M = -125$ into the expression $7M^{1/3}$.

c. Evaluate the expression $7M^{1/3}$, when $M = -125$.

Solution

a. The expression $M^{1/3}$ is the same as $\sqrt[3]{M}$. When M is a negative number, $\sqrt[3]{M}$ is negative. Because $7M^{1/3} = 7\sqrt[3]{M}$, this expression results in 7 times a negative number and, therefore, the result is negative.

b. To write the substitution, we need to remember that if we are applying an exponent to a negative number, the number must be written in parentheses.

$$7(-125)^{1/3}$$

c. $$7(-125)^{1/3}$$

$$= -35$$ This can be evaluated mentally or using a calculator.

In most of the previous examples, we evaluated the expressions without the use of a calculator because we were concentrating on learning what the notation means. There are many more examples of expressions for which the roots are not perfect, and a calculator is necessary for approximating the results. We shall see in the next example how to estimate expressions with reciprocal of integer exponents and radicals. Knowing these strategies can be helpful when we are estimating expressions or checking to see if an expression is entered correctly in a calculator.

Example 4

Estimate the value of each expression. Then use a calculator to approximate the value to the thousandths place.

a. $10^{1/2}$ b. $24^{1/3}$ c. $\sqrt[3]{3000}$

Solution a. We can write $10^{1/2}$ as $\sqrt{10}$. We want to estimate $\sqrt{10}$, so we begin by looking at perfect squares that we know are close to 10. We know that $3^2 = 9$ and $4^2 = 16$. Because $\sqrt{10}$ must be between $\sqrt{9}$ and $\sqrt{16}$, we know that $\sqrt{10}$ is between 3 and 4. Because 10 is closer to 9 than 16, $\sqrt{10}$ is probably closer to 3 than 4. So we estimate $10^{1/2}$ to be about 3.2. Now, using a calculator, we can either enter $10^{1/2}$ or $\sqrt{10}$ because these are equivalent expressions. In doing so we find that $10^{1/2} \approx 3.162$.

b. First, we rewrite $24^{1/3}$ as $\sqrt[3]{24}$. This time we want to estimate the cube root of 24, so we look for perfect cubes close to 24. We know that $2^3 = 8$ and $3^3 = 27$. Therefore $\sqrt[3]{24}$ is between 2 and 3 and is closer to 3. So we estimate $24^{1/3}$ to be about 2.8. Using a calculator, we find that $24^{1/3} \approx 2.884$.

c. We probably do not know exact perfect cubes close to 3000. One way to estimate $\sqrt[3]{3000}$ is to look at $10^3 = 1000$ and $20^3 = 8000$. (*Note:* We did not try to use 11^3, 12^3, and so on because these are not values that most of us can compute mentally.) Because 3000 is between $1000 = 10^3$ and $8000 = 20^3$, $\sqrt[3]{3000}$ is between 10 and 20. Because 3000 is closer to 1000, we estimate $\sqrt[3]{3000}$ to be about 13. Using a calculator, we see that $\sqrt[3]{3000} \approx 14.42$.

Example 5

Rewrite each expression that is in exponential form in radical form, and rewrite each expression that is in radical form in exponential form.

a. $2x^{1/4}$ b. $(2x)^{1/4}$ c. $(3.2x^2h)^{1/5}$ d. $25\sqrt{m}$ e. $\sqrt{25m}$ f. $7\sqrt[3]{a^2b^5}$

Solution a. The expression $2x^{1/4}$ is written in exponential form. The one-fourth power can be written as the fourth root and applies only to x. Therefore,

$$2x^{1/4} = 2\sqrt[4]{x}$$

b. The expression $(2x)^{1/4}$ is also written in exponential form but, in this expression, the exponent applies to the product $2x$. Therefore,

$$(2x)^{1/4} = \sqrt[4]{2x}$$

Notice that, when writing the expression in radical form, no parentheses are needed because the radical symbol groups the product of 2 and x.

c. Similarly,

$$(3.2x^2h)^{1/5} = \sqrt[5]{3.2x^2h}$$

d. The expression $25\sqrt{m}$ is written in radical form. A square root can be written as the one-half power. In this expression, the power should apply only to the base of m. Because a power is applied before a product, this expression can be written without parentheses. Therefore,

$$25\sqrt{m} = 25m^{1/2}$$

e. In the expression $\sqrt{25m}$ the radical is applied to the product of 25 and m. Therefore, when writing this expression in exponential form, we need to use parentheses to denote that the power applies to the product.

$$\sqrt{25m} = (25m)^{1/2}$$

f. In the expression $7\sqrt[3]{a^2b^5}$ the cube root can be written as a one-third power and applies to the product of a^2 and b^5, but not to the 7. Therefore,

$$7\sqrt[3]{a^2b^5} = 7(a^2b^5)^{1/3}$$

Recall that we created the definition of reciprocal of integer powers so that the previous properties of exponents would remain true, and they do. However, if the base is a negative number and the exponent is a fraction, the result is sometimes not a real number. For the properties to always hold true, we restrict the bases to positive real numbers. We restate these properties with their restrictions and then use them to simplify expressions with reciprocal of integer exponents.

Properties of Rational Exponents

Let x and y be positive real numbers. Let m and n be rational numbers. Then the following are true.

1. $x^m x^n = x^{m+n}$ When *multiplying powers of x,* add the exponents.

2. $\dfrac{x^m}{x^n} = x^{m-n}$ When *dividing powers of x,* subtract the exponents.

3. $(xy)^m = x^m y^m$ The *power of a product* can be simplified by applying the power to each factor.

4. $\left(\dfrac{x}{y}\right)^m = \dfrac{x^m}{y^m}$ The *power of a quotient* can be simplified by applying the power to each factor in the numerator and each factor in the denominator.

5. $(x^m)^n = x^{m*n}$ When simplifying a *power of a power of x,* multiply the exponents.

In Section 3.1, it was possible to simplify expressions by using the definition of exponents. For example, $x^3 * x^4$ can be thought of as the product of x used as a factor 3 times and x used as a factor 4 times. This can be simplified to x used as a factor 7 times. Notice that we did not use the properties of exponents. We just "counted" the exponents in our head. However, $x^{1/2} * x^{-1/3}$ cannot be simplified by this same process; we must rely on the properties of exponents to simplify this expression. So, $x^{1/2} * x^{-1/3} = x^{1/2 + -1/3} = x^{1/6}$.

Example 6

Simplify each expression. Express the results using only positive exponents. Assume that all variables are positive real numbers. Verify your results.

a. $\left(k^{1/3}\right)^{1/3}$ b. $5x^{-1/3}\left(8x^{1/4}\right)$ c. $\dfrac{m^{1/2}}{m^{-1/2}}$

Solution   a. $\left(k^{1/3}\right)^{1/3}$

$= k^{1/3*1/3}$ When raising a power to a power, multiply the exponents.

$= k^{1/9}$

VERIFY   Let $k = 10$

$\left(k^{1/3}\right)^{1/3} \overset{?}{=} k^{1/9}$ Is the original expression equivalent to our result?

$\left(10^{1/3}\right)^{1/3} \overset{?}{=} 10^{1/9}$ Substitute 10 for k.

$1.2915 = 1.2915$ ✔

We conclude that $\left(k^{1/3}\right)^{1/3}$ simplifies to $k^{1/9}$.

b. $5x^{-1/3}\left(8x^{1/4}\right)$

$= 5 * 8 * x^{-1/3+1/4}$ Multiply the numerical coefficients, and add the exponents.

$= 40x^{-1/12}$

$= \dfrac{40}{x^{1/12}}$ Rewrite $x^{-1/12}$ with a positive exponent. The exponent applies only to the x, so it is the only factor that moves to the denominator. Forty remains in the numerator.

VERIFY   Let $x = 100$

$5x^{-1/3}\left(8x^{1/4}\right) \overset{?}{=} \dfrac{40}{x^{1/12}}$

$5 * 100^{-1/3}\left(8 * 100^{1/4}\right) \overset{?}{=} \dfrac{40}{100^{1/12}}$

$27.25 = 27.25$ ✔

We conclude that $5x^{-1/3}\left(8x^{1/4}\right)$ simplifies to $\dfrac{40}{x^{1/12}}$.

c. $\dfrac{m^{1/2}}{m^{-1/2}}$

$= m^{1/2-{}^{-1/2}}$ When dividing powers of m, subtract the exponents.

$= m$

VERIFY   Let $m = 9$

$\dfrac{m^{1/2}}{m^{-1/2}} \overset{?}{=} m$

$\dfrac{9^{1/2}}{9^{-1/2}} \overset{?}{=} 9$

$9 = 9$ ✔

We conclude that $\dfrac{m^{1/2}}{m^{-1/2}}$ simplifies to m.

Example 7

Simplify the expression $\dfrac{\sqrt{64x}}{\sqrt{x}}$. Assume $x > 0$.

Solution   This expression is written in radical form, and we have no properties for simplifying radical expressions, so the first step is to rewrite the expression in exponential form.

$$\frac{\sqrt{64x}}{\sqrt{x}}$$

$$= \frac{(64x)^{1/2}}{x^{1/2}}$$

$$= \frac{64^{1/2}x^{1/2}}{x^{1/2}} \qquad \text{Each factor in the numerator is raised to the one-half power.}$$

$$= 8$$

VERIFY   Let $x = 5$

$$\frac{\sqrt{64x}}{\sqrt{x}} \overset{?}{=} 8$$

$$\frac{\sqrt{64 * 5}}{\sqrt{5}} \overset{?}{=} 8$$

$$8 = 8 \qquad \checkmark$$

We conclude that $\dfrac{\sqrt{64x}}{\sqrt{x}}$ simplifies to 8.

In Example 7, can you explain why x must be greater than zero?

In this section we discovered that exponents that are reciprocals of integers represent roots. We saw how to evaluate these expressions, being careful to decide when the result should be positive, negative, or not a real number. We were given two notations for rational exponents: **exponential notation** and **radical notation.** In the examples we found that it was easier to evaluate numerical expressions if they were written in radical notation. However, when simplifying algebraic expressions, we need to use exponential notation.

Problem Set 5.1

1. Translate each of the following English phrases into correct symbolic mathematics.
   a. the fifth root of y
   b. the one-third power of $^-27$
   c. the product of 5 and the one-half power of M
   d. the one-half power of the product of 5 and M
   e. the opposite of the cube root of 125
   f. the opposite of the one-third power of x
   g. the negative one-fifth power of y

2. Without using a calculator, evaluate the following radicals, if possible.
   a. $\sqrt{81}$
   b. $\sqrt[3]{-27}$
   c. $-\sqrt[4]{81}$
   d. $\sqrt[5]{32}$
   e. $\sqrt[4]{-16}$
   f. $-\sqrt[4]{16}$

3. Without using a calculator, evaluate the following numerical expressions, if possible.
   a. $49^{1/2}$
   b. $27^{1/3}$
   c. $64^{1/3}$
   d. $32^{1/5}$
   e. $\dfrac{1}{5^{-2}}$
   f. $25^{-1/2}$
   g. $-36^{1/2}$
   h. $(-27)^{-1/3}$

4. Without using a calculator, evaluate the following numerical expressions, if possible.
   a. $25^{1/2}$
   b. $(-27)^{1/3}$
   c. $(-4)^{-1/2}$
   d. $(-100)^{-1/2}$
   e. $-1^{1/10}$
   f. $(-1)^{1/10}$
   g. $(-1)^{1/9}$
   h. $(-1)^{-10}$
   i. $10{,}000^{1/4}$
   j. 10^{-4}
   k. $\left(\dfrac{1}{8}\right)^{-1/3}$

5. Rewrite each expression that is in exponential form in radical form, and rewrite each expression that is in radical form in exponential form. Assume that all variables are nonzero real numbers and that all quantities result in real numbers.
   a. $32x^{1/5}$
   b. $\sqrt{M+N}$
   c. $(32x)^{1/5}$
   d. $2\pi\sqrt[3]{R}$
   e. $\sqrt{\dfrac{L}{W}}$
   f. $-x^{1/5}$
   g. $(^-x)^{1/5}$
   h. $(p+q)^{-1/3}$
   i. $\dfrac{(w-2xy)^{1/2}}{5y}$

6. Estimate the value of each expression. Then use a calculator to approximate the value to the thousandths place.
   a. $27^{1/2}$
   b. $100^{1/4}$
   c. $\sqrt[3]{20{,}000}$

7. Without using a calculator, evaluate the following numerical expressions, if possible. If the problem is stated in fractional form, give the results as a fraction.
   a. $\left(\dfrac{1}{49}\right)^{1/2}$
   b. $(-1)^{1/7}$
   c. $\left(\dfrac{27}{8}\right)^{-1/3}$
   d. $81^{0.25}$
   e. $\left(\dfrac{-125}{8}\right)^{-1/3}$
   f. $(-1)^{1/4}$

8. For each of the following expressions, substitute the given value. You do *not* need to evaluate the expressions.

   a. $x^{1/2}$, when $x = 9$ **e.** $^{-}R^{1/2}$, when $R = {}^{-}1$

   b. $w^{1/3}$, when $w = {}^{-}27$ **f.** B^{-10}, when $B = {}^{-}10$

   c. $({}^{-}y)^{-1/4}$, when $y = {}^{-}10{,}000$ **g.** $({}^{-}M)^{1/3}$, when $M = {}^{-}125$

   d. $^{-}t^{1/2}$, when $t = 4$

9. Evaluate the expressions you wrote in the previous problem, if possible.

10. **a.** Complete the following table.

x	$x^{1/2}$	$x^{1/3}$	$x^{1/4}$	$x^{1/5}$
$^{-}4$				
$^{-}3$				
$^{-}2$				
$^{-}1$				
0				
1				
2				
3				
4			$\star$	

   b. Graph the following functions by plotting points from the table.

   $$y = x^{1/2} \qquad y = x^{1/3} \qquad y = x^{1/4} \qquad y = x^{1/5}$$

   c. What do you expect the graph of $y = x^{1/6}$ to look like? $y = x^{1/7}$? Be as specific as possible.

   d. If n is even, what does the graph of $y = x^{-1/n}$ look like? If n is odd, what does the graph of $y = x^{1/n}$ look like?

   e. What are reasonable inputs into the function $y = x^{1/n}$ when n is even? When n is odd?

11. Simplify the following expressions. Express the results using only positive exponents. Verify your results. Assume that all variables are positive real numbers.

   a. $6b^{1/2} * 9b^{-1/4}$ **d.** $(81R^{1/3})^{1/4}$

   b. $H^{1/2}(H^{-1/3})$ **e.** $\dfrac{x + x^{1/2}}{x^{1/2}}$

   c. $\dfrac{b * b^{-1/2}}{b^{1/2}}$

12. Simplify the following expressions. Express the results using only positive exponents. Verify your results. Assume that all variables are positive real numbers.

a. $\dfrac{B^{1/4}}{B^{1/3}}$

b. $(-27x^9 y^{1/2})^{1/3}$

c. $y^{1/2}(y^{-1/2} - y^{-1})$

d. $8m^{1/3} + (8m)^{1/3}$

e. $(z^{1/2} + 5)^2$

13. Simplify the following radicals. Verify your results. Assume that all variables are positive real numbers.

a. $\sqrt[3]{-8w^6}$

b. $\sqrt{\dfrac{25x^8}{36}}$

c. $\sqrt[5]{32a^5}$

14. Select *all* of the expressions that are equivalent to $x^{-1/3}$.

a. $x^{1/3}$　　**b.** $\dfrac{1}{x^{1/3}}$　　**c.** $-x^{1/3}$　　**d.** x^{-3}　　**e.** $\sqrt[3]{x}$　　**f.** $\dfrac{1}{\sqrt[3]{x}}$　　**g.** $\dfrac{1}{x^3}$

15. Always True?　First, try to decide whether each of the following equations is true for all values of the variable. Then substitute numerical values into the equation to verify your conjecture. Assume that all variables are nonzero.

a. $\sqrt{a^2 + b^2} = a + b$

b. $(a^3 b^3)^{1/3} = ab$

c. $(m^5 + 32)^{1/5} = m + 2$

d. $(a - b)^{-1/3} = \dfrac{1}{(a - b)^{1/3}},$　for all real values of a and b such that $a \neq b$.

16. a. Without calculating the value of the expression, decide whether $m^{-1/2}$ is positive, negative, or neither when $m = -324$. Explain.

b. Evaluate the expression $m^{-1/2}$ when $m = -324$ (if possible). Show your substitution.

17. a. Without calculating the value of the expression, decide whether $(-z)^{1/4}$ is positive, negative, or neither when $z = -1296$. Explain.

b. Evaluate the expression $(-z)^{1/4}$ when $z = -1296$ (if possible). Show your substitution.

18. a. Without calculating the value of the expression, decide whether $-y^{-1/2}$ is positive, negative, or neither when $y = -10$. Explain.

b. Evaluate the expression $-y^{-1/2}$ when $y = -10$ (if possible). Show your substitution.

19. a. Without calculating the value of the expression, decide whether $-R^{1/5}$ is positive, negative, or neither when $R = -100$. Explain.

b. Evaluate the expression $-R^{1/5}$ when $R = -100$ (if possible). Show your substitution.

20. Gravity.   The gravitational attraction between any two bodies is based on a formula which takes into account the mass of each of the bodies and the distance between their centers. The formula is

$$\text{gravitational force} = \frac{Gm_1 m_2}{r^2}$$

where the gravitational force is measured in meters per second squared,

G is a constant $= 6.67 * 10^{-11}$

m_1 and m_2 are the mass of the two objects in kilograms

and

r is the distance between the centers of the two objects in meters.

Consider an object of mass 56.8 kg standing on the surface of the earth. To calculate the force exerted by gravity on this object, we need to know the distance from the center of the earth to its surface. This is approximately $6.38 * 10^6$ meters. We also need to know the mass of the earth, this is about $5.98 * 10^{24}$ kg.

a. Substitute the known values into the formula.

b. *Without using a calculator,* estimate the value of this expression. You need to write out several steps to estimate this expression, but you should be able to round the numbers enough to do all of the arithmetic mentally.

c. Using your calculator, evaluate the expression from part a. How do your results compare?

5.2 Solving Power Equations

1. **a.** What do you expect $(x^3)^{1/3}$ to simplify to? Based on your answer, what do you expect the graph of $y = (x^3)^{1/3}$ to look like?

   b. Without simplifying, graph the function $y = (x^3)^{1/3}$ using your graphing calculator. What other function have you seen that has this same graph?

   c. Without simplifying, graph the function $y = (x^5)^{1/5}$ using your graphing calculator. What other function have you seen that has this same graph?

2. **a.** What do you expect $(x^2)^{1/2}$ to simplify to? Based on your answer, what do you expect the graph of $y = (x^2)^{1/2}$ to look like?

   b. Without simplifying, graph the function $y = (x^2)^{1/2}$ using your calculator. Does your graph look like what you expected? What other function have you seen that has this same graph?

   c. Without simplifying, graph the function $y = (x^4)^{1/4}$ using your calculator. What other function have you seen that has this same graph?

3. **a.** If n is a positive odd number, rewrite $(x^n)^{1/n}$ as an equivalent expression using your observations. Explain why your response always works when n is a positive odd number.

   b. If n is a positive even number, rewrite $(x^n)^{1/n}$ as an equivalent expression using your observations. Explain why your response always works when n is a positive even number.

4. **a.** Without simplifying, graph the function $y = (x^{1/2})^2$ using your graphing calculator. Describe what you see.

   b. Without simplifying, graph the function $y = (x^{1/4})^4$ using your graphing calculator. Describe what you see.

5. **a.** Without simplifying, graph the function $y = (x^{1/3})^3$ using your graphing calculator. What other function have you seen that has this same graph?

   b. Without simplifying, graph the function $y = (x^{1/5})^5$ using your graphing calculator. What other function have you seen that has this same graph?

6. **a.** If n is a positive even number, rewrite $(x^{1/n})^n$ as an equivalent expression using your observations. Explain why your response always works when n is a positive even number.

   b. If n is a positive odd number, rewrite $(x^{1/n})^n$ as an equivalent expression using your observations. Explain why your response always works when n is a positive odd number.

We know how to solve equations involving the operations of addition, subtraction, multiplication, and division by "undoing" the operations to isolate the variable. For example, to solve the equation $5x + 7 = 22$, we need to undo the operations of addition and multiplication as seen below.

$$5x + 7 = 22$$

$$5x + 7 - 7 = 22 - 7 \qquad \text{Subtract 7 from both sides to undo the addition.}$$

$$\frac{5x}{5} = \frac{15}{5} \qquad \text{Divide both sides by 5 to undo the multiplication.}$$

$$x = 3$$

Addition and subtraction are inverse operations, so these "undo" each other when we are solving an equation. Similarly, multiplication and division are inverse operations. In this section, we will look at how to solve power equations.

> Definition _____
>
> An equation in the form $ax^b + c = d$, where a, b, c, and d are constants, is called a **power equation.**

Notice that a power equation contains an exponent, and the variable is the base of the exponent. So, to solve a power equation, we need to "undo" the power. The inverse of a power is a root. For example, if we cube a number, $2^3 = 8$, and then take the cube root of the result, $\sqrt[3]{8} = 2$, we get the original number. The cube and cube root "undo" each other. Because any root can also be thought of as a reciprocal power, we can undo the exponent in an equation by applying the reciprocal power.

The properties of exponents tell us that $(x^n)^{1/n}$ is equal to x, but in the activities we found that this is not always true. If we restrict the graphs to cases when x is greater than or equal to zero it is true that $(x^n)^{1/n} = x$. When x is not restricted to the positive real numbers, we must pay careful attention to whether n is even or odd and to the order that the root and power are applied. We state our observations here.

> *Reciprocal Power Properties*
>
> If n *is an even whole* number, then
>
> - $(x^n)^{1/n} = |x|$, and
> - when $x \geq 0$, $(x^{1/n})^n = x$ and when $x < 0$ $(x^{1/n})^n$ is not a real number
>
> If n *is an odd whole* number, then
>
> - $(x^n)^{1/n} = x$, and
> - $(x^{1/n})^n = x$

We can see that when n is an odd whole number, we get what we expect when we simplify $(x^n)^{1/n}$ and $(x^{1/n})^n$. We also know that the properties of exponents apply when x is nonnegative, and our observations verify this. The cases that are different from what we expected occur when n is even and x is a negative number. Let's look at these cases carefully and see if we can justify our observations.

- If n is an even whole number and x is a negative number, then in the expression $(x^n)^{1/n}$, we are first raising a negative number to an even power. This results in a positive number. Next we take the root of this positive number, and the result is positive.

  For example, if $x = {}^-5$, we obtain

  $$[(-5)^2]^{1/2}$$
  $$= (25)^{1/2}$$
  $$= 5$$

  Notice we did not get x, we got $|x|$. If n is an even whole number, the expression $(x^n)^{1/n} = |x|$.

- If n is an even whole number and x is a negative number, then in the expression $(x^{1/n})^n$, the first step is to take an even root of a negative number. The even root of a negative is not real. Therefore, when x is negative this expression is not real. When an even root and then a power is applied, we must restrict the values for x. If n is an even whole number, then $\left(x^{1/n}\right)^n = x$ with the restriction that $x \geq 0$.

Notice that applying reciprocal powers in order undoes the powers and produces x *except* when x is raised to an even power followed by an even root. This is the one case when the result is something other than x. Although some other cases have restrictions on x, we will discover these as we check our solutions.

Even Power Followed by an Even Root

When x is raised to an even power followed by an even root, the result is the absolute value of x.

$$\left(x^n\right)^{1/n} = |x|$$

Example 1

Solve the following equations. Round approximate results to the thousandths place. Check your solutions.

a. $x^2 = 144$ b. $w^5 = 200$ c. $B^{1/3} = 4$

Solution   a.
$$x^2 = 144$$
$$\left(x^2\right)^{1/2} = 144^{1/2}$$ Raise both sides to the 1/2 power to undo the square.
$$|x| = 12$$ When x is raised to an even power followed by an even root, the result is the *absolute value* of x.

$$x = 12 \quad \text{or} \quad x = {}^-12$$

CHECK

$x = 12$	$x = {}^-12$
$x^2 = 144$	$x^2 = 144$
$12^2 \overset{?}{=} 144$	$({}^-12)^2 \overset{?}{=} 144$
$144 = 144$ ✔	$144 = 144$ ✔

Therefore, $x = 12$ or $x = {}^-12$ is the solution to the equation $x^2 = 144$.

b.
$$w^5 = 200$$
$$\left(w^5\right)^{1/5} = 200^{1/5}$$ Raise both sides to the 1/5 power to undo the 5th power.
$$w \approx 2.885$$ Odd reciprocal powers undo each other.

CHECK $w^5 \overset{?}{=} 200$
$$2.885^5 \overset{?}{=} 200$$
$$199.86 \approx 200 \qquad ✔$$

Therefore, $w \approx 2.885$ is the solution to the equation $w^5 = 200$.

c.
$$B^{1/3} = 4$$
$$\left(B^{1/3}\right)^3 = 4^3$$ Cube both sides to undo the 1/3 power.
$$B = 64$$

CHECK $B^{1/3} = 4$
$$64^{1/3} \overset{?}{=} 4$$
$$4 = 4 \qquad ✔$$

Therefore, $B = 64$ is the solution to the equation $B^{1/3} = 4$.

In Example 1, we saw how raising both sides of a power equation to the reciprocal power eliminates the power and allows us to solve the equation. When we apply this technique to solving a power equation, we must remember that the power we use applies to everything on both sides of the equation. Therefore, in more complex equations, we must first isolate the factor with the power before "undoing" the exponent. You will notice that the next example also contains equations with radicals. We know that any root can be written with exponents, so solving these equations will be similar to solving power equations.

Example 2

Solve the following equations. Round approximate solutions to the thousandths place. Check your solutions.

a. $5r^4 - 10 = 395$

b. $2\sqrt[3]{x + 5} = 22$

c. $\dfrac{M^{1/4}}{10} = {}^-2$

d. $^-2x = \sqrt{9x^2 - 45}$

Solution   a.

$$5r^4 - 10 = 395$$

$$5r^4 = 405 \qquad \text{Add 10 to both sides to isolate the term with the power.}$$

$$r^4 = 81 \qquad \text{Divide both sides by 5, to isolate the factor with the power.}$$

$$(r^4)^{1/4} = 81^{1/4} \qquad \text{Raise both sides to the 1/4 power to undo the 4th power.}$$

$$|r| = 3 \qquad \text{When } r \text{ is raised to an even power followed by an even root the result is the \textit{absolute value} of } r.$$

$$r = 3 \qquad \text{or} \qquad r = {}^-3$$

CHECK

$r = 3$	$r = {}^-3$
$5r^4 - 10 = 395$	$5r^4 - 10 = 395$
$5 * 3^4 - 10 \overset{?}{=} 395$	$5 * (^-3)^4 - 10 \overset{?}{=} 395$
$395 = 395$ ✔	$395 = 395$ ✔

Therefore, $r = 3$ or $r = {}^-3$ is the solution to the equation $5r^4 - 10 = 395$.

b. The first step is to rewrite the radical in exponential form. In this step, we are not applying a power to the entire side of the equation, we are just rewriting the expression.

$$2\sqrt[3]{x + 5} = 22$$

$$2(x + 5)^{1/3} = 22 \qquad \text{Rewrite in exponential form. Because we are just rewriting the root as an exponent, we do not raise the right-hand side to any power.}$$

$$(x + 5)^{1/3} = 11 \qquad \text{Divide both sides by 2.}$$

$$[(x + 5)^{1/3}]^3 = 11^3 \qquad \text{Cube both sides to the undo the one-third power.}$$

$$x + 5 = 1331$$

$$x = 1326$$

CHECK
$$2\sqrt[3]{x + 5} = 22$$
$$2\sqrt[3]{1326 + 5} \overset{?}{=} 22$$
$$22 = 22 \qquad ✔$$

Therefore, $x = 1326$ is the solution to the equation $2\sqrt[3]{x + 5} = 22$.

c. $\dfrac{M^{1/4}}{10} = {}^-2$

$M^{1/4} = {}^-20$ Multiply both sides by 10.

$\left(M^{1/4}\right)^4 = (-20)^4$ Raise both sides to the fourth power to undo the one-fourth power.

$M = 160{,}000$ Even though the powers are even, because M was raised to an even root followed by an even power it does *not* result in an absolute value.

CHECK $\dfrac{M^{1/4}}{10} = {}^-2$

$\dfrac{160{,}000^{1/4}}{10} \overset{?}{=} {}^-2$

$2 = {}^-2$ ✗

We see that $M = 160{,}000$ does not work in the original equation. Therefore, this equation has *no solution*. Whenever our solution does not work, it is a good idea to look back at the original equation to see if we can figure out why. In the equation $\dfrac{M^{1/4}}{10} = {}^-2$, we see that the numerator is the fourth root of M. The fourth root of a positive number is positive, and the fourth root of a negative number is not real. Therefore, $\dfrac{M^{1/4}}{10}$ cannot be negative.

d. $^-2x = \sqrt{9x^2 - 45}$

$^-2x = (9x^2 - 45)^{1/2}$ Rewrite in exponential form.

$(^-2x)^2 = \left[(9x^2 - 45)^{1/2}\right]^2$ Square both sides.

$4x^2 = 9x^2 - 45$ Simplify both sides of the equation.

$^-5x^2 = {}^-45$ Subtract $9x^2$ from both sides.

$x^2 = 9$ Divide both sides by $^-5$.

$(x^2)^{1/2} = 9^{1/2}$ Raise both sides to the one-half power.

$|x| = 3$ Because x is raised to an even power followed by an even root, the result is the absolute value of x.

$x = 3$ or $x = {}^-3$

CHECK

$x = 3$	$x = {}^-3$
$^-2x = \sqrt{9x^2 - 45}$	$^-2x = \sqrt{9x^2 - 45}$
$^-2 * 3 \overset{?}{=} \sqrt{9 * 3^2 - 45}$	$^-2 * {}^-3 \overset{?}{=} \sqrt{9 * (^-3)^2 - 45}$
$^-6 = 6$ ✗	$6 = 6$ ✔

We see that $x = {}^-3$ checks, but $x = 3$ does not. Therefore, $x = {}^-3$ is the only solution to the equation $^-2x = \sqrt{9x^2 - 45}$.

In Example 2, we saw a case in which our steps were correct algebraically; but when one of the final answers was substituted into the original equation it did *not* result in a true statement. When the process of solving an equation involves squaring both sides of the equation (or applying any even power) it is possible to create extraneous solutions. For example, if we start with a false statement like $^-6 = 6$ and square both sides, the statement becomes true. Let's look at this symbolically.

$^-6 = 6$ This is a *false* statement.

$(^-6)^2 = 6^2$ Square both sides.

$36 = 36$ Even though we began with a false statement, by squaring both sides, the statement is now *true*.

Extraneous solutions are solutions to an intermediate equation that *will not solve the original equation.* This means that checking your solutions to power equations is not only a good idea, it is a *necessary* step in the solution process.

Example 3

Chris and Terry have decided to invest $10,000 in a savings account for their daughter's college education. They would like their $10,000 to double in the next 10 years.

A formula for determining the amount in an account that is compounded annually is

$$A = P(1 + r)^t$$

where A = amount in the account

P = the original amount deposited

r = the interest rate as a decimal

t = number of years

Use the formula to determine the annual interest rate they need to double their $10,000 in ten years.

Solution   Chris and Terry want the amount in the account to double, so A is $20,000, P is $10,000, and t is 10 years. We can substitute these values into the formula and then solve for r.

$$20{,}000 = 10{,}000 * (1 + r)^{10}$$
$$2 = (1 + r)^{10}$$
$$2^{1/10} = [(1 + r)^{10}]^{1/10}$$
$$2^{1/10} = |1 + r|$$
$$2^{1/10} = 1 + r \qquad \text{or} \qquad -2^{1/10} = 1 + r$$
$$2^{1/10} - 1 = r \qquad \text{or} \qquad -2^{1/10} - 1 = r$$
$$0.07177 \approx r \qquad \text{or} \qquad -0.07177 \approx r$$

We are trying to find an interest rate, so we ignore the negative solution because it is meaningless. Most interest rates are rounded to the nearest tenth of a percent, so we round our answer to $r \approx 0.072$, or 7.2%.

$$\text{CHECK} \qquad 20{,}000 \stackrel{?}{=} 10{,}000 * (1 + 0.072)^{10}$$
$$20{,}000 \approx 20{,}042.31 \qquad ✔$$

Chris and Terry must invest their money in an account that pays an annual interest rate of about 7.2% per year or more if they want their money to double in 10 years.

In this section, we learned how to solve equations involving powers and roots using reciprocal exponents. We noted that reciprocal powers do not always merely cancel each other. In particular, if n is an even whole number, the nth power followed by the nth root results in the absolute value function, that is, $(x^n)^{1/n} = |x|$ when n is an even whole number. In the process of solving power equations, we may introduce **extraneous solutions.** This means that checking our potential solution is a crucial part of the problem solving process, not just a reassurance.

Problem
Set
5.2

1. Solve each equation. Check your solution. Round approximate numerical results to the thousandths place.

   a. $(4m)^3 = 1000$

   b. $5k^2 - 300 = 0$

   c. $10\sqrt{x} - 4 = 12$

   d. $(120 - y)^{1/5} = 3$

   e. $\dfrac{\sqrt[4]{R}}{8} = 20$

2. Solve each equation. If the equation has no solution, explain why.

   a. $x^3 = {}^-1$

   b. $x^{10} = {}^-1$

   c. $x^{12} = 1$

   d. $x^{1/7} = {}^-1$

   e. $x^{1/2} = 1$

   f. $x^{1/6} = {}^-1$

3. Based on your observations from the previous problem, determine the values of k for which the equation has solutions. (For example, does the equation have solutions when k is any real number, only if k is nonnegative, or other restrictions?)

   a. $x^4 = k$

   b. $x^3 = k$

   c. $x^{1/8} = k$

   d. $x^{1/5} = k$

4. Solve each equation. Check your solution. Round approximate numerical results to four significant digits.

   a. $\dfrac{(2p)^5}{3} = {}^-1$

   b. $4x^6 = 2916$

   c. $4\sqrt{x+2} - 5 = 15$

   d. $12 - 3(2x - 4) = 3x + 1$

   e. $625 = (3Q - 10)^4$

   f. $1000(1 + x)^{12} = 2000$

   g. $\dfrac{x+3}{8} = \dfrac{5x-1}{5}$

   h. $x^4 + 16 = 0$

5. Solve each equation. Check your solution. Round approximate numerical results to four significant digits.

   a. $x^3 + 1000 = 0$

   b. $(x + 1)^{1/3} - 8 = 0$

   c. $\dfrac{4}{x+1} + \dfrac{1}{x} = \dfrac{15}{2x}$

   d. $2\sqrt[4]{1-m} + 10 = 6$

   e. $34.5 = \dfrac{22t^3}{7}$

   f. $1000 + (3 - x)^{10} = 1978$

   g. $\sqrt{x^2 + 40} = 3x$

   h. ${}^-5x = \sqrt{2x^2 + 207}$

6. Solve each of the equations for the indicated variable.

   a. $K = (GM)^{1/3}$ for M

   b. $A = P(1 + r)^5$ for r

   c. $d = C_p v^2 A$ for A

   d. $V = \dfrac{4}{3}\pi r^3$ for r

   e. $T = 2\pi\sqrt{\dfrac{L}{g}}$ for L

7. Solve each of the equations for the indicated variable.

   a. $x - x_0 = \dfrac{1}{2}at^2$ for t

   b. $A = P_0 + \sqrt[3]{5.6MP}$ for M

   c. $A = P\left(1 + \dfrac{r}{k}\right)^{27}$ for r

   d. $A = \dfrac{R(1 - (1 + i)^{-n})}{i}$ for R

8. Roofing.   Many roofs are designed in the shape of a pyramid. The formula for the surface area of the lateral sides of a pyramid with a square base is

$$\text{Area} = 2s\sqrt{0.25s^2 + h^2}$$

where s = length of a side of the square base

$\qquad h$ = height from the base of the pyramid to the vertex

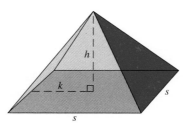

The base of a roof on a new manufactured home is to be 42.0 by 42.0 feet.

a. The minimum **pitch** for a roof (that is, the ratio of h to k in the figure) is 4 to 12. Find the minimum height of the roof for the manufactured home.

b. Suppose you are tiling the roof and the tile is sold in bundles that cover an area of 25 square feet. Use the previous formula to determine the maximum height you could build this roof if you have 80 bundles. Does this satisfy the minimum pitch rule?

9. Library Tax.   A random telephone poll of 400 registered voters determined that 55% would vote to support a tax measure for the library. The number 55% is an estimate and is usually reported with a plus-or-minus error. The formula for the error is

$$E = \pm 1.96\sqrt{\frac{p(1-p)}{n}}$$

where p = proportion favoring the tax, written as a decimal

$\qquad n$ = the number of people in the study

a. Determine the error for this study.

b. Solve the original formula for n.

c. Use your new formula from part b to find n, the number of people needed in this study if we want an error of at most ± 0.04. Assume that the proportion of voters supporting the library tax remains 55%.

d. Assume that a study reports support of the library tax to be 55%. You want to be assured that the tax will pass in the next election. What is the largest percent error that still means the tax would pass? Can you list any other considerations?

10. Braking Distance.   The formula for the distance it takes to stop a car traveling at a given speed is given in the formula

$$braking\ distance\ in\ feet = constant * (speed\ in\ mph)^2$$

where the constant is determined by the road conditions.

After leaving campus one day, another vehicle almost hits your car. You need to determine the speed of the other car. The skid mark left by the car measures 75 feet. By calling the local police department you find that the constant for the road condition in front of your school on this day is 0.037. What was the speed of the other car?

11. Population in Bend. Bend, Oregon, is a resort town in central Oregon that has been growing rapidly over the past decade. The population in Bend in 1988 was 18,970, and in 1998 was 32,250. The population in Bend can be modeled by the following formula, assuming that the growth rate is exponential,

$$A = P(1 + r)^t$$

where A = population after t years

       P = initial population

       r = annual rate of growth as a decimal

       t = time in years

   a. Use the formula to determine the annual rate of growth for Bend from 1988 to 1998.

   b. Determine the population in Bend in the year 2008, assuming the growth rate remains the same.

12. Determine the radius of a circle whose area is 1 square foot.

13. Find the length of the diagonal of a square whose area is 48 square inches.

14. Find the diameter of the sphere whose volume is 1000 cubic inches.

15. The Silo. A silo is formed from a right circular cylinder on top of a right circular cone. The height of the cylinder and the overall height of the silo are given in the following figure. The silo has a capacity of 144 cubic feet. Determine the diameter of the silo.

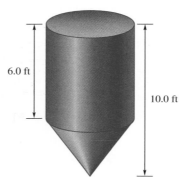

6.0 ft

10.0 ft

5.3 Rational Exponents

Activity Set 5.3

1. **a.** Evaluate the expression $\left(\sqrt{9}\right)^3$.

    b. Without simplifying first, rewrite the expression $\left(\sqrt{9}\right)^3$ in exponential form, and simplify this expression *using properties of exponents*.

    c. Without using a calculator, evaluate $9^{3/2}$. Explain how you did this.

    d. Without using a calculator, evaluate $25^{3/2}$.

2. **a.** Evaluate the expression $\left(\sqrt[3]{1000}\right)^2$.

    b. Without simplifying first, rewrite the expression $\left(\sqrt[3]{1000}\right)^2$ in exponential form, and simplify this expression *using properties of exponents*.

    c. Without using a calculator, evaluate $1000^{2/3}$. Explain how you did this.

    d. Without using a calculator, evaluate $125^{2/3}$.

    e. Without using a calculator, evaluate $81^{3/4}$.

3. Using your observations from Activities 1 and 2, write a definition for $x^{m/n}$.

Discussion 5.3

We already looked at exponents in the form $x^{1/n}$. Because the exponent is a fraction, these are called **rational exponents.** However, we restricted the definition to those rational numbers for which the numerator was 1. We now look at a more general definition for rational exponents. We want to define $x^{m/n}$, where m and n are positive integers, in such a way that the previous properties and definitions of exponents are still true. To do this, we start out by writing the fraction $\frac{m}{n}$ as a product, $\frac{m}{n} = \frac{1}{n} * m$. This means that $x^{m/n} = x^{1/n * m}$. Using the property that states that the power of power of x can be simplified by multiplying the exponents, $x^{1/n * m} = (x^{1/n})^m$. Next, we can rewrite $x^{1/n}$ in radical notation, $x^{1/n * m} = \left(\sqrt[n]{x}\right)^m$. We now have a definition for $x^{m/n}$ that satisfies all of our previous definitions and properties of exponents.

> #### Definition
> If x is a real number and $\frac{m}{n}$ is a rational number in reduced form, then $x^{m/n} = \left(\sqrt[n]{x}\right)^m$ provided $\sqrt[n]{x}$ is a real number.

We see from the definition that when evaluating an expression like $x^{m/n}$, we translate the denominator of the exponent to a root and the numerator to a power, in that order. We apply the root first because the definition includes the proviso that $\sqrt[n]{x}$ is a real number, and we want to check if the expression has meaning before we try to evaluate it. Taking the root first also makes mental arithmetic easier on those examples for which evaluating the root mentally makes sense. For example, if we are asked to evaluate $27^{2/3}$, it is much easier to evaluate $\sqrt[3]{27}$ than to try to evaluate 27^2 first!

Example 1

Without using a calculator, evaluate each expression.

a. $25^{3/2}$ b. $16^{5/4}$

Solution a.

$$25^{3/2}$$
$$= \left(\sqrt{25}\right)^3$$ Rewrite the rational exponent. Translate the denominator to a root and the numerator to a power in that order.
$$= 5^3$$
$$= 125$$

Therefore, $25^{3/2} = 125$.

b.

$$16^{5/4}$$
$$= \left(\sqrt[4]{16}\right)^5$$ Rewrite the rational exponent as a root followed by a power.
$$= 2^5$$
$$= 32$$

Therefore, $16^{5/4} = 32$.

The rules for evaluating expressions in which the base or the exponents are negative are no different when the exponent is rational. Recall that careful reading of the expressions is helpful in deciding whether the exponent is applied to a negative.

Example 2

Write out in words how you would read each expression. Then, without using a calculator, evaluate the expression.

a. $^-4^{3/2}$ b. $^-4^{-3/2}$ c. $(^-4)^{3/2}$ d. $^-27^{2/3}$ e. $^-27^{-2/3}$ f. $(^-27)^{2/3}$

Solution a. The expression $^-4^{3/2}$ is read as *the opposite of the three-halves power of four.*

$$^-4^{3/2}$$
$$= ^-\left(\sqrt{4}\right)^3$$ The power applies only to the 4, not to the negative sign. Rewrite the rational exponent as a root followed by a power.
$$= ^-(2)^3$$
$$= ^-8$$ The power applies only to the 2.

Therefore, $^-4^{3/2} = ^-8$.

b. The expression $^-4^{-3/2}$ is read as *the opposite of the negative three-halves power of four.* When we evaluate an expression with a negative fraction as an exponent, it is easier to rewrite the exponent in two steps. First rewrite the negative exponent as the reciprocal of the base raised to a positive exponent, then translate the fractional exponent into radical form.

$$^-4^{-3/2}$$

$$= -\frac{1}{4^{3/2}}$$ The power applies only to 4, not to the negative sign. First rewrite the negative exponent as the reciprocal.

$$= -\frac{1}{\left(\sqrt{4}\right)^3}$$ Rewrite the rational exponent.

$$= -\frac{1}{(2)^3}$$

$$= -\frac{1}{8}$$

Therefore, $^-4^{-3/2} = -\frac{1}{8}$.

c. The expression $(^-4)^{3/2}$ is read as *the three-halves power of* $^-4$.

$$(^-4)^{3/2}$$
$$= \left(\sqrt{^-4}\right)^3 \qquad \text{In this case the exponent applies to } ^-4.$$

The square root of a negative number is not real, so $\sqrt{^-4}$ is not real, and therefore $(^-4)^{3/2}$ is not a real number.

d. The expression $^-27^{2/3}$ is read as *the opposite of the two-thirds power of twenty-seven.*

$$^-27^{2/3}$$
$$= ^-\left(\sqrt[3]{27}\right)^2 \qquad \text{The power applies only to 27, not the negative sign. Rewrite the rational exponent according to the definition.}$$

$$= ^-(3)^2$$
$$= ^-9 \qquad \text{The power applies only to the 3.}$$

Therefore, $^-27^{2/3} = ^-9$.

e. The expression $^-27^{-2/3}$ is read as *the opposite of the negative two-thirds power of 27.*

$$^-27^{-2/3}$$

$$= ^-\frac{1}{27^{2/3}} \qquad \text{The power applies only to 27, not to the negative sign. Rewrite the negative exponent as the reciprocal.}$$

$$= ^-\frac{1}{\left(\sqrt[3]{27}\right)^2} \qquad \text{Rewrite the rational exponent.}$$

$$= ^-\frac{1}{(3)^2}$$

$$= ^-\frac{1}{9}$$

Therefore, $^-27^{-2/3} = ^-\frac{1}{9}$.

f. The expression $(^-27)^{2/3}$ is read as *the two-thirds power of* $^-27$.

$$(^-27)^{2/3}$$
$$= \left(\sqrt[3]{^-27}\right)^2 \qquad \text{The power applies to } ^-27.$$
$$= (^-3)^2 \qquad \text{The cube root of a negative number is negative.}$$
$$= 9 \qquad \text{The power applies to } ^-3.$$

Therefore, $(^-27)^{2/3} = 9$.

You may have noticed in Example 2 that when the exponent is a negative fraction two steps are needed to rewrite it. First we rewrite the negative exponent as the reciprocal of the base raised to a positive exponent, then we translate the fractional exponent to an expression written in radical notation.

Example 3

a. If R is a negative number, is the expression $-R^{4/3}$ positive, negative, or not a real number?

b. Show your substitution of $R = -125$ into the expression $-R^{4/3}$.

c. Evaluate the expression $-R^{4/3}$, when $R = -125$.

Solution

a. In the expression $-R^{4/3}$, the power applies only to the R, not the negative sign. We know that $R^{4/3}$ means the same as $\left(\sqrt[3]{R}\right)^4$. If R is negative then the cube root of R is also negative. The fourth power of a negative number is positive. Therefore, $R^{4/3}$ is positive and $-R^{4/3}$ is negative. Let's look at this again, somewhat symbolically.

$$-R^{4/3}$$
$$= -(\text{neg \#})^{4/3} \qquad \text{Represent the negative number } R \text{ with "neg \#."}$$
$$= -\left(\sqrt[3]{\text{neg \#}}\right)^4 \qquad \text{Write in radical form.}$$
$$= -(\text{neg \#})^4 \qquad \text{The cube root of a negative number is negative.}$$
$$= -(\text{pos \#}) \qquad \text{The fourth power of a negative number is positive.}$$
$$= \text{neg \#} \qquad \text{The opposite of a positive number is negative.}$$

b. When substituting a negative number into an expression with exponents, we must remember to write the negative number in parentheses. Substituting $R = -125$ into $-R^{4/3}$ is

$$-(-125)^{4/3}$$

c. $\qquad -(-125)^{4/3}$

$$= -625 \qquad \text{This can be evaluated mentally or using a calculator.}$$

We are now ready to simplify expressions with rational exponents. However, there is nothing new here. We defined rational exponents so that all of our previous properties and definitions remained true. We restate the properties here, but remember, they are no different from those we worked with previously.

Properties of Exponents

Let x and y be positive real numbers. Let m and n be rational numbers. Then the following are true.

1. $x^m x^n = x^{m+n}$ When *multiplying powers of x*, add the exponents.

2. $\dfrac{x^m}{x^n} = x^{m-n}$ When *dividing powers of x*, subtract the exponents.

3. $(xy)^m = x^m y^m$ The *power of a product* can be simplified by applying the power to each factor.

4. $\left(\dfrac{x}{y}\right)^m = \dfrac{x^m}{y^m}$ The *power of a quotient* can be simplified by applying the power to each factor in the numerator and each factor in the denominator.

6. $(x^m)^n = x^{m*n}$ When simplifying a *power of a power of x*, multiply the exponents.

Example 4

Simplify the following expressions. Verify your results. Assume that all variables are positive real numbers.

a. $(4x^{2/3})(5x^{1/3})$ b. $(1024w^2)^{1/5}$

Solution   a. $(4x^{2/3})(5x^{1/3})$

$= 20x^{2/3+1/3}$ Multiply the numerical factors, and add the exponents.

$= 20x$

VERIFY   Let $x = 8$

$(4x^{2/3})(5x^{1/3}) = 20x$

$(4 * 8^{2/3})(5 * 8^{1/3}) \overset{?}{=} 20 * 8$

$160 = 160$ ✔

We conclude that $(4x^{2/3})(5x^{1/3})$ simplifies to $20x$.

b. $(1024w^2)^{1/5}$

$= 1024^{1/5}(w^2)^{1/5}$ Raise each factor to the one-fifth power.

$= 4w^{2/5}$ Evaluate the numerical coefficient, and multiply the exponents on w.

VERIFY   Let $w = 3$

$(1024w^2)^{1/5} = 4w^{2/5}$

$(1024 * 3^2)^{1/5} \overset{?}{=} 4 * 3^{2/5}$

$6.207 = 6.207$ ✔

We conclude that $(1024w^2)^{1/5}$ simplifies to $4w^{2/5}$.

In the previous section, we learned that equations that include terms or factors raised to a power can be solved by isolating these terms or factors and then "undoing" the power by raising both sides of the equation to the reciprocal power. Because equations that include roots can be written in exponential form the same process applied. This same process for solving equations with powers or roots applies here as well.

Example 5

Solve the following equations. Check your solutions. Round approximate solutions to four significant digits.

a. $2x^{3/2} - 100 = 28$ b. $\left(\sqrt{10y + 25}\right)^5 = 7500$

Solution   a. First isolate the factor that contains the exponent.

$2x^{3/2} - 100 = 28$

$2x^{3/2} = 128$ Add 100 to both sides.

$x^{3/2} = 64$ Divide both sides by 2.

$(x^{3/2})^{2/3} = 64^{2/3}$ To undo the power, we raise both sides to the reciprocal power.

$x = 16$

CHECK

$2 * 16^{3/2} - 100 \overset{?}{=} 28$

$28 = 28$ ✔

Therefore, $x = 28$ is the solution to $2x^{3/2} - 100 = 28$.

238   Chapter 5   Exponents

b. Our first step is to rewrite the left side of the equation in exponential form.

$$\left(\sqrt{10y+25}\right)^5 = 7500$$

$$(10y+25)^{5/2} = 7500$$

We are just *rewriting* the radical, so we do not apply any power to the other side of the equation.

$$\left[(10y+25)^{5/2}\right]^{2/5} = 7500^{2/5}$$

Because the factor that was raised to an exponent was already isolated, we raise both sides to the reciprocal power.

$$10y + 25 \approx 35.48$$

$$10y \approx 10.48$$

$$y \approx 1.048$$

CHECK

$$\left(\sqrt{10*1.048+25}\right)^5 \stackrel{?}{=} 7500$$

$$7498 \approx 7500 \qquad ✔$$

Therefore $y \approx 1.048$ is the solution for $\left(\sqrt{10y+25}\right)^5 = 7500$.

In this section, we learned that **rational exponents** represent a combination of an integer root followed by an integer power. When evaluating expressions with rational exponents, we explored when the base can be negative and when it cannot. Being able to easily determine when an expression is real or not, and when the result is positive or negative, is helpful as we work with expressions with rational exponents. Once again, we found that the properties of exponents hold as long as the base is positive. We also extended solving equations to those that include rational exponents.

Problem
Set
5.3

1. Translate each of the following English phrases into correct symbolic mathematics.

   a. the four-fifths power of twice x

   b. the fifth power of the square root of y

   c. the opposite of the cube root of the fourth power of x

   d. the opposite of the four-thirds power of x

   e. the three-fourths power of the difference of x and 3

   f. the product of π and the cube root of the square of r

   g. the product of π and the three-halves power of r

2. Rewrite each expression that is in exponential form in radical form and rewrite each expression that is in radical form in exponential form. Assume that all variables are nonzero real numbers and that all quantities result in real numbers.

   a. $x^{5/4}$

   b. $\left(\sqrt{m}\right)^3$

   c. $25\left(\sqrt[3]{T}\right)^4$

   d. $\left(\sqrt[3]{25T}\right)^4$

   e. $8y^{5/2}$

   f. $(8y)^{5/2}$

   g. $7\left(\sqrt[3]{cd}\right)^5$

   h. $\sqrt[3]{x^2}$

   i. $\left(\dfrac{3V}{4}\right)^{2/3}$

   j. $\dfrac{-3x^{1/2}}{y}$

3. Rewrite each expression that is in exponential form in radical form and rewrite each expression that is in radical form in exponential form. Assume that all variables are nonzero real numbers and that all quantities result in real numbers.

   a. $-\sqrt[3]{x^5}$

   b. $\sqrt[3]{(^-y)^5}$

   c. $\dfrac{\sqrt{n^3}}{x}$

   d. $(16x)^{3/2}$

   e. $16x^{3/2}$

   f. $(^-x)^{5/2}$

   g. $-x^{5/2}$

   h. $2\pi\sqrt{V^3}$

   i. $\left(\sqrt{\dfrac{c}{d}}\right)^3$

4. Without using a calculator, evaluate the following numerical expressions, if possible.

   a. $25^{3/2}$

   b. $(^-27)^{2/3}$

   c. $(^-36)^{3/2}$

   d. $-9^{3/2}$

   e. $1000^{-2/3}$

   f. $\left(\dfrac{1}{8}\right)^{-2/3}$

   g. $(^-32)^{3/5}$

5. Without using a calculator, evaluate the following numerical expressions, if possible.

   a. $^-10^{-4}$

   b. $\left(\dfrac{16}{81}\right)^{3/4}$

   c. $\left(\dfrac{^-1}{27}\right)^{1/3}$

   d. $\left(\dfrac{1}{9}\right)^{-2}$

   e. $(^-1)^{7/4}$

   f. $(^-1)^{4/7}$

   g. $(16)^{-3/4}$

6. Without using a calculator, simplify the following radicals. (*Hint:* Pay close attention to the results from part a and b as you simplify each of the expressions because we are *not* assuming that all variables represent strictly positive numbers.)

   a. $\sqrt[4]{(^-1)^4}$

   b. $\sqrt[3]{(^-1)^3}$

   c. $\sqrt[4]{x^4}$

   d. $\sqrt[3]{x^3}$

7. Without using a calculator, simplify the following radicals. (*Hint:* Pay close attention to the results from the previous problem as you simplify each of the expressions because we are *not* assuming that all variables represent strictly positive numbers.)

   a. $\sqrt{x^2}$

   b. $\sqrt[4]{16y^4}$

   c. $\sqrt[3]{\dfrac{27}{B^3}}$

   d. $\left(2\sqrt[3]{x}\right)^3$

8. Select *all* of the expressions that are equivalent to $x^{-3/4}$.

   a. $x^{4/3}$

   b. $\dfrac{1}{x^{4/3}}$

   c. $\dfrac{1}{x^{3/4}}$

   d. $\dfrac{1}{\left(\sqrt[3]{x}\right)^4}$

   e. $(^-x)^{3/4}$

   f. $\dfrac{1}{\left(\sqrt[4]{x}\right)^3}$

9. a. Without calculating the value of the expression, decide whether $m^{4/3}$ is positive, negative, or neither when $m = ^-216$. Explain.

   b. If possible, evaluate the expression $m^{4/3}$ when $m = ^-216$. Show your substitution.

10. a. Without calculating the value of the expression, decide whether $z^{-2/3}$ is positive, negative, or neither when $z = ^-8000$. Explain.

    b. If possible, evaluate the expression $z^{-2/3}$ when $z = ^-8000$. Show your substitution.

11. a. Without calculating the value of the expression, decide whether $-y^{3/2}$ is positive, negative, or neither when $y = -2500$. Explain.

   b. If possible, evaluate the expression $-y^{3/2}$ when $y = -2500$. Show your substitution.

12. a. Without calculating the value of the expression, decide whether $(-R)^{3/4}$ is positive, negative, or neither when $R = -100$. Explain.

   b. If possible, evaluate the expression $(-R)^{3/4}$ when $R = -100$. Show your substitution.

13. Which of the following expressions are equivalent to $x^{-4/5} * x$?

   a. $2x^{-4/5}$

   b. $x^{1/5}$

   c. $x^{-1/5}$

   d. $\sqrt[5]{x}$

   e. $\dfrac{1}{\sqrt[5]{x}}$

14. Rewrite the following radical expressions in exponential form and simplify the exponential expressions. Round approximate numbers to the hundredths place. Assume that all variables represent positive quantities.

   a. $\sqrt[3]{125x^5y^3}$ **b.** $\sqrt{10x^8}$ **c.** $5x\sqrt[3]{729x^4y^6}$

15. a. Complete the following tables.

x	$x^{1/2}$	$(x^{1/2})^3$
-4		
-3		
-2		
-1		
0		
1		
2		
3		
4		

x	$x^{1/3}$	$(x^{1/3})^2$
-4		
-3		
-2		
-1		
0		
1		
2		
3		
4		

x	$x^{1/4}$	$(x^{1/4})^5$
-4		
-3		
-2		
-1		
0		
1		
2		
3		
4		

x	$x^{1/5}$	$(x^{1/5})^4$
-4		
-3		
-2		
-1		
0		
1		
2		
3		
4		

   b. Based on your tables, what do you expect the graphs of the following functions to look like?

$$y = x^{3/2} \qquad y = x^{2/3} \qquad y = x^{5/4} \qquad y = x^{4/5}$$

   c. Graph the four functions in part b using your calculator. Are the graphs what you expect? Graph the four functions in part a using your graphing calculator and entering them as they are written in the table. Are the graphs what you expect? What conclusions can you draw?

   d. If m is odd and n is even, what does the graph of $y = x^{m/n}$ look like? If m is even and n is odd, what does the graph of $y = x^{m/n}$ look like?

   e. Under what conditions is a positive real number a reasonable input into the function $y = x^{n/m}$? Under what conditions is a negative real number a reasonable input into the function $y = x^{n/m}$?

16. Simplify each expression below. Express the results using only positive exponents. Numerical coefficients should be written exactly or rounded to three decimal places. Verify your results. Assume that all variables are positive real numbers.

    a. $x^{2/3} * x$

    b. $\dfrac{b^{3/5}}{b^2}$

    c. $5x^{-2/3}$

    d. $x^{1/2} * (x^{3/4} - x^{1/2}) + 5x^{5/4}$

    e. $(2.5w^{1/3})^3$

    f. $(x^{3/2} - y)^2$

    g. $\dfrac{R + R^{1/2}}{R^{-1/2}}$

    h. $\dfrac{15t^{1/4}}{6t^{-1/2}}$

17. Solve each equation. Check your solution. Round approximate numerical results to the thousandths place.

    a. $10w^{3/2} = 80$

    b. $\left(\sqrt{5y}\right)^3 = 4$

    c. $(B + 5)^{3/4} = {}^-2$

    d. $\sqrt{x^2 - 2.4^2} = 7.3$

18. Solve each equation. Check your solution. Round approximate numerical results to the thousandths place.

    a. ${}^-1215 = 0.0016x^{5/3}$

    b. ${}^-10 = (x^2 - 64)^{0.25}$

    c. $375 = 500(1 - r)^{3/2}$

    d. $2.01 = \sqrt[3]{\pi M^2}$

    e. $(100 - x^4)^{1/2} = {}^-2x^2$

19. **Growth of Bacteria.**   A single bacterium divides every 3 hours to produce two complete bacteria. If we find that the colony has 500 bacteria at 8:00 A.M., complete the following table of values for this problem.

Time (in hours) After 8:00 A.M.	0	3	6	9	. . . 24
Number in Colony					

    a. Graph the data in your table.

    b. Write an equation for the number of bacteria in the colony in terms of the time.

    c. If we had checked the colony earlier, how many bacteria would we have found at 5:00 A.M.? At 7:00 A.M.?

    d. At what time would we have expected to find only one bacterium in the colony?

20. **Growth of Cholera.**   A single cholera bacterium divides every 0.5 hours to produce two complete bacteria. If we find that the colony has 500 bacteria at 8:00 A.M., complete the following table of values for this problem.

Time (in hours) After 8:00 A.M.	0	0.5	1	3	. . . 24
Number in Colony					

    a. Graph the data in your table.

    b. Write an equation for the number of bacteria in the colony in terms of the time.

    c. If we had checked the colony earlier how many bacteria would we have found at 7:30 A.M.? At 7:00 A.M.? At 5:00 A.M.?

    d. How many bacteria would we expect to find in the colony at 8:10 A.M.?

    e. At what time would we have expected to find only one bacterium in the colony?

21. Write an equation for any bacteria that divides every k hours. Use A for amount in the colony at any given time, t for time in hours, and A_0 for the initial amount in the colony. (*Hint:* Check your formula with data from the two previous growth problems.)

22. Old Growth Problem.   The following formula approximates the volume (V) in *hundreds of board feet* in a tree that is t years old.

$$V = (5.86 * 10^{-4})t^{2.5}$$

a. Predict the number of board feet in a tree that is 150 years old.

b. Make a table of values for the volume. Let t range from 0 to 200 years. Include at least eight values in your table.

c. Draw a graph by plotting the points from your table and connecting the points with a smooth curve.

d. How many years will it take a tree to reach a volume of 5000 board feet?

23. The Power Line Revisited.   A power company needs to string a power line between two buildings. One building is located on the west side of a river, and the other building is located 3000 feet downriver on the opposite shore. The river is 500 feet wide. The power line must run underground on the shore and underwater across the river. The cost to run the line underwater is $50 per foot, and the cost to run the line underground is $40 per foot.

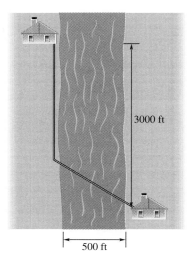

3000 ft

500 ft

a. How much does it cost to run 500 feet of wire underground and the rest underwater?

b. How much does it cost to run 1000 feet of wire underground and the rest underwater?

c. Make a table for the total cost to run the power line between the two buildings in terms of the length of wire run underground for several lengths. (What are the minimum and maximum values for length of wire run underground? Make sure that these are included in your table.)

d. Graph the data from your table.

e. Write an equation for the total cost to run the power line between the two buildings in terms of the amount of wire run underground.

24. Wind Resistance.   The power needed to overcome wind resistance in a vehicle is determined by the frontal surface area of a car and the speed the automobile is traveling.

$$\text{Horsepower} = \frac{s^3 A}{150,000}$$

where s = speed in miles per hour

A = frontal surface area of the vehicle in square feet

Determine the speed of the vehicle if the power needed to overcome wind resistance is 230 horsepower and the frontal surface area is 70.0 square feet. Assume all measurements are to three significant digits.

Chapter Five Summary

We know that a *positive exponent* is used as a shorthand notation for *repeated multiplication*. For example, x^4 means $x * x * x * x$. A *negative exponent* is used as a shorthand notation for *repeated division*. That is, x^{-4} represents $\frac{1}{x^4}$ or, equivalently, $\frac{1}{x * x * x * x}$. In this chapter, we extended our study of exponents to include fractional exponents.

- $x^{1/n} = \sqrt[n]{x}$ provided $\sqrt[n]{x}$ is a real number
- $x^{m/n} = \left(\sqrt[n]{x}\right)^m$ provided $\sqrt[n]{x}$ is a real number

---

Notice that each of these definitions includes the proviso $\sqrt[n]{x}$ is a real number. This is because even roots of negative numbers are not real numbers. This means that a negative number raised to a fractional power in which the denominator is an even number is not real. However, we need to be careful when reading an expression that the negative number is actually being raised to the power.

The following key ideas are useful when working with exponents.

- Remember that exponents apply to the item that is to the exponent's immediate left. For example, $-5^{1/2} = -\sqrt{5}$ and $(-5)^{1/3} = \sqrt[3]{-5}$.
- When substituting a negative number for a variable that is raised to a power, include parentheses around the negative number, and place the exponent just to the right of the parentheses.
- When evaluating numerical expressions that include fractional exponents, it is usually easier to rewrite the expression in radical form first.
- When simplifying algebraic expressions involving fractional exponents, you should *not* rewrite the exponent in radical form. Instead, apply the properties of exponents using the exponents given. Then, verify your result numerically.
- Remember that *no* properties of exponents apply to a sum or difference. The only way to rewrite a power of a sum or difference is to write out the meaning of the exponent. Then perform the resulting multiplication, if possible. For example,

$$(x + 5)^2$$
$$= (x + 5)(x + 5)$$
$$= x^2 + 10x + 25$$

In Section 5.2, we learned how to use the properties of reciprocal powers to solve power equations, that is, an equation in the form $ax^b + c = d$, where a, b, c, and d are constants.

Strategy

Solving Power Equations in One Variable

1. If the equation is written using a radical, first rewrite the radical as a power. Then isolate the power. For example, in the equation $2(x + 1)^{1/2} + 4 = 10$, we isolate the power by isolating the expression $(x + 1)^{1/2}$.

2. Apply the reciprocal power to both sides. Recall that an even power followed by an even root simplifies to an absolute value.

3. Continue solving the equation using methods we previously learned.

4. Check your solution in the original equation.

Because solving power equations may introduce extraneous solutions, checking the results is a crucial part of the solving process, not just for reassurance.

Chapter Six

Exponential Functions

In Chapter 6, we will examine exponential functions numerically, graphically, and algebraically. We will learn how to recognize these functions from any of the three models. This type of function will make it possible for us to model situations that involve exponential growth or decay.

6.1 Exponential Functions from a Numerical Perspective

Activity Set 6.1

Previously in this text we studied linear functions, specific absolute value functions, and power functions.

- A linear function can be written in the form $y = mx + b$, where m and b are constants.
- We studied absolute value functions of the form $y = |mx + b|$, where m and b are constants.
- A power function is of the form $y = ax^b$, where a and b are constants, with $a \neq 0$.

In this chapter, we will begin to study a new class of functions called exponential functions.

> ### Definition
>
> An **exponential function** is any function that can be written in the form $y = ab^x$, where a and b are constants, with $a \neq 0$, $b > 0$, and $b \neq 1$.

You may have noticed that a power function and an exponential function appear similar. Notice that the variable occurs in the base of a power function and the variable occurs in the exponent of an exponential function.

1. Complete each table. List all of the observations you can about the relationship between the table of values and the corresponding function.

x	$y = 3^x$
0	
1	
2	
3	
4	
5	

x	$y = 6 * 2^x$
0	
1	
2	
3	
4	
5	

x	$y = 4 * 5^x$
0	
1	
2	
3	
4	
5	

2. Based on your observations from Activity 1, what do you expect to see in a table of values for the function $y = 10 * 3^x$? Verify your conjecture by making a table.

3. What do you expect to see in a table of values for the function $y = 32 * \left(\frac{1}{2}\right)^x$? Verify your conjecture by making a table.

4. a. For each table, compute differences between successive output values.

  b. Compute ratios between successive output values.

  c. Based on your observations from parts a and b, identify each table of values as linear or exponential.

  d. Write an equation that models the data. Check your equations.

TABLE A

x	y	Difference	Ratio
0	2		
		4	3
1	6		
2	10		
3	14		
4	18		

TABLE B

x	y	Difference	Ratio
0	5		
		10	3
1	15		
2	45		
3	135		
4	405		

TABLE C

x	y	Difference	Ratio
0	625		
1	250		
2	100		
3	40		
4	16		

5. a. For each table, determine whether the data is linear or exponential.

  b. Write an equation that models the data. Check your equations.

TABLE A

x	y
0	4
1	12
2	36
3	108
4	324

TABLE B

x	y
10	20
11	15
12	10
13	5
14	0

TABLE C

x	y
0	8
5	16
10	32
15	64
20	128

TABLE D

x	y
0	64
1	48
2	36
3	27
4	20.25

6. a. Complete each table.

  b. List any observations you have about the output values that correspond to negative input values.

TABLE A

x	$y = 2 * 5^x$
−3	
−2	
−1	
0	
1	
2	
3	

TABLE B

x	$y = 10 * \left(\frac{1}{2}\right)^x$
−3	
−2	
−1	
0	
1	
2	
3	

Discussion

6.1

As mentioned in the activities, we will be discussing the exponential function in this chapter.

> #### Definition
> An **exponential function** is any function that can be written in the form $y = ab^x$, where a and b are constants, with $a \neq 0$, $b > 0$, and $b \neq 1$.

The form of the exponential function $y = ab^x$ and the form of the power function $y = ax^b$ are very similar. It is important to have a clear understanding of the differences between these two. In the exponential function, the variable occurs in the exponent. In the power equation, the variable is the base of an exponent. For example, $y = 5 * 2^x$ is an exponential function, and $y = 5 * x^2$ is a power function.

In the activities, you may have found that you can identify an exponential function from a table of values. If the *ratio* of successive outputs is constant for a constant change in input, then the function is exponential. Let's see why this is reasonable. Consider the function $y = 4 * 5^x$ from Activity 1. To determine the output value for successive rows in the table, we increase the value of the exponent by 1. Because exponents are a shorthand notation for repeated multiplication, we multiply by one more factor of 5 each time. Therefore, the ratio of successive outputs is a constant 5, the base of the exponent. In addition, notice that the output corresponding to $x = 0$ is 4, which corresponds to the coefficient a in the general form $y = ab^x$.

x	$y = 4 * 5^x$
0	$4 * 5^0 = 4$
1	$4 * 5^1 = 20$
2	$4 * 5^2 = 4 * 5 * 5 = 100$
3	$4 * 5^3 = 4 * 5 * 5 * 5 = 500$
4	$4 * 5^4 = 4 * 5 * 5 * 5 * 5 = 2500$
5	$4 * 5^5 = 4 * 5 * 5 * 5 * 5 * 5 = 12{,}500$

We can also look at this symbolically. Suppose we input t into our function. Then the output is $4 * 5^t$. The next input following t is $t + 1$. This produces an output of $4 * 5^{t+1}$. We can take the ratio of these successive outputs algebraically.

$$\frac{4 * 5^{t+1}}{4 * 5^t}$$

$$= \frac{5^{t+1}}{5^t} \qquad \text{The common factor of 4 cancels.}$$

$$= 5^{(t+1)-t} \qquad \text{Subtract the exponents.}$$

$$= 5 \qquad \text{The ratio of any two outputs for } y = 4 * 5^x \text{ is 5.}$$

We can use this information to identify tables of values as exponential and write the equation that is modeled by the exponential data.

Example 1

For each part, determine whether the table of data is linear or exponential. Then write the equation that is modeled by the data. Check your equation.

a.

x	y
10	15
20	13
30	11
40	9
50	7

b.

x	y
0	100
2	40
4	16
6	6.4
8	2.56

Solution

a. To determine whether a table is linear, we need to find the successive *differences* in output values. If these differences are constant for a constant change in input, then the table is linear. If the *ratios* of successive output values are constant for a constant change in input, then the data is exponential. We start by computing both successive differences and ratios.

x	y	Difference	Ratio
10	15		
		-2	$\frac{13}{15} \approx 0.87$
20	13		
		-2	$\frac{11}{13} \approx 0.85$
30	11		
		-2	≈ 0.82
40	9		
		-2	≈ 0.78
50	7		

Although the ratios are close, we can see that the table is linear because the differences are constant. To write the equation of a line, we first need to determine the slope. Because the change in output is -2 for a change in input of 10, the slope is $\frac{-2}{10} = -0.2$. Next, we can write the equation for the line using the slope–intercept equation.

$y = -0.2x + b$	Substitute the slope into the equation $y = mx + b$.
$15 = -0.2 * 10 + b$	Substitute a data point into the equation.
$15 = -2 + b$	Simplify the equation.
$17 = b$	Solve for b.

The equation for the line modeled by the data is $y = -0.2x + 17$. We can also find the vertical intercept by extending our table to include $x = 0$.

To check this equation, we need to substitute values for x from the table. We can do this by making a table using our equation.

x	y = −0.2x + 17
10	15
20	13
30	11
40	9
50	7

We can see that our equation checks. Therefore, $y = -0.2x + 17$ models the data from the table.

b. Again we need to look at differences and ratios.

x	y	Difference	Ratio
0	100		
		$^-60$	$\frac{40}{100} = 0.4$
2	40		
		$^-24$	$\frac{16}{40} = 0.4$
4	16		
		$^-9.6$	0.4
6	6.4		
		$^-3.84$	0.4
8	2.56		

In this table, we can see that the differences are not constant but the ratios are. Therefore, this data represents an exponential function. The constant ratio is the base in our equation, and the output corresponding to $x = 0$ is our coefficient. Therefore, let's try $y = 100 * 0.4^x$. To check we can substitute a few values from our table.

x	$y = 100 * 0.4^x$
0	100
2	16
4	
6	
8	

We can see immediately from the second value in our table that our equation is incorrect. The input values increase by 2's. Because we know the ratio of successive outputs is 0.4, we know that each successive output can be found by multiplying the previous output by 0.4. Let's try this in our table.

x	$y = 100 * 0.4^x$
0	100
2	$100 * 0.4$
4	$100 * 0.4 * 0.4 \qquad\qquad = 100 * 0.4^2$
6	$100 * 0.4 * 0.4 * 0.4 \qquad = 100 * 0.4^3$
8	$100 * 0.4 * 0.4 * 0.4 * 0.4 = 100 * 0.4^4$

From this table, we can see that the only part of the output that changes is the exponent. When the input is 2, the exponent is 1. When the input is 4, the exponent is 2. When the input is 6, the exponent is 3. From this we can see that the exponent is the input divided by 2. Therefore, the equation for this table is $y = 100 * 0.4^{x/2}$. Let's check this equation.

x	$y = 100 * 0.4^{x/2}$
0	100
2	40
4	16
6	6.4
8	2.56

Our equation checks. Therefore, the equation for this table is $y = 100 * 0.4^{x/2}$.

All of the tables of exponential functions that we encountered so far have included the coordinates corresponding to $x = 0$. How do we write the equation if the table does not include this value?

Example 2

Write the equation for the exponential function modeled in the following table.

x	y
6	1,125
7	5,625
8	28,125
9	140,625

Solution

We are already told that this table is exponential. However, we still need to compute successive ratios to determine the base of our exponential equation. In this case, $\frac{5625}{1125} = 5$ is our ratio. Therefore, we know our equation looks like $y = a * 5^x$. Because our table does not include the value corresponding to $x = 0$, we need to do some work to find the value of a. We have two choices. We can extend the table to include this value by dividing by 5, or we can substitute a point from the table into our equation and solve for a. Both of these are good strategies, although sometimes extending the table is extremely time-consuming. Let's use the second strategy and substitute the point (6, 1125) from our table.

$$y = a * 5^x$$
$$1125 = a * 5^6 \qquad \text{Substitute the point (6, 1125).}$$
$$1125 = 15{,}625a \qquad \text{Evaluate } 5^6.$$
$$\frac{1125}{15{,}625} = a \qquad \text{Solve for } a.$$
$$0.072 = a$$

Now, we can substitute this value into our equation to obtain $y = 0.072 * 5^x$. Let's check our equation.

x	y = 0.072 * 5^x
6	1125
7	5625
8	28,125
9	140,625

Our equation checks. Therefore, the equation that models the data is $y = 0.072 * 5^x$.

Throughout this text we used tables as tools to solve applications. We often found it beneficial to write expressions in our tables rather than just results. These expressions allowed us to generalize patterns and write equations. Although this is still an excellent strategy, we occasionally start with a table and do not have enough information to write expressions in our tables. With the experience we now have, we know of two instances where we can write the equations from the data. If successive output values have a constant difference for a constant change in input, we can write a linear function for the data. If the ratio of successive output values is constant for a constant change in input, we can write an exponential function that models the data.

Example 3

Wade and Sam put $1000 into a bank account two years ago. The account compounds interest quarterly. They do not remember the interest rate on the account, but they have the following data from their records.

    a. Write an equation that models the situation.

    b. Determine annual interest for their account.

Number of Months	Balance in Account
0	$1000.00
3	1011.25
6	1022.63
9	1034.13
12	1045.77
15	1057.53
18	1069.43
21	1081.46
24	1093.60

Solution   a. Let's see if the table represents a linear function or an exponential function. If it is not one of these we may be stuck. To decide if it's linear we need to look at successive differences. The first four differences are 11.25, 11.38, 11.50, and 11.64. Although the differences are close, we can see that they are increasing. Therefore, the data are probably not linear.

Next let's look at successive ratios. The first four ratios are 1.01125, ≈ 1.011253, ≈ 1.011246, and ≈ 1.011256. These ratios are very close. Therefore, we conclude that the table is exponential with a factor of about 1.01125.

If we let m represent the number of months and B represent the account balance, then the equation for this situation is

$$B = 1000 * 1.01125^{m/3}$$

Can you explain why the exponent is $m/3$ instead of m?

    b. From the equation and the data, we can see that every three months we multiply the old balance by 1.01125. This means that the interest rate for the three months is 0.01125, or 1.125%. To determine the annual interest rate, we need to multiply the interest rate for one quarter of the year by 4. This gives us $4 * 1.125\% = 4.5\%$. Therefore, the annual interest rate is 4.5%.

In this section we explored exponential functions numerically. An exponential function can be written in the form $y = ab^x$, where $a \neq 0$, $b > 0$, and $b \neq 1$. We found that a table of values for which the ratio of successive output values is constant for a constant change in input values represents an exponential function. The base for this function is the constant ratio, and the coefficient is the output value when the input is zero. In some situations we do not know the output when the input is zero. In these instances we can substitute a known value from our table into the form of the exponential equation and solve for the coefficient.

Problem
Set
6.1

1. Identify each of the following equations as representing a linear function, a power function, an exponential function, or other.

   a. $y = 3x + 1$ **d.** $y = 4.5 * x^4$ **g.** $x - y = 3.4$

   b. $y = 3x^2$ **e.** $y = \dfrac{2}{x} + 7$ **h.** $y = x^2 + 3^x$

   c. $y = 1.75^x$ **f.** $y = 8 * 2^x$

2. Determine whether each of the following tables represents a linear function, an exponential function, or neither. Explain.

a.

Input	Output
0	2
1	12
2	72
3	432
4	2592

b.

Input	Output
10	42
20	84
30	126
50	210
100	420

c.

Input	Output
0	300.00
5	180.00
10	108.00
15	64.80
20	38.88

d.

Input	Output
1	2
2	5
3	10
4	17
5	26

e.

Input	Output
6	10
7	50
8	250
10	6,250
12	156,250

f.

Input	Output
0	$^-15$
2	$^-12$
4	$^-9$
6	$^-6$
8	$^-3$

3. Each table below represents an exponential function. Write an equation that models the data. Check your equation.

a.

x	y
0	30
1	60
2	120
3	240
4	480

b.

t	A
0	2025
1	675
2	225
3	75
4	25

c.

x	y
0	0.25
10	0.50
20	1.00
30	2.00
40	4.00

d.

m	C
4	100.000
5	120.000
6	144.000
7	172.800
8	207.360

4. The following table represents an exponential function. The output values are rounded to the nearest hundredth. Write an equation that models the data in the table. Check your equation.

t	B
0	500.00
0.5	510.00
1	520.20
1.5	530.60
2	541.22

5. For each table, write an equation that models the data. Check your equation.

a.

x	y
0	0.5
1	3
2	18
3	108
4	648

b.

m	P
0	10,000
15	5,000
30	2,500
45	1,250
60	625

c.

t	C
20	22
25	64
30	106
50	274
100	694

d.

x	y
0	150
2	90
4	54
6	32.40
8	19.44

6. Write an equation that models the data in the table. Check your equation.

t	0	1	2	4	6
B	11	33	99	891	8019

7. Solve the following equations. Check your solutions.

a. $27 = 62 - x^4$

b. $3x^3 - 17 = 82$

c. $2\sqrt[5]{x} - 2 = 10$

d. $\sqrt[4]{3 - x} + 10 = 12$

e. $3 + (4x + 1)^{1/2} = 0$

f. $1375 = 550(1 + r)^5$

g. $-3x = \sqrt{4x^2 + 180}$

h. $A = P\left(1 + \dfrac{r}{k}\right)^9$ for r

8. Marty's Allowance Proposal.   Jerry's daughter Marty has asked him to consider a change in her allowance. She has proposed that he pay her 2¢ this week and that he double the amount each week. It sounds good for this week but Jerry is suspicious! How much would Jerry pay in week 5? Week 15? Week 25? Week 50? Should Jerry accept this proposal?

9. The Bouncing Ball.   A ball is dropped from a height of 36 ft. At each bounce it reaches a height that is three quarters of the previous height.

   a. How high does the ball bounce after hitting the ground the first time?

   b. Create a table of values for the height of the ball in terms of the number of times it has hit the ground.

   c. Write an equation for the height of the ball in terms of the number of times it has hit the ground.

   d. How many bounces must the ball make before it rebounds less than 1 foot?

10. Radioactive Decay.   Polonium-210 is a radioactive element with a half-life of 140 days. This means that half of the polonium decays every 140 days.

  a. If 400 mg is in the initial sample, complete the table.

Number of Days	Amount of Polonium Remaining (mg)
0	400
140	
280	
420	
700	

  b. Write an equation for the amount of polonium-210 that remains after n days.

  c. How many days does it take for the amount of remaining polonium-210 to be less than 1 mg?

11. Suburb 1.   The following chart shows the population in a certain suburb from 1980 to 1990.

  a. Write an equation that models the data.

  b. Assuming that the trend indicated by the table continues, predict the population in suburb 1 for the year 2010.

  c. When does your equation predict the population of suburb 1 will be greater than 30,000?

Year	Suburb 1
1980	12,690
1982	13,071
1984	13,463
1986	13,867
1988	14,283
1990	14,711

12. Suburb 2.   The following chart shows the population in a certain suburb from 1980 to 1990.

  a. Write an equation that models the data.

  b. Assuming that the trend indicated by the table continues, predict the population in suburb 2 for the year 2010.

  c. When does your equation predict the population of suburb 2 will be greater than 30,000?

Year	Suburb 2
1980	23,112
1982	23,693
1984	24,274
1986	24,855
1988	25,436
1990	26,017

6.2 Graphing Exponential Functions

In Chapter 2, we studied linear functions. We found that linear functions can be written in slope–intercept form as $y = mx + b$, where m and b are constants. Furthermore, we found that m represents the slope of the line and $(0, b)$ is the vertical intercept. Because m and b are arbitrary constants whose values affect the graph of the linear function, we call m and b **parameters** of the equation.

One of our goals in this section is to determine how the parameters a and b affect the graphs of exponential functions of the form $y = a * b^x$.

1. A set of four tables follows representing exponential functions.

   a. For each table, write an equation that models the data. Check your equation.

   b. For each table, draw a graph by plotting the points. Identify the vertical intercept of each graph.

   c. All of the functions you wrote in part a were of the form $y = a * b^x$. List all of your observations about the relationship between the parameters of the equation and the graph.

TABLE A	
x	y
−2	$\frac{1}{4}$
−1	$\frac{1}{2}$
0	1
1	2
2	4
3	8

TABLE B	
x	y
−2	0.24
−1	1.2
0	6
1	30
2	150

TABLE C	
x	y
−1	384
0	96
1	24
2	6
3	1.5

TABLE D	
x	y
−2	25
−1	5
0	1
1	0.2
2	0.04

2. Based on your observations about the effects of the parameters on the graph of the equation of $y = a * b^x$, predict what the graph of each function will look like. Verify your predictions by graphing each function on your calculator.

   a. $y = 3 * 2^x$

   b. $y = 4^x$

   c. $y = 50 * \left(\frac{1}{3}\right)^x$

3. a. Complete the following table of values. Express each entry as an integer or a fraction rather than as a decimal.

   b. Which of the expressions in your table are equivalent? Explain.

   c. Predict how the graph of $y = 5^{-x}$ will look. Verify your prediction by graphing.

x	3^x	$\left(\frac{1}{3}\right)^x$	3^{-x}
3			
2			
1			
0			
−1			
−2			

4. Which of the following functions are equivalent to $y = 2^{-x}$?

   a. $y = \left(\frac{1}{2}\right)^{1/x}$ **b.** $y = 2^{1/x}$ **c.** $y = \left(\frac{1}{2}\right)^{-x}$ **d.** $y = \frac{1}{2^x}$ **e.** $y = \left(\frac{1}{2}\right)^{x}$

5. Predict how the graph of each function will look. Verify your predictions by graphing each function on your calculator.

   a. $y = 1.5 * 6^x$ **b.** $y = 500 * 4^{-x}$

6. **a.** Describe the graph of $y = a * b^x$ when $a > 0$ and $b > 1$.

   b. Describe the graph of $y = a * b^x$ when $a > 0$ and $0 < b < 1$.

   c. Describe the graph of $y = a * b^{-x}$ when $a > 0$ and $b > 1$.

7. **a.** Complete the table of values for $y = (-2)^x$.

   b. Explain why $(-2)^{1/2}$ is undefined and $(-2)^{1/5}$ is defined.

   c. Plot the points from your table. Is it reasonable to connect the points with a smooth curve? Explain.

   d. Is $(-2)^{8/5}$ defined? Explain.

x	$(-2)^x$
2	
1	
$\frac{1}{2}$	
$\frac{1}{5}$	
0	
$-\frac{1}{5}$	
$-\frac{1}{2}$	

8. Graph each set of three functions. List all of the observations you can make about the relationships between the parameters and the graphs.

$$\begin{cases} y = 2^x \\ y = 2^{3x} \\ y = 2^{x/2} \end{cases} \qquad \begin{cases} y = 0.5^x \\ y = 0.5^{3x} \\ y = 0.5^{x/2} \end{cases} \qquad \begin{cases} y = 10^{-x} \\ y = 10^{-3x} \\ y = 10^{-x/2} \end{cases}$$

Discussion
6.2

In Activity 1, you wrote the equations and plotted the graphs of four exponential functions. Table A was modeled by the equation $y = 2^x$, table B by $y = 6 * 5^x$, table C by $y = 96 * 0.25^x$, and table D by $y = 0.2^x$. Let's look at the graphs of these functions and the effects of the parameters on the graph of the function $y = a * b^x$.

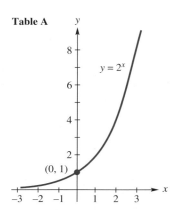

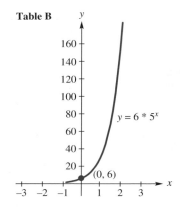

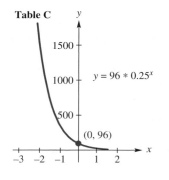

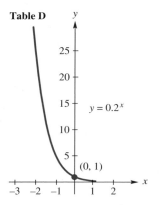

The graphs of tables A and B both rise from left to right. The output values increase slowly at first and then increase more and more rapidly. Because many exponential functions exhibit this rapid growth, this characteristic is called **exponential growth.** This reference often occurs in discussions of population growth and bacteria growth.

The graphs of tables C and D both fall from left to right, a pattern often referred to as **exponential decay.** Many chemical substances decay in a manner that can be modeled by a decreasing exponential function.

In the activities, you may have made the following observations.

- The vertical intercept of $y = a * b^x$ is (0, a).
- When $b > 1$, the graph of $y = a * b^x$ rises from left to right.
- When $0 < b < 1$, the graph of $y = a * b^x$ falls from left to right.
- All of the graphs of $y = a * b^x$ get very close to the horizontal axis and may touch it.

Now we examine each of these observations to see if they are, in fact, true. Let's look first at the observation that the vertical intercept of the equation $y = a * b^x$ appears to be $(0, a)$. To determine a vertical intercept algebraically, we set $x = 0$ and solve for y.

$y = a * b^x$

$y = a * b^0$ Let $x = 0$ to find the vertical intercept.

$y = a * 1$ We know that $b^0 = 1$, provided $b \neq 0$.

$y = a$

We see that our observation is correct. The graph of $y = a * b^x$ has a vertical intercept of $(0, a)$.

You might have noticed that for equations that modeled the data from tables A and B, the parameter b was greater than 1. In the equations that modeled the data from tables C and D, the parameter b was between 0 and 1. We made the observation that when $b > 1$, the function $y = a * b^x$ represents exponential growth, and when $0 < b < 1$, the function $y = a * b^x$ represents exponential decay. In fact, these observations are true. Let's look at why this makes sense.

Consider the function $y = 2^x$, where the base is larger than one. As x increases by 1, the output is twice as large as the previous output. This results in the output values increasing as the input increases. When the base is smaller than 1, such as in the function $y = 96 * 0.25^x$, we multiply by a factor smaller than 1. This results in each output being smaller than the previous output. In this particular example, each output is one quarter of the previous output because $0.25 = \frac{1}{4}$. Therefore, the graph of $y = 96 * 0.25^x$ decreases as x increases.

When $b > 1$, we multiply by a factor that makes the function increase. When $0 < b < 1$, we multiply by a factor that makes the function decrease.

One of the observations that you might have made about the graphs of exponential functions is that all of the graphs of $y = a * b^x$ get very close to the horizontal axis and might touch it. In fact, the graph never touches the axis. A graph touches or crosses the horizontal axis when $y = 0$. For example, to find where $y = 2^x$ might touch the horizontal axis, we need to solve the equation $2^x = 0$. There is no power to which we can raise a positive number to get 0. Therefore, there is no input value for which $a * b^x$ can equal zero. Therefore, the graph of $y = a * b^x$ gets close to the horizontal axis on the left (in the case of growth) or on the right (in the case of decay) but never touches or crosses the axis. It gets so close that it appears to touch the axis. Be careful when you draw graphs of exponential functions to show clearly that they do not touch or cross. This type of behavior is called **asymptotic behavior.** We say that the graph of $y = a * b^x$ has a horizontal asymptote of $y = 0$.

To summarize:

Graphs of $y = a * b^x$ for $a > 0$, $b > 0$, and $b \neq 1$

- The function $y = a * b^x$ with $b > 1$ graphs as exponential growth with a vertical intercept at $(0, a)$. The horizontal asymptote is the line $y = 0$.

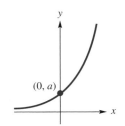

- The function $y = a * b^x$ with $0 < b < 1$ graphs as exponential decay with a vertical intercept at $(0, a)$. The horizontal asymptote is the line $y = 0$.

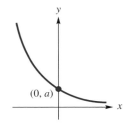

In the activities you looked at the function $y = 2^{-x}$. Also, in Section 6.1, you wrote models that had exponents such as $2x$ or $\frac{x}{2}$. These two observations suggest that we need to consider exponential functions in which the exponent is something other than just x. For that reason, we will now expand our definition of exponential functions.

> Definition _____
>
> Functions of the form $y = a * b^{kx}$, with $a \neq 0$, $b > 0$, $b \neq 1$, and $k \neq 0$, are called **exponential functions.**

You might have noticed that our definition of exponential functions says $a \neq 0$. The coefficient a can be negative. However, in most applications the value of a is positive. For this reason we restrict our explorations and observations to positive values of a.

We need to determine the effect of the parameter k on the graph of $y = a * b^{kx}$. Consider the function $y = 3^{-x}$. We know from the properties of exponents that

$$3^{-x} = \frac{1}{3^x} = \left(\frac{1}{3}\right)^x$$

We already know what the graph of $y = \left(\frac{1}{3}\right)^x$ looks like. Because $y = \left(\frac{1}{3}\right)^x$ graphs as exponential decay with a vertical intercept of $(0, 1)$, so does $y = 3^{-x}$.

Example 1

Predict how the graph of each of the following functions will look. Verify your prediction.

    a. $y = 5 * 2^x$

    b. $y = 100 * 0.46^x$

    c. $y = 24 * 3^{-x}$

Solution   a. We know that $y = 5 * 2^x$ is an exponential function. Therefore, the graph either rises or falls exponentially.

- Because the base 2 is greater than 1 and the coefficient of the exponent 1 is positive, we predict that the graph of $y = 5 * 2^x$ rises exponentially.
- Because $a = 5$, the vertical intercept is $(0, 5)$.

Therefore, we predict that the graph looks like the following.

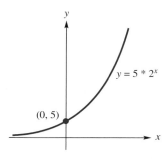

In selecting a window to view the graph of $y = 5 * 2^x$, we consider the vertical intercept of $(0, 5)$ and the fact that the graph grows rapidly because each output is twice as large as the previous one. If our first window is not a good choice, we need to play with the settings to get a more reasonable picture.

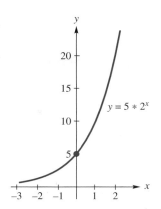

We can see that our predictions about the graph of $y = 5 * 2^x$ were correct.

b. The function $y = 100 * 0.46^x$ also fits the definition of an exponential function.

- Because the base 0.46 is between 0 and 1 and the coefficient of the exponent is positive, we predict that the graph of $y = 100 * 0.46^x$ falls exponentially.

- Because $a = 100$, the vertical intercept is (0, 100).

Therefore, we predict that the graph looks like the following.

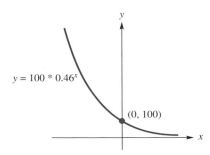

Considering that the vertical intercept is (0, 100) and the graph falls exponentially from there, we probably are most interested in outputs less than 100. To get some perspective, we might choose a maximum output of 200. We are also probably more interested in positive inputs, so we extend the positive side of the x-axis farther than the negative.

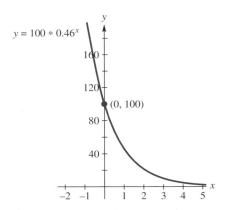

We can see that our predictions about the graph of $y = 100 * 0.46^x$ were correct.

c. The function $y = 24 * 3^{-x}$ also fits the definition of an exponential function.

   • Because the base 3 is greater than 1 and the coefficient of the exponent is $^-1$, we predict that the graph of $y = 24 * 3^{-x}$ falls exponentially.

   • Because $a = 24$, the vertical intercept is $(0, 24)$.

   Therefore, we predict that the graph looks like the following.

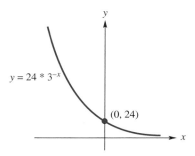

Again, because we are looking at exponential decay, we are most interested in outputs less than the vertical intercept of 24 and in positive inputs. With those observations, we might choose the following window.

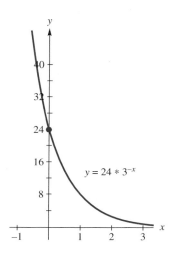

We can see that our predictions about the graph of $y = 24 * 3^{-x}$ were correct.

Earlier in this section, we looked at the graphs of exponential functions of the form $y = a * b^{kx}$, where the parameter k had values of 1 or $^-1$. Whether k was positive or negative, together with the value of the base b, determined if a function represented exponential growth or decay.

In Activity 8, you looked at the set of functions $y = 2x$, $y = 2^{3x}$, and $y = 2^{x/2}$. The graphs of these functions help us as we examine further effects of the parameter k on the graph of the function $y = a * b^{kx}$. We know that these graphs should all have a vertical intercept of $(0, 1)$ and all exhibit exponential growth.

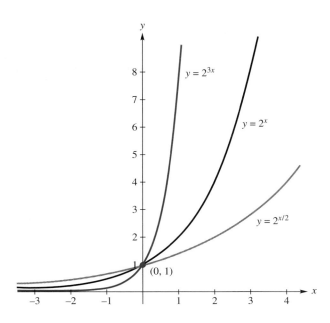

In the function $y = 2^{x/2}$, the parameter k has a value of $\frac{1}{2}$. The effect of the $\frac{1}{2}$ is to make the graph rise less rapidly. This makes sense. In the function $y = 2^x$, the output is multiplied by a factor of 2 for every 1-unit increase in the input. In the function $y = 2^{x/2}$, it takes a 2-unit increase in the input to affect the output to the same degree. In the function $y = 2^{3x}$, k is equal to 3. The effect of the 3 is to make the graph grow more rapidly. Every time x increases by 1, we multiply by three more factors of 2. Notice that when the value of k remains positive, the graph rises exponentially.

In Activity 8, the graphs of $y = 10^{-x}$, $y = 10^{-3x}$, and $y = 10^{-x/2}$ all fall exponentially. We have seen two effects of the parameter k. When it is negative and the base greater than 1, the graph falls rather than rises. This is important because it affects whether the graph represents growth or decay. When k has a value other than 1 or $^-1$, it affects how rapidly the graph rises or falls. This is a more subtle point and not as important in predicting the general shape of a graph.

Graphs of Exponential Functions of the Form $y = a * b^{kx}$, Where $a > 0$

- The function $y = a * b^{kx}$ with $b > 1$ and $k > 0$ graphs as exponential growth with a vertical intercept at $(0, a)$. The horizontal asymptote is $y = 0$.

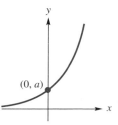

- The function $y = a * b^{kx}$ with $b > 1$ and $k < 0$ graphs as exponential decay with a vertical intercept at $(0, a)$. The horizontal asymptote is $y = 0$.

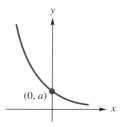

- The function $y = a * b^{kx}$ with $0 < b < 1$ and $k > 0$ graphs as exponential decay with a vertical intercept at $(0, a)$. The horizontal asymptote is $y = 0$.

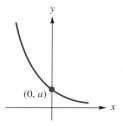

You may have noticed that we did not consider the case when $0 < b < 1$ and k is negative, as in the function $y = \left(\frac{1}{2}\right)^{-x}$. We choose instead to rewrite this as $y = 2^x$, and we already know what this graph looks like.

Example 2

Annual Compounding of Interest.   How long does April have to wait for a $1000 investment to double if it is in an account that is earning 6.5% annual interest, compounded annually?

Solution   Let's first build a table of values for the first few years. In doing this, we can think carefully about how we generate each of the outputs based on the input, as we did in creating this type of model earlier in the course. Instead of this approach, which can be difficult, we apply the techniques you learned in Section 6.1.

At the end of each year, we multiply the previous balance by 0.065 and add it to the balance. Let t represent the number of years the money has been in the account and B represent the account balance.

t	B ($)
0	1000
1	1065
2	1134.23
3	1207.95

Because multiplication was used to calculate the table entries, we might suspect we are more likely to find a common ratio than a common difference.

Looking at the ratios of successive balances, we get

$$\frac{1065}{1000} = 1.065, \qquad \frac{1134.23}{1065} \approx 1.065, \qquad \text{and} \qquad \frac{1207.95}{1134.23} \approx 1.065$$

This indicates that we can model this data with an exponential function with a base of 1.065. Notice that this base represents 100% + 6.5%.

Because the constant ratio is 1.065 and the output is $1000 for year 0, we know that one possibility for our function is $B = 1000 * 1.065^t$. Calculating the outputs for the first three years using this model and rounding to cents gives us exactly the values in our previous table.

Next we want to use our model to decide when the account balance reaches $2000. In other words, we need to solve the following equation.

$$2000 = 1000 * 1.065^t$$

We do not yet have an algebraic method to solve this equation. For now, we must rely on numerical or graphical methods. In Section 10.2, we will learn an analytic method to solve this type of equation.

Let's first look at a graphical solution to $2000 = 1000 * 1.065^t$. To do this, we can graph $B_1 = 2000$ and $B_2 = 1000 * 1.065^t$. Applying what we know about each of these graphs, we can make the following observations.

- The graph of $B_1 = 2000$ is a horizontal line with a vertical intercept of (0, 2000).
- The graph of $B_2 = 1000 * 1.065^t$ is an exponential function that rises and has a vertical intercept of (0, 1000).

In graphing these functions, we need to consider what window to use. We know that our outputs are all positive, and we need to see values somewhat larger than 2000. Let's use a minimum vertical value of $^-100$ and a maximum vertical value of 3500. We cannot be as certain what the appropriate input values are. We know that we are interested only in positive values and that we are talking about a number of years. Let's start with a minimum input of $^-5$ and a maximum value of 25. If we cannot see the intersection of the two graphs in this window, we need to adjust it. (We chose to start at $^-100$ and $^-5$ rather than at 0 so we can see the axes and tic marks on the calculator.) Graphing these two functions with the window we described gives us the following graph.

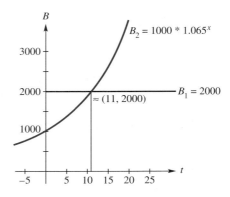

We can see that the two graphs intersect when x is approximately 11.

We can also solve the problem numerically.

x	$1000(1.065)^x$
10	≈ 1877.14
11	≈ 1999.15
12	≈ 2129.10

From this table we see that April's money comes within 85¢ of doubling after 11 years.

In Example 2, we developed a model to calculate compound interest when the interest is being compounded annually. Following is a formula that can be used for a much wider number of compounding situations.

Compound Interest Formula

The following formula gives the accumulated amount in an account when interest is being compounded.

$$A = P\left(1 + \frac{r}{n}\right)^{nt}$$

where A = the accumulated amount

$\qquad P$ = the principal, that is, the original amount

$\qquad r$ = the annual interest rate as a decimal

$\qquad n$ = the number of compounding periods per year

$\qquad t$ = the number of years

Next we explore how the number of compounding periods in a year affects the accumulated amount.

Suppose we deposit \$1 at a 100% annual interest rate. We look at compounding it in various ways, for instance, quarterly, monthly, daily. Substituting the known values of our situation into the compound interest formula gives us

$$A = 1 * \left(1 + \frac{1}{n}\right)^{n*1} \qquad \text{or} \qquad A = \left(1 + \frac{1}{n}\right)^{n}$$

The following table shows the values of A for various values of n.

Although we seem to do better with each new value of n up through 365, beyond that, only slow improvement occurs. If our amount is rounded to dollars and cents, it appears that \$2.72 is the best we can expect from our investment.

The value that we approached for $\left(1 + \frac{1}{n}\right)^{n}$ as n got larger has a special significance in mathematics. This same number occurs in many naturally occurring growth situations. This number, like π, is an irrational number. It is represented by the symbol e and is

Compounded	n	$\left(1 + \frac{1}{n}\right)^{n}$
Annually	1	2
Semiannually	2	2.25
Quarterly	4	≈ 2.44141
Monthly	12	≈ 2.61304
Daily	365	≈ 2.71457
Hourly	8,760	≈ 2.71813
Every minute	525,600	≈ 2.71828
Every second	31,536,000	≈ 2.71828

sometimes called the **natural base.** To see an approximation of the value of e to several decimal places, find the e^{x} button on your calculator and calculate e^{1}. You should get $e \approx 2.7182818285$.

$\mathsf{Definition}$ _____

The **natural base** is an irrational constant. It is represented by the symbol **e,** and its value is approximately 2.71828.

Because of the natural occurrence of this constant e in growth situations, the exponential function $y = e^{x}$ is important and is used often in mathematics. This function is referred to as the natural exponential function.

$\mathsf{Definition}$ _____

The function $y = e^{x}$ is called the **natural exponential function.**

Because the expression $\left(1 + \frac{1}{n}\right)^{n}$ approached the number e as n got very large, the compound interest formula presented earlier takes on the form $A = Pe^{rt}$ for **interest compounded continuously.**

Continuous Compound Interest Formula

The formula $A = Pe^{rt}$ can be used to calculate the amount accumulated in an account in which interest is being compounded continuously and

P = the principal

r = the annual interest rate as a decimal

t = the time in years

Example 3

Compounding Continuously.   In Example 2, we found it would take April approximately 11 years to double her money if she earned 6.5% annual interest compounded annually. If her account instead earned 6.5% annual interest compounded continuously, how long would it take for her $1000 to grow to $2000?

Solution   Using $A = Pe^{rt}$, we have the equation $2000 = 1000 * e^{0.065t}$. The following table allows us to solve this problem numerically.

t	$1000 * e^{0.065t}$
8	≈ 1682.03
9	≈ 1794.99
10	≈ 1915.54
11	≈ 2044.19
12	≈ 2181.47

This show us that April's money doubles sometime during the tenth year. Because the compounding happens continuously, it seems reasonable to get a more precise answer. Let's look at the balance at the end of each month. We can accomplish this by setting the increment on our table to $\frac{1}{12}$ of a year between years 10 and 11.

t	$1000 * e^{0.065t}$
10	1915.54
10.083	1925.94
10.167	1936.41
10.25	1946.92
10.333	1957.50
10.417	1968.13
10.5	1978.82
10.583	1989.57
10.667	2000.37
10.75	2011.24

We can see from our new table that the account has a balance of $2000 by the end of August of the tenth year.

We can also solve this problem graphically.

In this section, we explored **exponential functions** graphically. Functions of the form $y = a * b^{kx}$ with $a > 0$, $b > 0$, $b \neq 1$, and $k \neq 0$ have possible inputs of all real numbers and outputs equal to all positive real numbers. We now know how a, b, and k affect the graphs of exponential functions. The **natural base e** was introduced. We looked at applications involving exponential functions and solved them both graphically and numerically. Later in the text, we will solve exponential equations analytically.

Problem Set 6.2

1. Identify each of the following equations as representing a linear function, a power function, an exponential function, or other.

   a. $y = 3 * 4^x$

   b. $y = 3 * x^4$

   c. $y = 5 - x$

   d. $y = \dfrac{2}{x + 1}$

   e. $y = 4.6(0.05)^{5x}$

   f. $x + 3y = 6.2$

2. Predict how the graph of each of the following exponential functions will look. Verify your predictions by graphing.

   a. $y = 2.5^x$

   b. $y = 0.15^x$

   c. $y = 3.1^{-x}$

   d. $y = e^x$

3. Predict how the graph of each of the following functions will look. Verify your predictions by graphing.

   a. $y = 5 * 3.4^x$

   b. $y = 10\left(\dfrac{1}{2}\right)^x$

   c. $y = 28 * 0.75^x$

   d. $y = 0.56 * 3^{2x}$

   e. $y + 3x = 6$

   f. $y = 4 * 3^{-x}$

4. Predict how the graph of each of the following functions will look. Verify your predictions by graphing.

   a. $y = 12 * 0.85^{3x}$

   b. $\dfrac{x + y}{5} = -1$

   c. $y = \dfrac{4}{5} * 3^x$

   d. $y = 3\left(\dfrac{4}{5}\right)^x$

   e. $y = 1000 * e^x$

   f. $y = 0.4 * 5^{-x}$

5. Rewrite each expression in the form ax^n, where a is the numerical coefficient and n is either a negative or positive integer. For example, we can rewrite $\dfrac{2}{3^x}$ as $2 * 3^{-x}$.

   a. $\dfrac{7}{2^x}$

   b. $\dfrac{2}{5^{-x}}$

   c. $\dfrac{1}{2 * 3^x}$

   d. $\dfrac{8}{4 * 10^x}$

   e. $\dfrac{1}{5e^{2x}}$

6. Using your results from the previous problem, predict how the graph of each of the following functions will look. Verify your predictions by graphing.

   a. $y = \dfrac{7}{2^x}$

   b. $y = \dfrac{2}{5^{-x}}$

   c. $y = \dfrac{1}{2 * 3^x}$

   d. $y = \dfrac{8}{4 * 10^x}$

   e. $y = \dfrac{1}{5e^{2x}}$

7. Predict how the graph of each of the following functions will look. Verify your predictions by graphing.

   a. $y = 100 * 2^x$

   b. $y = \dfrac{10}{4^x}$

   c. $y = \dfrac{100}{2 * 3^x}$

   d. $y = 28 * 0.45^{2x}$

8. Earning Interest.   Calculate the accumulated amount in each of the following situations.

   a. $2500 is deposited at 6.25% annual interest compounded quarterly for 12 years.

   b. $1200 is deposited at 8.5% annual interest compounded monthly for 5 years.

   c. $1200 is deposited at 7% annual interest compounded continuously for 6 years.

9. How long does it take $2500 to reach a value of $8000 if it is earning 6.5% annual interest compounded continuously?

10. Passenger Car Problem.   The number of cars in the United States can be modeled by the equation

$$N = 29(1.036)^t$$

where N is the number of millions of passenger cars in the United States and t is the number of years after 1940.

  a. Graph the equation for values of t from 0 to 65.

  b. How many passenger cars does the model predict for the year 2000?

  c. In what year was the number of passenger cars double the number of cars in 1940? In what year would the number of passenger cars double again?

11. Endangered Birds.   An equation used to predict the population of a bird that is on the endangered species list is

$$\text{predicted population} = \text{current population} * 9^{-(\text{years}/9)}$$

This equation predicts the number of surviving birds over the next 15 years if no measures are taken to protect the population.

  a. If the current population is about 4000 birds, sketch a graph of the bird population over the period the model covers.

  b. What is the predicted population five years from now?

  c. It has been determined that if the population drops below 150 birds, it may not be possible to avoid extinction without removing the birds from the wild and raising them in captivity. Will the population fall below 150 in the next 15 years? If yes, when?

12. Growth of Bacteria.   A colony of bacteria being grown in a jug of apple cider is known to triple every 4 hours. We know that the number of bacteria at 8 A.M. was 600.

  a. What is the population of the bacteria colony at 12 noon?

  b. What was the population of the bacteria colony at 4 A.M.?

  c. Write an equation for the number of bacteria in the colony in terms of the number of hours after 8 A.M. Check your responses in parts a and b in your equation.

  d. Our lab manual suggests that when the population reaches about 10,000, it will run out of food to continue growing. Approximately when will this occur?

  e. Graph this problem situation. Take into account what will happen after the population reaches 10,000. Clearly state the assumptions you are making to explain your graph's behavior.

13. How Many Bacteria?   Bacteria are known to grow exponentially. A culture of a particular bacterium was begun at 2 P.M. on Tuesday. An hour later, the number of bacteria in the culture was 800. After another 30 minutes, the population was 1600.

  a. What was the original population of the bacteria colony? Explain how you decided.

  b. Write an equation for the number of bacteria in terms of the number of hours after the culture started to grow.

  c. At approximately what time did the population reach 25,000?

Chapter Six Summary

In this chapter we were introduced to exponential functions.

> ### Definition
> Functions of the form $y = a * b^{kx}$, with $a \neq 0$, $b > 0$, $b \neq 1$, and $k \neq 0$, are called **exponential functions.**

We learned to recognize exponential functions from the three different models.

- In the **equation of an exponential function,** the variable occurs in the exponent. For example, $y = 4 * 5^x$ is an exponential function. In contrast, the function $y = 4 * x^5$ is a power function because the variable is the base rather than the exponent.
- In a **table of values of an exponential function,** the ratio of successive outputs is constant for a constant change in inputs.
- The **graph of an exponential function** of the form $y = a * b^{kx}$ appears as exponential growth or decay. The horizontal asymptote is the x-axis and the vertical intercept is $(0, a)$.

With the knowledge we now have, there are two instances where we can easily write the equations from the numerical data. If there is a constant difference between successive output values for a constant change in input, we can write a linear function for the data. If the ratio of successive output values is constant for a constant change in input, we can write an exponential function that models the data.

Strategy

> **Writing the Equation of an Exponential Function from a Table of Values**
>
> 1. Compute the ratios of successive outputs. If these ratios are constant for a constant change in input, then the table represents an exponential function.
> 2. The constant ratio that you identified is the base b of the exponential function $y = a * b^{kx}$.
> 3. If the inputs that gave you the constant ratio differed by 1, the value of k is 1. If the values of the inputs that gave you the constant ratio differed by c, the value of k is $\frac{1}{c}$.
> 4. The coefficient a is the value of the function for an input of 0. If you know the output for an input of 0, then you know the value of the coefficient a.
> 5. If you do not know the output when the input is 0, substitute a known ordered pair into the equation $y = a * b^{kx}$. Because b and k are now known, you can solve for a.

Previously, we found that we could predict the behavior of the graphs of linear functions $y = mx + b$ from the parameters m and b. The coefficient m is the slope of the line, and the constant b is the vertical intercept of the line.

We can use the parameters a, b, and k to predict the shape of an exponential function of the form $y = a * b^{kx}$ as summarized in the following box.

Graphs of Exponential Functions of the Form $y = a * b^{kx}$, where $a > 0$

- The function $y = a * b^{kx}$ with $b > 1$ and $k > 0$ graphs as exponential growth with a vertical intercept at $(0, a)$. The horizontal asymptote is $y = 0$.

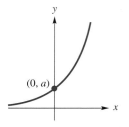

- The function $y = a * b^{kx}$ with $b > 1$ and $k < 0$ graphs as exponential decay with a vertical intercept at $(0, a)$. The horizontal asymptote is $y = 0$.

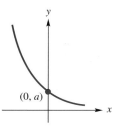

- The function $y = a * b^{kx}$ with $0 < b < 1$ and $k > 0$ graphs as exponential decay with a vertical intercept at $(0, a)$. The horizontal asymptote is $y = 0$.

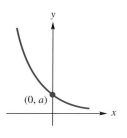

In looking at applications of exponential functions, we were introduced to the **natural base e** and the **natural exponential function $y = e^x$**. The natural base e is an irrational constant and its value is approximately 2.71828.

We know how to solve exponential equations graphically and numerically. In Chapter 10 we will learn to solve these equations analytically.

Chapter
Review

4-6

1. Solve each inequality algebraically. Graph your solution on a number line. Check your solution. (Be sure to check the border point, a point that lies in the solution region, and one that does not lie in the solution region.)

   a. $8 - 4x \geq 32$ 　　　　**b.** $^-3x - 10 < x + 10$ 　　**c.** $\frac{5}{8} > \frac{5}{16}R$

2. **a.** *Use the given graph* to solve the following equation.

   $$\frac{4(30 - 20x)}{3} + 560 = 8(5 - x) + 4(7x - 35)$$

   b. *Use the given graph* to solve the following system of equations.

   $$\begin{cases} y = \dfrac{4(30 - 20x)}{3} + 560 \\ y = 8(5 - x) + 4(7x - 35) \end{cases}$$

   c. *Use the given graph* to solve the inequality.

   $$\frac{4(30 - 20x)}{3} + 560 > 8(5 - x) + 4(7x - 35)$$

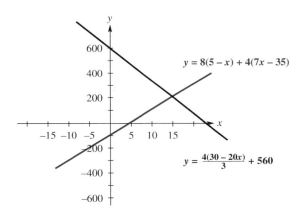

3. Solve each inequality algebraically. Graph your solution on a number line. Check your solution. (Be sure to check the border point, a point that lies in the solution region, and one that does not lie in the solution region.)

   a. $\dfrac{3}{4}x - 10 \leq ^-42$

   b. $24.3 - 6.2p > 4.2p + 102.3$

   c. $44 - 8M \leq 4(3M + 2)$

   d. $\dfrac{7x - 10}{5} + \dfrac{x - 5}{10} \geq x + \dfrac{7}{4}$

4. Solve each inequality graphically. Include a sketch of your two-dimensional graph. Report your solution using inequality notation. Graph your solution on a number line. Check your solution.

   a. $5(4 - 2k) + 11k \leq 100$

   b. $6 + \dfrac{x}{2} \geq 4(x - 4)$

   c. $\dfrac{x}{2} - \dfrac{10x - 25}{10} > 3(x + 3) - (x + 14)$

5. Determine whether each of the following tables represents a linear function. For each table that is linear, write the equation of the line. Check your equation.

a.

x	y
4	1
5	−5
6	−11
7	−17
8	−23

b.

m	P
2	3
4	6
6	12
8	24
10	48

c.

x	y
3	−69
4	−44
5	−19
6	6
8	56

d.

t	D
2	19
4	24
6	29
10	39
20	64

6. Hurricane Floyd.   On Tuesday, September 14, 1999, the eye of Hurricane Floyd was approaching the coast of South Carolina. At 5:00 P.M., the eye of the hurricane was about 480 miles from Charleston, SC; by 11:00 P.M. it was 415 miles from Charleston.

a. Write the equation to predict how far the eye of the hurricane is from Charleston. Assume that Hurricane Floyd travels at a constant rate.

b. What is the slope of your equation? What does it mean in this situation?

c. What is the vertical intercept of your equation? What does it mean in this situation?

d. Based on your equation, when did the eye of Hurricane Floyd hit Charleston?

e. Hurricane Floyd's diameter was about 600 miles. Use your equation to determine when the edge of Hurricane Floyd hit Charleston.

7. Write an inequality or compound inequality to describe each of the following graphs.

a.
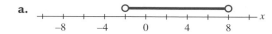

b.
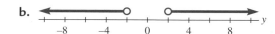

c.

d.
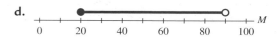

8. *Use the given graph* to solve the following inequality.

$$\frac{5}{8}x + 25 \le \frac{-5(10 - 2x)}{4} + \frac{105}{2}$$

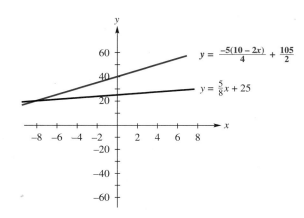

9. Without using your calculator, determine whether each function graphs as a line or a V.

a. $y = -x$

b. $y = |x - 5|$

c. $y = 6.7 + 3.4x$

d. $y = |3 - 2x|$

e. $y = \left|5x + \frac{2}{3}\right|$

f. $y = \frac{6x - 7}{2}$

g. $y = 10$

10. Predict how the graph of each function will look. Then graph the function on your calculator to verify your prediction.

a. $y = |x - 6|$

b. $y = x$

c. $y = |5 + 2x|$

d. $y = |x + 10|$

e. $y = |30 - 2x|$

f. $y = 0.4x - 55$

11. Solve each absolute value equation or inequality graphically (include a sketch of the two-coordinate graph). Write your solution using equality or inequality notation. Graph your solution on a number line.

a. $|w - 2| \le 8$

b. $\left|\frac{x}{2} - 12\right| \ge 12$

c. $|32 + 4x| = 22$

d. $|0.5m + 2| > 9$

e. $41 - 3x < 17$

f. $|k - 1| < {}^-5$

12. For each compound inequality, graph the solution on a number line.

a. $8 < x$

b. $R \le {}^-3$ and $R \ge {}^-6$

c. $k < {}^-4$ or $k \ge 3$

d. ${}^-10 \le x \le 25$

e. $x > 0$ and $x > 3.5$

f. $6 < m$ or $m < {}^-1$

13. Nutcracker.   The student activities club is offering tickets to the *Nutcracker* ballet at a reduced price. The matinee performance tickets are $8.50, and the evening performance tickets are $11.00. In all, 98 tickets were sold and they collected $923. How many matinee tickets were sold?

14. For each of the following graphs write the equation for the line, and check your equation.

a.

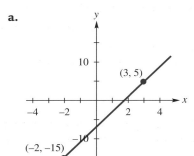

b.

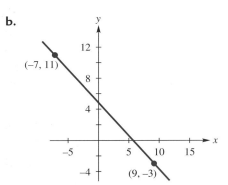

15. Without using a calculator, evaluate the following numerical expressions, if possible.

a. $^-10^0$

b. $\left(\dfrac{81}{16}\right)^{3/4}$

c. $\left(\dfrac{81}{16}\right)^{-3/4}$

d. $\left(\dfrac{-1}{2}\right)^{-3}$

e. $(^-1)^{5/2}$

f. $(^-4)^{-2}$

g. $(^-8)^{-4/3}$

16. Without using a calculator, evaluate the following numerical expressions, if possible.

a. $8^{2/3}$

b. $(^-8)^{2/3}$

c. $^-8^{2/3}$

d. $^-25^{3/2}$

e. 5^{-2}

f. $\left(\dfrac{4}{9}\right)^{-3/2}$

g. $(^-3)^{-4}$

17. Simplify each expression. Express the results using only positive exponents. Verify your results. Assume that all variables are positive real numbers.

a. $x^{5/2} * x$

b. $\dfrac{m^3}{m^{1/2}}$

c. $\left(25x^2\right)^{-1/2}$

d. $\dfrac{p^{1/5}}{p^{-1/5}}$

e. $27(w^3)^{2/3}$

f. $(a^{5/2} + b)^2$

g. $x * (3x^{1/4} + x^{3/4}) + x^{5/4}$

h. $\dfrac{4x + x^{1/4}}{2x^{-3/4}}$

18. Oil Production.   Between 1900 and 1970, the world production of oil doubled about every 10 years. Approximately 150 million barrels of oil were produced in 1900.

  a. Write an equation for the amount of oil produced in terms of the number of years after 1900.

  b. Use your equation to predict the amount of oil produced in 1930 and in 1955.

  c. After 1970, the increase in the production of oil slowed dramatically. The world production of oil in 1988 was 21,338 million barrels. What does your equation predict for 1988?

19. Use the given graph to approximate the solution to each absolute value equation or inequality. Write your solution using equality or inequality notation. Graph your solution on a number line.

  a. $|15x + 37.5| = 67$ **b.** $|15x + 37.5| \leq 67$

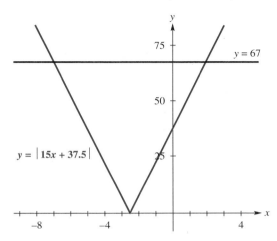

20. Solve each absolute value equation or inequality graphically (include a sketch of the two-coordinate graph). Write your solution using equality or inequality notation. Graph your solution on a number line.

  a. $|x - 7| < 3$ **b.** $|x + 3| > 15$ **c.** $|x| \leq 7.5$ **d.** $|17 - x| = 15$

21. Solve the following inequality either graphically or algebraically. Report your solution using inequality notation. Graph your solution on a number line. Check your solution.

  $$12.4x + 11.2 > 14.5x + 84.7$$

22. Solve each of the following absolute value inequalities graphically. Graph the solution on a number line. Include a sketch of the two-coordinate graph that you use to solve the inequality.

  a. $|7.5x + 15| \leq 2.5x + 30$

  b. $|0.25x - 12| > 8.2 - 0.5x$

  c. $|x - 10| > {}^{-}2 + \frac{2}{3}x$

23. For each part, sketch the graph and write the equation of the line satisfying the given conditions. Label two points on the line.

   a. With a slope of 0.25 and going through the point $(3, {}^-9)$

   b. Passing through the points $({}^-5, 32)$ and $(15, 28)$

   c. Passing through the point $({}^-5, 2)$ and parallel to $x + 2y = 14$

   d. Passing through the point $({}^-1, 12)$ and perpendicular to $y = {}^-0.5x + 5$

   e. A horizontal line passing through the point $(3, 4)$

24. The Boussinesq Equation.   The Boussinesq equation can be used to determine the vertical pressure caused by a concentrated load placed on the soil. The equation is

   $$q = \frac{3Q}{2\pi z^2[1 + (r/z)^2]^{5/2}}$$

   where q = approximate vertical stress at depth z in pounds per square foot

   Q = concentrated load in pounds

   z = depth in feet

   r = horizontal distance in feet from the point of application of Q to the point at which q is desired

   a. Determine the vertical stress exerted by a 250-ton load at a point 20 feet deep and 16 feet horizontally from the point of the load. See the picture.

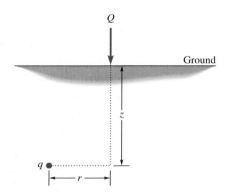

   b. If the vertical stress at a point 20 feet deep and 16 feet horizontally from the point of the load should not exceed 250 pounds per square foot, what is the maximum concentrated load that can be placed on the soil?

25. Solve for the indicated variable. Verify your solutions.

   a. $\dfrac{m + 4}{3p} = m,$ for m

   b. $\dfrac{m + 4}{3p} = m,$ for p

   c. $\sqrt{4x - 3} = 5y,$ for y

   d. $\sqrt{4x - 3} = 5y,$ for x

   e. $7k(a - 5) + 3 = ak,$ for k

   f. $\dfrac{B + 1}{A} = A,$ for A

   g. $\dfrac{4}{n + 2} + x = 9,$ for n

   h. $\dfrac{2}{k} + 5a = \dfrac{a}{k},$ for k

26. Identify each of the following equations as representing a linear function, a power function, or other.

   a. $y = 3.5x - 7.8$

   b. $y = {}^-4$

   c. $y = 2.5x^5$

   d. $y = 2.5 * 4^x$

   e. $3x + 2y = 15$

   f. $y = 2 + \dfrac{1}{x}$

   g. $y = 6 * x^{3/2}$

27. Solve each equation. Check your solution. Round approximate numerical results to four decimal places.

   a. $3p^5 = {}^-96$

   b. $16m^4 = 625$

   c. $\sqrt{2x - 5} + 6 = 9$

   d. $4(k + 2)^{1/3} = 20$

   e. $24 - 6(2x - 4) = 2(3x + 1)$

28. Solve each equation. Check your solution. Round approximate numerical results to four decimal places.

   a. $1000(1 + t)^8 = 5000$

   b. $\sqrt{58 + 6x} + 15 = x + 18$

   c. $\dfrac{x + 18}{3x} = \dfrac{x + 2}{6}$

   d. $p^2 + 16 = 0$

29. Inheritance.   When Benni was 18 years old she inherited \$2000 from an aunt she had never met. Benni was busy at college at the time and just invested the money in a fund that has been earning 8% annual interest compounded annually.

   a. Write an equation for the amount of money in Benni's account in terms of the number of years since it was invested.

   b. Benni is now 50 years old. How much money does she have in this account?

   c. Most of us do not invest in one large sum; instead, we deposit money into savings monthly. A formula to determine the amount in an account is

$$A = P * \left[\frac{\left(1 + \dfrac{r}{12}\right)^{12t} - 1}{\dfrac{r}{12}} \right]$$

      where P is the amount invested each month

            r is the annual interest rate

            t is the number of years

      If we use 8% as the annual interest rate, how much do we need to deposit monthly to earn as much as Benni in 32 years?

30. Solve each equation. Check your solution. Round approximate numerical results to four decimal places.

   a. $(28 - x)^{1/4} = 7$

   b. $\sqrt{5R^2 + 11} = {}^-4R$

   c. $\dfrac{10}{x - 3} + \dfrac{28}{x} = \dfrac{98}{x}$

   d. $\sqrt[4]{x} + 1 = 10$

   e. $24.8m^4 = 248$

   f. $(2t - 7)^2 = 81$

31. Solve each equation for the indicated variable.

    a. $k^2 + p^2 = m^2$, for p **d.** $P = T_0 m^3 A$, for A

    b. $A = 2000(1 + r)^5$, for r **e.** $P = T_0 m^3 A$, for m

    c. $\dfrac{1 + B}{A} = \dfrac{5B}{A}$, for B **f.** $\sqrt{5L + B} = K$, for L

32. Identify each of the following equations as representing a linear function, a power function, an exponential function, or other.

    a. $y = 2.5 * x^5$ **d.** $y = x^3 + 4^x$ **g.** $x + y = 25$

    b. $y = 2 - 3.2x$ **e.** $y = \dfrac{2x + 1}{x}$ **h.** $y = \dfrac{5^x}{4}$

    c. $y = 5^x$ **f.** $y = 8 * x^{3/2}$ **i.** $x = {}^{-}y$

33. Predict how the graph of each of the following functions will look. Verify your predictions by graphing.

    a. $y = 2 * 4^x$ **c.** $y = 15 * 0.45^x$ **e.** $y = 3^{2x}$

    b. $y = 25\left(\dfrac{1}{2}\right)^x$ **d.** $y = 3x - 7$ **f.** $y = 42.5 * 2^{-x}$

34. Hood to Coast.   The Hood-to-Coast relay is the largest relay race in the world. One thousand teams with 12 runners each run from Mt. Hood in Oregon to the Pacific Ocean. The relay is 195 miles long. Each runner runs three legs, which are 3.9 to 8.2 miles long. Matt has put together a team of friends that he ran with in high school and college. They predict that their average pace will be 8 minutes per mile. Matt's mom, Laurie, is on a team with friends from work. Her team predicts their average pace will be 10 minutes per mile. Because of the number of participants in this race, they stagger the start. Laurie's team is scheduled to start at 11:30 A.M. on Friday. Matt's team will start at 4:00 P.M. on Friday.

    a. Write an equation for the distance traveled by Matt's team. Write an equation for the distance traveled by Laurie's team. (*Hint:* You need to convert their pace in minutes per mile to a rate in miles per hour.)

    b. Graph both of your equations on a single set of axes.

    c. How far has Matt's team run by midnight Friday? How far has Laurie's team run by midnight Friday?

    d. Does Matt's team ever catch Laurie's team? If yes, at what time (and day) do they meet?

    f. At what time (and day) does Matt's team reach the Pacific Ocean? Laurie's team?

35. Determine whether each of the following tables represents a linear function, an exponential function, or neither. Explain.

a.

Input	Output
0	5.5
1	11
2	22
3	44
4	88

b.

Input	Output
5	7.5
10	20
15	32.5
20	45
25	57.5

c.

Input	Output
1	⁻12
2	⁻48
3	⁻192
4	⁻768
5	⁻3072

d.

Input	Output
1	24
2	12
3	8
4	6
5	4.8

36. For each table, write an equation that models the data. Check your equation.

a.

x	y
0	150
1	300
2	600
3	1200
4	2400

b.

m	P
2	5
4	25
6	125
8	625
10	3125

c.

x	y
3	108
4	324
5	972
6	2,916
8	26,244

d.

t	D
0	−22
2	−17
4	−12
6	−7
8	−2

37. For each table, write an equation that models the data. Check your equation.

a.

t	C
3	21
6	17
9	13
12	9

b.

n	B
5	102.4
6	204.8
7	409.6
8	819.2

c.

m	P
0	4096
1	2560
2	1600
3	1000

d.

n	B
10	−104
12	−124.8
14	−145.6
16	−166.4

38. The Puddle.   The area of a large puddle is evaporating at the rate of 25% each hour. At 12:00 noon the diameter of the puddle was 3.02 meters.

a. Complete the table below. (*Hint:* If 25% evaporates in 1 hour, how much of the puddle remains?)

b. Write an equation for the area of the puddle in terms of the number of hours after 12:00 noon.

c. How large is the puddle at midnight?

d. Predict when the puddle is less than 100 square centimeters.

Hours After 12:00 Noon	0	1	2	3
Area of Puddle (cm²)				

39. Predict how the graph of each of the following functions will look. Verify your predictions by graphing.

a. $y = 0.75^{3x}$

b. $y = 25$

c. $y = \dfrac{4}{5^x}$

d. $y = \left(\dfrac{4}{5}\right)^x$

e. $y = 500 * e^x$

f. $y = 0.4 * 10^{-x}$

40. Earth Density.   The density of the earth can be measured using an instrument called a torsion balance. The density of the earth is not uniform. Knowing the mean and standard deviation is important in answering questions about the earth's composition. The following density measurements are expressed in grams per cubic centimeter. Determine the mean and standard deviation for these measurements.

5.50   5.57   5.42   5.61   4.88   5.62   4.07

41. Table Salt. A machine that packages table salt historically has had a mean of 737 grams per box and a standard deviation of 5 grams. Last week the following samples were taken.

Sample Number	1	2	3	4	5	6	7	8	9	10
Weight (grams)	739	741	720	749	746	754	748	746	740	753

a. Plot the data from the table. Graph "sample number" on the horizontal axis and weight on the vertical axis. Draw a horizontal line through the process mean of 737 grams. Compute the mean plus three standard deviations and the mean minus three standard deviations. Draw a dotted horizontal line through each of these values.

b. A production process is considered "in-control" if all of the data falls inside the two dotted lines on your graph. Is this process in control?

c. Use the ten sample weights to compute a mean and standard deviation for these ten samples.

Chapter Seven

Quadratic Functions

In Chapters 2 and 4, we learned about linear functions. We learned how to recognize when an equation is linear. We learned about the relationship between the parameters of a linear equation and its graph. These relationships allowed us to graph linear equations more efficiently. In Chapter 6, we studied exponential functions. Again we studied the relationship between the parameters of an exponential equation and its graph. In this chapter we will study quadratic functions and equations. We will learn how to graph quadratic functions in two variables, how to solve quadratic equations in one variable, and the relationships between the one-variable equation and the two-variable function.

7.1 The Graphs of Quadratic Functions

Activity Set 7.1

Before we begin to study quadratic functions, we need to define a line of symmetry. We will discuss this definition in the context of absolute value functions, with which you are already familiar.

> **Definition** _____
>
> A **line of symmetry** is a line about which a figure is symmetrical.

We often think of a line of symmetry of a graph as a line about which the graph produces a mirror image of itself. For example, if we consider the graph of the absolute value function $y = |x - 3|$, the portion of the graph that lies to the left of 3 is a mirror image of the portion that is to the right of 3. Therefore, the line of symmetry for this graph is the vertical line that passes through the vertex of the V. The equation for this line of symmetry is $x = 3$.

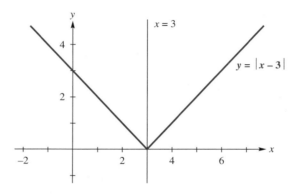

The line of symmetry for this graph is $x = 3$.

1. Write the equation for the line of symmetry for each absolute value function graphed here.

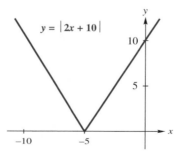
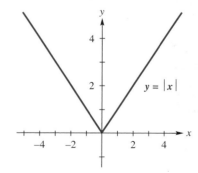

2. Graph each of the following equations using your graphing calculator in its standard graphing window. Draw a sketch of each graph for future reference.

$$y = x^2 \qquad\qquad y = x^2 + 5$$
$$y = x^2 + 1 \qquad\quad y = x^2 - 6$$

a. How are these graphs alike and how are they different? Be as specific as possible.

b. Determine the vertical intercept for each graph.

c. Write the equation of the line of symmetry for each graph.

d. Describe how the graph of $y = x^2 + c$ looks. Be as specific as possible.

3. Graph each of the following equations using your graphing calculator in its standard graphing window. Draw a sketch of each graph for future reference.

$$y = x^2 \qquad\qquad y = 2x^2 \qquad\qquad y = 0.5x^2$$
$$y = {}^-x^2 \qquad\qquad y = {}^-2x^2 \qquad\qquad y = {}^-0.5x^2$$

   a. How are these graphs alike and how are they different? Be as specific as possible.
   b. Determine the vertical intercept for each graph.
   c. What is the equation of the line of symmetry?
   d. Describe how the graph of $y = ax^2$ looks. Be as specific as possible.

4. Based on your observations about the effects of the parameters a and c on the equation $y = ax^2 + c$, predict how the graph of each function will look. Verify your predictions by graphing each function on your calculator.

$$y = {}^-0.25x^2 \qquad\qquad y = {}^-2x^2 + 8$$
$$y = x^2 + 10 \qquad\qquad y = x^2 - 25$$

5. For each function in Activity 4, determine the horizontal intercepts algebraically.

6. Using your calculator, graph each of the following equations in its standard graphing window.

$$y = x^2 + 3x \qquad\qquad y = 2x^2 - 7x$$
$$y = x^2 - 4x \qquad\qquad y = {}^-x^2 + 5x$$

   a. How are these graphs alike and how are they different? Be as specific as possible.
   b. For each graph, find the horizontal intercepts algebraically. (*Hint:* You will need the **zero-product principle,** which states that for real numbers A and B, if $A * B = 0$, then $A = 0$ or $B = 0$. In other words, if the product of two factors is zero, then at least one of the factors must equal zero.)
   c. Write the equation of the line of symmetry for each graph.
   d. Complete the following table using your results from parts b and c.

Quadratic Function	Horizontal Intercepts	Line of Symmetry
$y = x^2 + 3x$		
$y = x^2 - 4x$		
$y = 2x^2 - 7x$		
$y = {}^-x^2 + 5x$		
$y = ax^2 + bx$		

   e. What is the relationship between the horizontal intercepts and the coefficients of the second-degree term and the first-degree term?
   f. What is the relationship between the horizontal intercepts and the line of symmetry?
   g. What is the relationship between the equation of the line of symmetry and the coefficients?

7. Graph each of the following boxed triples of functions using your graphing calculator in its standard graphing window.

$$y = x^2 + 3x$$
$$y = x^2 + 3x + 2$$
$$y = x^2 + 3x - 5$$

$$y = 2x^2 - 7x$$
$$y = 2x^2 - 7x + 4$$
$$y = 2x^2 - 7x - 7$$

a. Within each triple, how are the graphs alike and how are they different? Be as specific as possible.

b. Write the equation of the line of symmetry for each triple.

c. Based on your previous results, what is the equation of the line of symmetry for the graph of the equation $y = ax^2 + bx + c$?

Discussion 7.1

In Chapter 2, we studied linear functions. We found that linear functions can be written in slope–intercept form as $y = mx + b$, where m and b are constants. Furthermore, we found that m represents the slope of the line and $(0, b)$ is the vertical intercept. Because m and b are arbitrary constants whose values affect the graph of the linear function, we call m and b **parameters** of the equation.

In this section we will look at a new set of functions called quadratic functions. Any function of the form $y = ax^2 + bx + c$, where a, b, and c are constants with $a \neq 0$, is a **quadratic function.** A quadratic function graphs as a U-shape, called a **parabola.** Our goal in this section is to determine how the parameters a, b, and c of the function $y = ax^2 + bx + c$ affect its graph.

It is important to note that the graphs of many other functions have a general U-shape; however, it is only the quadratic function whose U-shaped graph is called a parabola. This parabolic shape has special reflective properties that not all U-shaped figures have.

Before we begin to discuss the effects of the parameters we need to introduce some terminology. The expression $ax^2 + bx + c$ has three terms; ax^2, bx, and c. The term ax^2 is referred to as the **second-degree term** because the power of x is 2. The term bx is referred to as the **first-degree term** because the power of x is 1. The term c is referred to as the **constant term** because c is a constant.

To determine the effects of the parameters of the quadratic function on its graph, we look at these in pieces. First, in the activities, we saw that the graph of $y = x^2$ is a parabola. This parabola is symmetrical about the y-axis. Therefore, the equation for the line of symmetry for $y = x^2$ is $x = 0$. Can you explain why $y = x^2$ is symmetrical? Consider the results of substituting $x = 2$ and $x = {}^-2$ into this equation. What happens?

The low or high point of a parabola is called the **vertex.** The vertex of $y = x^2$ is at $(0, 0)$ as seen on the following graph. Remember that the window you choose affects the apparent steepness of the graph. Therefore, if you graph this function in a different window than the one shown here, its appearance may look slightly different.

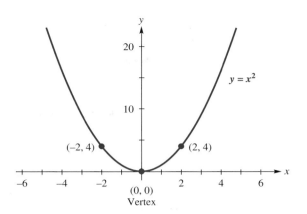

Next, let's look at the effect of adding a constant term to the function $y = x^2$. For example, in the activities, we compared the four functions $y = x^2$, $y = x^2 + 1$, $y = x^2 + 5$, and $y = x^2 - 6$. Here we look at three of these: $y = x^2$, $y = x^2 + 5$, and $y = x^2 - 6$. It appears that the constant term corresponds to the vertical intercept, and the line of symmetry remains $x = 0$.

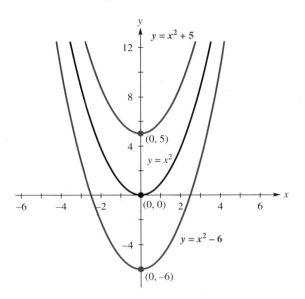

To see why the constant term gives us the vertical intercept, let's find the vertical intercept of $y = x^2 + 5$ algebraically. Remember to find the vertical intercept, we substitute $x = 0$ into the equation and find the value of y.

$$y = 0^2 + 5$$
$$y = 5$$

Therefore, the vertical intercept of $y = x^2 + 5$ is $(0, 5)$. Similarly, the vertical intercept for $y = x^2 - 6$ is $(0, {}^-6)$.

From this, we know that $y = x^2 + c$ graphs as a parabola opening up with a vertical intercept at $(0, c)$. Another way to think of the effect of the constant term is to think of it as shifting the parabola vertically according to the value of c.

Next, let's consider how the coefficient of the second-degree term affects the graph of the quadratic function. In the activities, we compared the graphs of $y = x^2$, $y = {}^-x^2$, $y = 2x^2$, $y = {}^-2x^2$, $y = 0.5x^2$, and $y = {}^-0.5x^2$. You may have found that the sign of the coefficient affects the direction in which the parabola opens. If the coefficient of x^2 is positive, the parabola opens up, and if the coefficient of x^2 is negative, the parabola opens down.

In addition, the magnitude of the coefficient affects the steepness of the parabola. A quadratic function in which the coefficient of x^2 is larger than 1 or smaller than $^-1$ graphs as a parabola that is "steeper" than $y = x^2$. A quadratic function in which the coefficient of x^2 is between $^-1$ and 1 graphs as a parabola that is less steep than $y = x^2$.

Notice that in all of the parabolas that we discussed so far, the line of symmetry has remained $x = 0$. In other words, the parabola has not moved horizontally yet.

Putting the two pieces we have discussed together, we know that $y = ax^2 + c$ graphs as a parabola. If a is positive, the parabola opens up. If a is negative, the parabola opens down. The vertical intercept is at $(0, c)$. The parabola has not shifted horizontally; therefore, the line of symmetry is the y-axis whose equation is $x = 0$.

Example 1

Predict how the graph of each function will look. Determine the horizontal intercepts algebraically. Verify your prediction by graphing the function on your calculator.

   a. $y = {}^-3x^2 + 75$ b. $y = 0.4x^2 - 15$

Solution   a. $y = {}^-3x^2 + 75$

Because the coefficient of x^2 is $^-3$, the parabola opens down and is steeper than if the coefficient were $^-1$. The vertical intercept is at $(0, 75)$, and the equation for the line of symmetry is $x = 0$. Furthermore, because the parabola is not shifted horizontally, the vertical intercept is also the vertex of the parabola. With this information, we can draw a rough sketch.

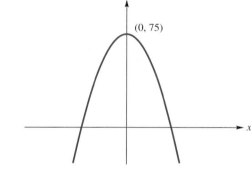

To determine the horizontal intercepts, we substitute $y = 0$ and solve for x.

$$0 = {}^-3x^2 + 75$$
$$3x^2 = 75$$
$$x^2 = 25$$
$$(x^2)^{1/2} = 25^{1/2}$$
$$|x| = 5$$
$$x = \pm 5$$

Therefore, the horizontal intercepts are $(^-5, 0)$ and $(5, 0)$. With this we can graph $y = {}^-3x^2 + 75$ in a reasonable window to verify our conjecture.

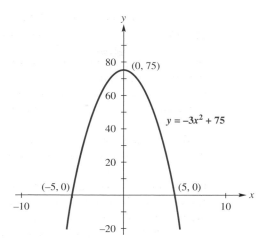

b. $y = 0.4x^2 - 15$

Because the coefficient of x^2 is 0.4, this function graphs as a parabola that opens up. It is less steep than $y = x^2$, the vertical intercept is at $(0, ^-15)$, and the line of symmetry is $x = 0$. Again, because this parabola is not shifted horizontally, the vertical intercept is the vertex of the parabola. A rough sketch of this function is shown in the figure.

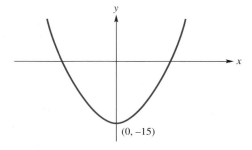

To find the horizontal intercepts, we substitute $y = 0$ into the equation.

$$0 = 0.4x^2 - 15$$
$$15 = 0.4x^2$$
$$37.5 = x^2$$
$$37.5^{1/2} = (x^2)^{1/2}$$
$$6.12 \approx |x|$$
$$x \approx \pm 6.12$$

The horizontal intercepts of $y = 0.4x^2 - 15$ are $\approx(^-6.12, 0)$ and $\approx(6.12, 0)$. With this we can graph $y = 0.4x^2 - 15$ in a reasonable window to verify our conjecture.

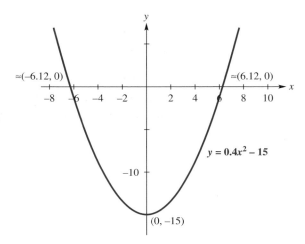

In Example 1, both parabolas had two horizontal intercepts. Do all parabolas have two? Can you draw a parabola with one? None?

So far, we looked at the effects of the parameters a and c in the quadratic function $y = ax^2 + bx + c$. In all of these cases, b was equal to zero. Because all of the previous quadratic functions that we looked at were not shifted horizontally, we might guess that when b is not equal to zero the parabola shifts horizontally, and, thus, the line of symmetry is not the y-axis.

In Activity 6, we investigated the graphs of $y = x^2 + 3x$, $y = x^2 - 4x$, $y = 2x^2 - 7x$, and $y = -x^2 + 5x$. Each of these functions graphed as a parabola with two horizontal intercepts, one of which was at $(0, 0)$. Knowing the horizontal intercepts allowed us to determine the line of symmetry. Let's look at the quadratic function $y = 2x^2 - 7x$ in detail to see how this worked. Because the coefficient of x^2 is positive, the parabola opens up. The vertical intercept is at $(0, 0)$, because the constant term is 0. As we did in the activities, let's find the horizontal intercepts by substituting $y = 0$ into our equation.

$$0 = 2x^2 - 7x$$

To solve this equation, we factor the right-hand side and then use the **zero-product principle.**

$$0 = x(2x - 7) \qquad \text{Factor } x \text{ out of the right-hand side.}$$
$$x = 0 \quad \text{or} \quad 2x - 7 = 0 \qquad \text{Set each factor equal to zero.}$$
$$x = 0 \quad \text{or} \quad x = 3.5$$

The horizontal intercepts of $y = 2x^2 - 7x$ are at $(0, 0)$ and $(3.5, 0)$. A graph of $y = 2x^2 - 7x$ is shown in the following figure.

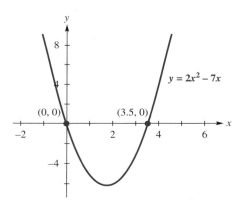

Then, by the symmetry of the graph, we know that the line of symmetry must be exactly halfway between the horizontal intercepts. Because one of the intercepts is at $(0, 0)$, the line of symmetry is half of 3.5, that is, $x = \frac{1}{2} * 3.5$, or $x = 1.75$.

Let's see if we can generalize this to the function $y = ax^2 + bx$. First, we need to find the horizontal intercepts.

$$0 = ax^2 + bx$$
$$0 = x(ax + b) \qquad \text{Factor } x \text{ out of the right-hand side.}$$
$$x = 0 \quad \text{or} \quad ax + b = 0 \qquad \text{Set each factor equal to zero.}$$
$$x = 0 \quad \text{or} \quad x = \frac{-b}{a}$$

The horizontal intercepts of $y = ax^2 + bx$ are at $(0, 0)$ and $\left(\frac{-b}{a}, 0\right)$. Then, the line of symmetry is halfway between these intercepts at $x = \frac{1}{2} * \frac{-b}{a}$, which simplifies to $x = \frac{-b}{2a}$.

Finally, we need to consider the case when all three terms are present. As we did in the activities, we add a constant to $y = 2x^2 - 7x$. Let's graph the following three quadratic functions and see what observations we can make.

$$y = 2x^2 - 7x$$
$$y = 2x^2 - 7x + 4$$
$$y = 2x^2 - 7x - 7$$

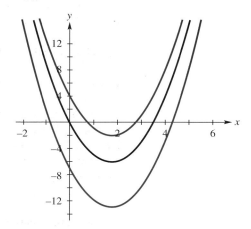

Let's list the observations we can make based on these graphs.

• All of the graphs open up.
• Each graph is a vertical shift of the other graphs.
• All of the parabolas have the same line of symmetry.
• The constant term in the equation corresponds to the vertical intercept.
• The parabolas have different horizontal intercepts.

Based on these observations, the graph of $y = ax^2 + bx + c$ is the graph of $y = ax^2 + bx$ shifted vertically c units. This means the graph of $y = ax^2 + bx + c$ has the same line of symmetry as the graph of $y = ax^2 + bx$, which is the vertical line $x = \frac{-b}{2a}$. However, the vertical and horizontal intercepts of these functions are different.

Notice that we did not find the horizontal intercepts for $y = 2x^2 - 7x + 4$ and $y = 2x^2 - 7x - 7$. At this point, we can approximate the coordinates of the horizontal intercepts using the graph, but we do not have an algebraic technique when all three terms are present. We will learn how to solve equations of the form $0 = ax^2 + bx + c$ in the next section.

Let's put these conclusions together with our previous results. The function $y = ax^2 + bx + c$, where $a \neq 0$, graphs as a parabola. If a is positive the parabola opens up, and if a is negative, the parabola opens down. The vertical intercept is the point $(0, c)$. The equation for the line of symmetry is $x = \frac{-b}{2a}$. Because the vertex lies on the line of symmetry, the x-coordinate of the vertex is also $\frac{-b}{2a}$.

Graphs of Quadratic Functions

The graph of the quadratic function $y = ax^2 + bx + c$, where $a \neq 0$, is a parabola.

• If $a > 0$, then the parabola opens up.
• If $a < 0$, then the parabola opens down.
• The vertical intercept is the point $(0, c)$.
• The equation of the line of symmetry is $x = \frac{-b}{2a}$.
• The x-coordinate of the vertex is $x = \frac{-b}{2a}$.

Example 2

Graph the quadratic function $y = {}^{-}2x^2 - 100x + 55$. Label the vertical intercept and vertex on the graph.

Solution   If we graph this in the standard window on a graphing calculator we obtain a graph similar to the following.

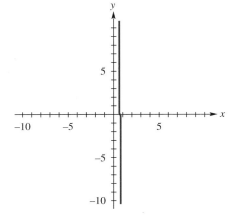

However, based on our experience in the activities and our discussion in this section, we know that this is not the complete graph. The graph of a quadratic function is a parabola. Therefore, we need to use the information we discovered about the effects of the parameters to determine a window that shows a complete graph.

We know that the graph of $y = {}^{-}2x^2 - 100x + 55$ is a downward parabola because the coefficient of x^2 is negative. The vertical intercept is at $(0, 55)$ because the constant term is 55. The line of symmetry is

$$x = \frac{-b}{2a}$$

$$x = \frac{--100}{2 * {}^{-}2} \qquad \text{Substitute the values for } a \text{ and } b.$$

$$x = {}^{-}25$$

With this information, we can make a rough sketch as seen in the following figure.

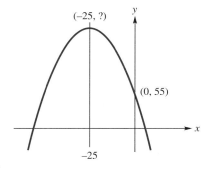

To determine the scale for the graph, it is helpful to know the coordinates of the vertex. We know the horizontal coordinate of the vertex is $^{-}25$, so we can substitute $^{-}25$ into the equation for x and solve for y.

$$y = {}^{-}2x^2 - 100x + 55$$

$$y = {}^{-}2 * ({}^{-}25)^2 - 100 * {}^{-}25 + 55 \qquad \text{Substitute the value of the } x\text{-coordinate of the vertex to determine the } y\text{-coordinate.}$$

$$y = 1305$$

Therefore, the vertex is the point $({}^{-}25, 1305)$.

If we put all of this information together, we can choose a reasonable window to view the complete graph. By the symmetry of the graph, we know we at least need to double the line of symmetry. Because the line of symmetry is $x = {}^-25$, a reasonable horizontal range could be $[{}^-60, 10]$. The vertical intercept is $(0, 55)$, and the vertex is $({}^-25, 1305)$; therefore, a reasonable vertical range is $[{}^-500, 1500]$. With this window we obtain the following graph.

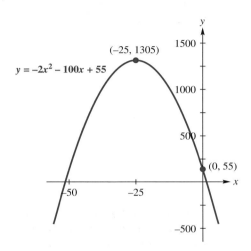

In the previous example, we could have spent time with our calculator, using trial and error to find a complete graph. Given the information we now know about parabolas, trial and error is a poor choice of strategies for graphing a quadratic function. Instead, we should use our knowledge about the effects of the parameters of the equation $y = ax^2 + bx + c$ on its graph.

Also, there are applications where the exact coordinates of the vertex must be found. Technology can approximate these but will not always provide us with exact values. Here again it is important for us to use relationships between the equation of a parabola and its graph.

Example 3

The Profit.   A business's profit for selling suits is given by the equation $P = {}^-200 + 210t - t^2$, where P is the profit in dollars and t is the number of suits sold.

    a. Graph the function.

    b. How much profit is made if no suits are sold?

    c. How many suits should be sold to make the most profit?

    d. How many suits should be sold to keep the profit above $6000?

    e. How many suits should be sold to keep the profit above $12,000?

Solution   a. The equation $P = {}^-200 + 210t - t^2$ is a quadratic function that will graph as a parabola. The coefficient of the second-degree term is negative, which means the parabola will open down. The vertical intercept is $(0, {}^-200)$. We can use the equation $x = \frac{{}^-b}{2a}$ to find the line of symmetry, except in this function the independent variable is t.

$$t = \frac{{}^-210}{2 * {}^-1}$$

$$t = 105$$

With this we can sketch a rough graph.

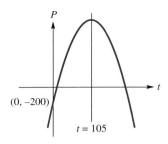

To determine the scale for our graph, we will need to know the coordinates of the vertex. We know that the horizontal coordinate of the vertex is 105, so we can substitute 105 into the equation for t and solve for P.

$$P = {}^-200 + 210t - t^2$$

$$P = {}^-200 + 210 * 105 - 105^2 \quad \text{Substitute } t = 105 \text{ to determine the } P\text{-coordinate of the vertex.}$$

$$P = 10{,}825$$

Therefore, the vertex is the point (105, 10,825).

If we put all of this information together, we can choose a reasonable window to view the complete graph. The line of symmetry is $t = 105$, so the horizontal range can be $[^-20, 250]$. The vertical intercept is $(0, {}^-200)$, and the vertex is (105, 10,825). Therefore, the vertical range can be $[^-1000, 11{,}000]$. With this window we obtain the following graph.

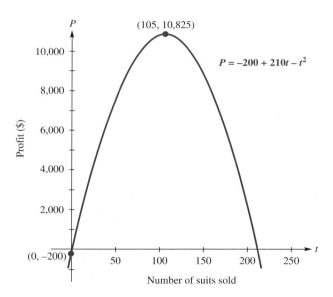

b. If no suits are sold, then t is zero.

$$P = {}^-200 + 210 * 0 - 0^2$$

$$P = {}^-200$$

The profit is $^-\$200$ if no suits are sold, that is, a loss of $200. Notice that this point corresponds to the vertical intercept.

c. Because the highest profit occurs at the vertex of the parabola, the business realizes the largest profit if it sells 105 suits.

d. To determine the number of suits that must be sold to keep the profit above $6000, we first find when the profit is equal to $6000. This corresponds to solving the following equation.

$$6000 = {}^-200 + 210t - t^2$$

We do not have an algebraic technique to solve this equation. Therefore, we use the graph to answer this question. We graph the line $P = 6000$ together with the graph of the quadratic function.

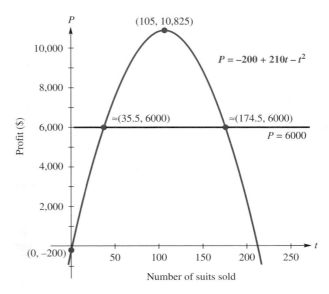

We can see that the two graphs intersect in two points, $\approx(35.5, 6000)$ and $\approx(174.5, 6000)$, and that values of t between 35.5 and 174.5 produce a profit that is greater than 6000. Because we cannot sell a partial suit, we must determine the endpoints of the interval. We can do this by evaluating P at values near the intersection points.

t	P
35	5925
36	6064

t	P
174	6064
175	5925

From the graph and the table, we can see that if we sell more than 35 suits and fewer than 175 suits we make a profit of at least $6000. We can also say that we need to sell at least 36 suits and no more than 174 suits.

We found the border points of our interval by making a table. We can also use the graph. For example, consider the left intersection point. Because the t-coordinate is $t \approx 35.5$ and we know that the profit is larger for values between 35.5 and 174.5, we can conclude that when $t = 36$ the profit is more than $6000 and when $t = 35$ the profit is less than $6000.

e. We know that the highest profit occurs at the vertex, which is (105, 10,825). This means that the maximum profit is $10,825; therefore, it is not possible to make a profit of $12,000.

In this section we looked at functions of the form $y = ax^2 + bx + c$, where $a \neq 0$. These functions are called **quadratic functions** and graph as **parabolas.** A parabola is a U-shaped graph that has either a low point or a high point called the **vertex.** The parabola is an upward parabola with a low point if a is positive and a downward parabola with a high point if a is negative. The vertex always occurs on the **line of symmetry,** which is the line about which the graph is symmetrical. The equation of the line of symmetry is $x = \dfrac{-b}{2a}$.

We looked at two special cases in which the quadratic function contained only two terms: $y = ax^2 + c$ and $y = ax^2 + bx$. In both cases, we can algebraically determine the horizontal intercepts, if there are any, by substituting $y = 0$ into the equation. In the case of the quadratic function $y = ax^2 + c$, the parabola is not shifted horizontally. This means the line of symmetry of $y = ax^2 + c$ is the y-axis, and the vertical intercept is also the vertex of the parabola. Because the vertical intercept is the vertex, this particular quadratic is relatively easy to graph.

Remember, at this point, when all three terms are present we cannot algebraically determine the horizontal intercepts.

Problem Set 7.1

1. Without using your calculator, predict how the graph of each function will look. Verify your prediction by graphing the function on your calculator.

    a. $y = {}^-5x^2 + 15$ **c.** $y = 42 - 4x^2$

    b. $y = x^2 + 20$ **d.** $y = -\frac{3}{2}x^2 - 70$

2. Without using a calculator, determine whether each equation graphs as a line, a parabola, or a V-shaped graph.

    a. $y = 25 - 4x^2$ **e.** $x = 42$

    b. $y = |2x + 15|$ **f.** $y = 2x + 15$

    c. $y = 5x - 2x + 12$ **g.** $y = x(5 - 0.25x)$

    d. $y = 6x^2 + 32x$ **h.** $y = 3(2x + 5) - x$

3. For each function in Problem 2, graph the function by considering the parameters. Verify your prediction by graphing the function on your calculator.

4. **a.** Draw a parabola that has no horizontal intercepts.

    b. Write the equation of a parabola that has no horizontal intercepts.

    c. Draw a parabola that has one horizontal intercept.

5. For each quadratic function, determine the equation of the line of symmetry, the vertex, and the vertical intercept. Use these to draw a graph.

    a. $y = {}^-4x^2 + 484$ **g.** $y = 1280 + 100x + 5x^2$

    b. $y = 0.5x^2 - 8x + 17$ **h.** $y = \frac{x}{2}(18 + 2x)$

    c. $y = 10x^2 + 25$ **i.** $y = {}^-9.8x^2 + 49.6x$

    d. $y = 46x - 3x^2$ **j.** $y = \frac{1}{2}x^2 - 20x + 180$

    e. $y = x^2 - 15x + 60$ **k.** $y = 3.1x^2 + 247.2x + 5172.0$

    f. $y = x(10 - x)$

6. Without using a calculator, determine which of the following graphs *could be* the graph of $y = 1.2 * 10^{-2}x^2 + 2.3 * 10^{-4}x$. Explain.

a.

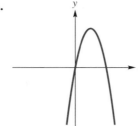

b.

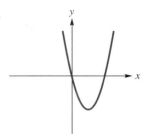

c.

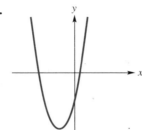

d.

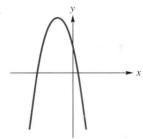

e.

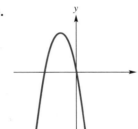

f.

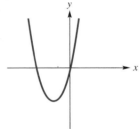

7. The Arrow.   An arrow is shot vertically into the air. The height (in feet) of the arrow after t seconds is given by $h = {}^{-}16t^2 + 112t$.

  a. Graph this quadratic function.

  b. Determine the maximum height the arrow attains.

  c. At what time is the arrow at its maximum height?

  d. Determine the time when the arrow hits the ground.

8. The Largest Rectangular Pen with a Border. Chris has bought 50 feet of fencing to build a rectangular pen. Chris is going to build the fence along the side of a hay barn. Therefore, the barn can serve as one side of the pen.

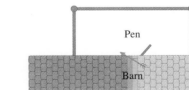

  a. Write an equation for the area of the pen as a function of the length of the side opposite the barn.

  b. Graph your function.

  c. Determine the dimensions of the pen that yield the largest area.

9. The Tip Top Tennis Balls manufacturing company makes and sells tennis balls. The profit (in dollars) the company receives in producing and selling x cases (12 cans per case, 3 balls per can) of tennis balls in one week is given by $P = -0.51x^2 + 102x - 1500$. All of the following questions refer to a one-week period; only full cases of balls are sold.

  a. Graph this function.

  b. If no tennis balls are made or sold, what is the profit?

  c. How many cases of tennis balls need to be made and sold to produce the maximum profit?

  d. What is the maximum profit?

  e. Use your graph to determine the number of cases of tennis balls that must be made and sold for the profit to be greater than zero.

  f. Use your graph to determine the number of cases of tennis balls that must be made and sold to keep the profit above $2000.

  g. Use your graph to determine the number of cases of tennis balls that must be made and sold to keep the profit above $10,000.

10. Minimum Sound Intensity.   A scientist is doing experiments with sound. A sound must have a minimum intensity or the subject cannot hear it. This intensity is referred to as the threshold of hearing. Suppose that the following equation gives a description of the threshold of hearing for a range of frequencies from 500 to 4000 Hz for a certain subject.

$$I = (3.713 * 10^{-6})f^2 - (2.1 * 10^{-2})f + 34.5714$$

where I = the intensity measured in decibels and

f = the frequency measured in hertz (Hz).

  a. Sketch a graph of the quadratic function.

  b. Determine the frequency for which the intensity is minimum.

11. The following table represents the input and output values for a specific quadratic function. Use the table to answer the following questions.

  a. Determine the vertical intercept of this quadratic function.

  b. Estimate the horizontal intercepts of this quadratic function.

  c. Does this quadratic function graph as an upward parabola or a downward parabola? Explain.

  d. An upward parabola has a minimum value, and a downward parabola has a maximum value. Estimate the maximum or minimum value of this quadratic function. Describe how you used the table to arrive at your result.

Input	Output
−6	6.9866
−5	4.3286
−4	2.0966
−3	0.2906
−2	−1.0894
−1	−2.0434
0	−2.5714
1	−2.6734
2	−2.3494
3	−1.5994
4	−0.4234
5	1.1786
6	3.2066
7	5.6606

12. Pressure Washer.   Your company just purchased a pressure washer for $3500. Maintenance, detergent, and fuel to operate the pressure washer costs about 57¢ an hour. The operator of the pressure washer is paid $14.50 per hour. You plan to charge customers $37 an hour for the pressure washing service.

   a. Write an equation for the cost to operate the pressure washer in terms of the number of hours it is in use. Include the original cost of the equipment in your equation.

   b. Write an equation for your company's income earned from the pressure washing service in terms of the number of hours it is in service.

   c. What is your cost to operate the pressure washer for 100 hours? What is your income for operating the pressure washer for 100 hours? After 100 hours, is your profit greater than zero?

   d. Graph the two equations on one set of coordinates axes. Be sure your graph clearly shows the intersection of the graphs of the two equations.

   e. At what point do you break even? That is, where is the cost equal to the income?

13. Write an expression for the shaded area in the following figure in terms of k. Assume that the shaded portion on the right-hand side is a semicircle.

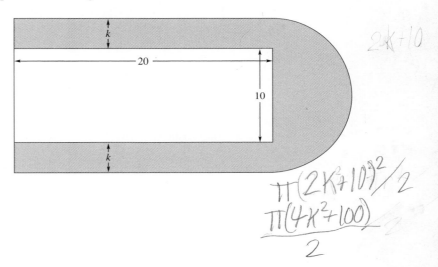

$$2k + 10$$

$$\frac{\pi(2k + 10)^2}{2}$$

$$\frac{\pi(4k^2 + 100)}{2}$$

7.2 Solving Quadratic Equations

Activity Set 7.2

1. For each function shown here do the following.

   a. Determine the horizontal intercepts *algebraically*.

   b. Using your calculator, graph the function and verify that you have the correct horizontal intercepts.

$$y = x^2 - 7x$$
$$y = x^2 - 4$$
$$y = 2x^2 + 10x$$
$$y = -3x^2 + 27$$

2. a. Describe in words the process used to solve $0 = ax^2 + bx$, for x.

   b. Describe in words the process used to solve $0 = ax^2 + c$, for x.

3. a. Describe the relationship between the function $y = x^2 - 7x$ and the equation $0 = x^2 - 7x$.

   b. Describe the relationship between the function $y = x^2 - 4$ and the equation $0 = x^2 - 4$.

   c. Describe the relationship between the function $y = ax^2 + bx + c$ and the equation $0 = ax^2 + bx + c$.

4. Use the zero-product principle to solve the quadratic equation $0 = (x - 2)(x + 10)$. Check your solution. (*Hint:* Do not multiply the factors.)

5. The equation $0 = ax^2 + bx + c$, with $a \neq 0$, is called the **standard form** of a quadratic equation.

> **The Quadratic Formula**
> The solution to the quadratic equation $0 = ax^2 + bx + c$, with $a \neq 0$, is
> $$x = \frac{-b + \sqrt{b^2 - 4ac}}{2a} \quad \text{or} \quad x = \frac{-b - \sqrt{b^2 - 4ac}}{2a}$$

   a. The quadratic equation $0 = (x - 2)(x + 10)$ is not written in standard form. Rewrite the equation in standard form, and identify the coefficients a, b, and c.

   b. Use the quadratic formula to determine the solution to the equation $0 = (x - 2)(x + 10)$. Does this solution agree with what you found in Activity 4?

6. a. Identify the coefficients a, b, and c in the equation $0 = 4x^2 - 20x - 96$.

   b. Use the quadratic formula to determine the solutions to the equation $0 = 4x^2 - 20x - 96$. Check your solutions.

   c. Based on your solution to part b, how do you expect the graph of $y = 4x^2 - 20x - 96$ to look? Graph this function to verify your conjecture.

7. Solve each equation using the quadratic formula. Check your solutions.

   a. $5x^2 - 31x - 190 = 0$

   b. $8x^2 + 14x = 15$

   c. $25x - 12 = 3.1x^2$

8. a. Predict how the graph of $y = x^2 + 4$ will look. Graph this function to verify your conjecture.

  b. Based on your graph, what is the solution to the equation $0 = x^2 + 4$?

9. We know three different algebraic methods to solve quadratic equations: factoring, using a square root, and the quadratic formula. For each quadratic equation, decide which of the three methods you think is "most efficient."

  a. $0 = 8x - x^2$ **e.** $25k^2 = 330k - 648$

  b. $12x = 0.5x^2$ **f.** $(t + 7)^2 - 36 = 0$

  c. $0 = 6x^2 + 23x - 18$ **g.** $x^2 + 5 = 0$

  d. $3x^2 = 75$ **h.** $0.7m^2 = 0.2m + 0.03$

10. Solve each quadratic equation in Activity 9 using the most efficient algebraic method.

11. Graph $y = 12x$ and $y = 0.5x^2$. How do the graphs of these two functions relate to your solution to Activity 10b?

Discussion 7.2

In Section 7.1, we saw that the graph of any quadratic function $y = ax^2 + bx + c$, with $a \neq 0$, is a parabola. We were able to algebraically find the vertex and vertical intercept of any parabola. Knowing the vertex and the vertical intercept of a parabola allowed us to readily determine an appropriate window in which to graph the function.

To find the horizontal intercepts of the function $y = ax^2 + bx + c$, we need to solve the corresponding equation $0 = ax^2 + bx + c$. Although these two equations are closely related and look very similar, we need to understand some important differences between them.

The quadratic equation $0 = ax^2 + bx + c$ has at most two solutions. These solutions are values of x that make the equation true, and these solutions correspond to the horizontal intercepts of the graph of the function $y = ax^2 + bx + c$. In contrast, the function $y = ax^2 + bx + c$ has an infinite number of solutions. The solutions to the function are ordered pairs of numbers that graph as a parabola.

Our focus in this section is on solving equations of the form $0 = ax^2 + bx + c$. When only two of the three terms are present in this quadratic equation, we can use algebraic techniques that we already learned to solve the equation. We will begin by reviewing these algebraic techniques. Then, we will present a method that will work when all three terms are present.

Solving Quadratic Equations of the Form $ax^2 + c = 0$

In Chapter 5, we solved equations in which the variable was raised to a power. The equation $ax^2 + c = 0$ is a special case of this. To solve this type of equation, we need to isolate the square and then undo the square by raising both sides to the $\frac{1}{2}$ power or equivalently taking the square root of both sides.

Example 1

Solve each quadratic equation algebraically. Check your solutions.

a. $4x^2 - 100 = 0$ b. $2x^2 + 42 = 0$

Solution   a. $4x^2 - 100 = 0$

$4x^2 = 100$ Isolate the squared term by first adding 100.

$x^2 = 25$ Then divide by 4.

$(x^2)^{1/2} = 25^{1/2}$ Raise both sides to the $\frac{1}{2}$ power.

$|x| = 5$ The $\frac{1}{2}$ power of the square is the absolute value.

$x = \pm 5$

To check, we must check both values.

CHECK

$x = 5$ $x = {}^-5$

$4 * 5^2 - 100 \overset{?}{=} 0$ $4 * ({}^-5)^2 - 100 \overset{?}{=} 0$

$0 = 0$ ✔ $0 = 0$ ✔

Therefore, $x = 5$ or $x = {}^-5$ is the solution to the equation $4x^2 - 100 = 0$.

b. $2x^2 + 42 = 0$

$2x^2 = {}^-42$ Isolate the squared term by first subtracting 42.

$x^2 = {}^-21$ Then divide by 2.

Because the square of a real number can never equal a negative number, we conclude that this equation has no real solutions. Because we have no solutions to check, we may want to look at this problem graphically to give us confidence in our conclusion. The solutions to the equation $2x^2 + 42 = 0$ are the horizontal intercepts of the graph of $y = 2x^2 + 42$. Let's list what we know about the graph of $y = 2x^2 + 42$.

- The graph of $y = 2x^2 + 42$ is a parabola.
- The parabola opens up because the coefficient of x^2 is positive.
- The line of symmetry is $x = 0$. When the term bx is missing, we know the graph is not shifted horizontally and, therefore, the line of symmetry remains the y-axis.
- The vertex and vertical intercept are both (0, 42).

With this information we can sketch a rough graph of the function.

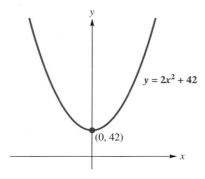

From the graph, we can clearly see that the function $y = 2x^2 + 42$ has no horizontal intercepts. This confirms our conclusion that the equation $2x^2 + 42 = 0$ has no real solutions.

Solving Quadratic Equations of the Form $ax^2 + bx = 0$

In equations of the form $ax^2 + bx = 0$, the left-hand side of the equation has a common factor of x. This allows us to factor the side containing the common factor and use the zero-product principle to solve the equation.

Example 2   Solve each quadratic equation algebraically. Check your solutions.

a. $0 = 4m^2 - 34m$ b. $58x + 2.5x^2 = 0$

Solution   a. $0 = 4m^2 - 34m$

$0 = m(4m - 34)$ Factor out m on the right-hand side.

$m = 0$ or $4m - 34 = 0$ Set each factor equal to zero.

$m = 0$ or $4m = 34$ Solve each of the resulting equations.

$m = 0$ or $m = 8.5$

CHECK

$m = 0$ $m = 8.5$

$0 \overset{?}{=} 4 * 0^2 - 34 * 0$ $0 \overset{?}{=} 4 * 8.5^2 - 34 * 8.5$

$0 = 0$ ✔ $0 = 0$ ✔

The solution to $0 = 4m^2 - 34m$ is $m = 0$ or $m = 8.5$.

b. $58x + 2.5x^2 = 0$

$x(58 + 2.5x) = 0$ Factor out x on the right-hand side.

$x = 0$ or $58 + 2.5x = 0$ Set each factor equal to zero.

$x = 0$ or $2.5x = {}^-58$ Solve each of the resulting equations.

$x = 0$ or $x = {}^-23.2$

CHECK

$x = 0$ $x = {}^-23.2$

$58 * 0 + 2.5 * 0^2 \overset{?}{=} 0$ $58 * {}^-23.2 + 2.5 * ({}^-23.2)^2 \overset{?}{=} 0$

$0 = 0$ ✔ $0 = 0$ ✔

The solution to $58x + 2.5x^2 = 0$ is $x = 0$ or $x = {}^-23.2$.

Example 3   Use the solution to the equation in Example 2b to sketch a graph of the quadratic function $y = 58x + 2.5x^2$.

Solution   Because the solution to the equation $58x + 2.5x^2$ is $x = 0$ or $x = {}^-23.2$, we know that the horizontal intercepts of the graph of $y = 58x + 2.5x^2$ are $(0, 0)$ and $({}^-23.2, 0)$. We also know that the function $y = 58x + 2.5x^2$ graphs as a parabola that opens up because the coefficient of x^2 is positive. With this we can sketch a rough graph.

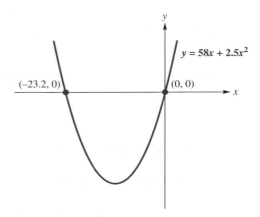

Notice that this sketch does not include a scale. To determine a reasonable scale we need to find the vertex. We can use the formula $x = \frac{-b}{2a}$ to find the line of symmetry or we know that the line of symmetry is halfway between the horizontal intercepts. Using either method, we find that the line of symmetry is $x = {}^-11.6$. Then, the y-coordinate can be found by substituting $x = {}^-11.6$ into the equation.

$$y = 58x + 2.5x^2$$
$$y = 58 * {}^-11.6 + 2.5 * ({}^-11.6)^2$$
$$y = {}^-336.4$$

The vertex is $({}^-11.6, {}^-336.4)$. Therefore, a reasonable window might be $[{}^-30, 10]$ by $[{}^-400, 200]$.

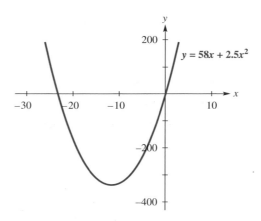

Solving Quadratic Equations of the Form $ax^2 + bx + c = 0$

Until now we were not able to solve quadratic equations that contained all three terms. However, there is a formula that solves *any* quadratic equation. It is called the **quadratic formula** and is stated in the following box. Those students wishing to see a derivation of the formula should talk to their instructor.

> *The Quadratic Formula*
>
> The equation $0 = ax^2 + bx + c$, $a \neq 0$, is called the standard form of a quadratic equation.
>
> The solution to the quadratic equation $0 = ax^2 + bx + c$, $a \neq 0$, is
>
> $$x = \frac{-b + \sqrt{b^2 - 4ac}}{2a} \qquad \text{or} \qquad x = \frac{-b - \sqrt{b^2 - 4ac}}{2a}$$

This solution is often written as

$$x = \frac{-b \pm \sqrt{b^2 - 4ac}}{2a}$$

The symbol $\pm$ is read as "plus or minus." The use of the symbol $\pm$ is a shorthand notation. We should always be aware that it represents two expressions.

Example 4

Solve each quadratic equation using the quadratic formula. Check your solutions.

a. $0 = 2x^2 - 47x + 140$ b. $15.6p + 36 = 1.2p^2$

Solution

a. To use the quadratic formula we must identify a, b, and c in the standard form of the quadratic equation $0 = 2x^2 - 47x + 140$. In this case,

$a = 2$

$b = {}^-47$

$c = 140$

The quadratic formula then says that the solution to the equation $0 = 2x^2 - 47x + 140$ is

> $$x = \frac{-b + \sqrt{b^2 - 4ac}}{2a}$$
> $$x = \frac{47 + \sqrt{(-47)^2 - 4 * 2 * 140}}{2 * 2}$$
> $$x = 20$$

or

> $$x = \frac{-b - \sqrt{b^2 - 4ac}}{2a}$$
> $$x = \frac{47 - \sqrt{(-47)^2 - 4 * 2 * 140}}{2 * 2}$$
> $$x = 3.5$$

CHECK

$x = 20$

$0 \overset{?}{=} 2 * 20^2 - 47 * 20 + 140$

$0 = 0$ ✔

$x = 3.5$

$0 \overset{?}{=} 2 * 3.5^2 - 47 * 3.5 + 140$

$0 = 0$ ✔

The solution to $0 = 2x^2 - 47x + 140$ is $x = 20$ or $x = 3.5$.

b. This second equation, $15.6p + 36 = 1.2p^2$, is not in standard form, so we must first put it in standard form. This can be done by moving all of the terms to the left-hand side or all to the right-hand side. These equations will look different but will have the same solutions. If we move everything to the left, the standard form is

$$^-1.2p^2 + 15.6p + 36 = 0$$

Then, $a = {}^-1.2$, $b = 15.6$, and $c = 36$. Using the quadratic formula, we obtain the following solution. Notice that in this case we are using the shorthand notation $\pm$, and we are solving for the variable p.

$$p = \frac{^-b \pm \sqrt{b^2 - 4ac}}{2a}$$

$$p = \frac{^-15.6 \pm \sqrt{15.6^2 - 4 * {}^-1.2 * 36}}{2 * {}^-1.2}$$

$$p = {}^-2 \qquad \text{or} \qquad p = 15$$

To check, we must substitute our solution in the original equation.

CHECK

$p = {}^-2$ $\qquad\qquad\qquad\qquad$ $p = 15$

$15.6 * {}^-2 + 36 \overset{?}{=} 1.2 * ({}^-2)^2$ $\qquad$ $15.6 * 15 + 36 \overset{?}{=} 1.2 * 15^2$

$\qquad\qquad 4.8 = 4.8$ $\quad$ ✔ $\qquad\qquad\qquad\qquad 270 = 270$ $\quad$ ✔

The solution to $15.6p + 36 = 1.2p^2$ is $p = {}^-2$ or $p = 15$.

In Example 4, if we had moved all of the terms to the right-hand side, we would have obtained $0 = 1.2p^2 - 15.6p - 36$. In this case, $a = 1.2$, $b = {}^-15.6$, and $c = {}^-36$. These values are just the opposite of the values we used, but if you substitute them into the quadratic formula you obtain the same solutions.

The most difficult part about using the quadratic formula is entering the expression correctly in your calculator. Here is a list of key items that may improve your chances of success.

- If b is a negative number, you need parentheses to substitute into the term b^2. Be sure that the power of 2 is outside these parentheses.
- When entering the expression on your calculator, you must group the following.

  the numerator

  the denominator

  the radicand

In Example 4, we used the quadratic formula to solve two different quadratic equations, both of which contained three terms. However, the quadratic formula solves *all* quadratic equations. In Example 2b, we solved the quadratic equation $0 = 4x^2 - 34x$ by factoring and using the zero-product principle and obtained the solution $x = 0$ or $x = 8.5$. Let's re-solve this same problem using the quadratic formula and compare the processes.

To use the quadratic formula, we need to identify a, b, and c. In this case, $a = 4$, $b = {}^-34$, and $c = 0$. Then, the quadratic formula gives the following solutions.

$$x = \frac{34 \pm \sqrt{(-34)^2 - 4 * 4 * 0}}{2 * 4}$$ The $^-34$ is put in parentheses with the exponent outside these parentheses.

$$x = 8.5 \qquad \text{or} \qquad x = 0$$

As we can see, the quadratic formula yields the same solution. We have several different algebraic techniques for solving quadratic equations, so we need to consider when to use each method. Because none of the methods yields a "better" solution, we need to think about their efficiency. In the case of the equation $0 = 4x^2 - 34x$, it is probably more efficient to solve the equation by factoring rather than using the quadratic formula because the quadratic formula can be difficult to enter on a calculator.

Choosing an Efficient Algebraic Method

To summarize, we discussed three different algebraic techniques for solving quadratic equations.

- **Factoring** and using the zero-product principle
- Isolating the square and taking the **square root** of both sides
- The **quadratic formula**

For the purposes of this discussion we refer to these as the factoring method, the square root method, and the quadratic formula. The quadratic formula solves *all* quadratic equations. However, it is not always the most efficient method. Both the factoring method and square root method are very efficient but work only on special forms of the quadratic equation. In the next example, we will look at several quadratic equations and choose the most efficient algebraic method.

Example 5

Choose the most efficient method to solve each quadratic equation.

a. $8x^2 = 42$

d. $(2x + 10)(x - 12) = 0$

b. $0 = 5t^2 + 42t - 81$

e. $(m - 7)^2 = 64$

c. $7x = 10x^2$

Solution　Again, the quadratic formula solves all of these equations, but our goal is to choose the most efficient method.

a. The first quadratic equation, $8x^2 = 42$, contains only two terms and x occurs only to the second power. Therefore, we could use the *square root method* to solve the equation.

b. The equation $0 = 5t^2 + 42t - 81$ contains all three terms and does not factor. Therefore, the *quadratic formula* is the only algebraic method we have to solve the equation.

c. The equation $7x = 10x^2$ contains two terms. Because each of the terms contains the variable *x,* we can solve this equation by *factoring* and using the zero-product principle.

d. The equation $(2x + 10)(x - 12) = 0$ is in a form different from many we have seen. We could multiply the two factors, resulting in a quadratic equation containing three terms. In this case we would then use the quadratic formula.

Alternatively, we could use the *zero-product principle,* because the left-hand side is already in *factored* form. Let's carry out this process to see how it looks.

$$(2x + 10)(x - 12) = 0$$

$2x + 10 = 0$	or	$x - 12 = 0$	Set each factor equal to zero.
$2x = {}^-10$	or	$x = 12$	Solve both of the equations.
$x = {}^-5$	or	$x = 12$	

If you check these values you will find that they both work. Therefore, the solution to $(2x + 10)(x - 12) = 0$ is $x = {}^-5$ or $x = 12$.

e. In Chapter 5 we solved equations of this form. Because the square is isolated, we can use the *square root method* to solve this equation.

$$(m - 7)^2 = 64$$
$$((m - 7)^2)^{1/2} = 64^{1/2} \qquad \text{Raise both sides to the } \tfrac{1}{2} \text{ power.}$$
$$|m - 7| = 8 \qquad \text{The } \tfrac{1}{2} \text{ power of the square is the absolute value.}$$

$$m - 7 = 8 \qquad \text{or} \qquad m - 7 = {}^-8$$
$$m = 15 \qquad \text{or} \qquad m = {}^-1$$

Again, you can check to see that both of the values are solutions. Therefore, the solution to $(m - 7)^2 = 64$ is $m = 15$ or $m = {}^-1$.

We can also solve this problem by rewriting the equation in standard form and using the quadratic formula. Which method is more efficient is somewhat a matter of opinion. People with strong algebraic skills may find the square root method more efficient, whereas people with strong calculator skills may prefer the quadratic formula. There is no single right answer to this question.

Example 6

Everett is planning a sidewalk around two sides of his house as shown in the following figure. Everett's friend Susan has just redone the siding on her house and in the process removed a brick facade. Everett is thinking of using the bricks from Susan's house to construct his sidewalk. There are 1500 bricks, and each brick together with the mortar take up a rectangular space that is 3 in. × 8 in. If Everett uses all of the bricks, how wide would the sidewalk be?

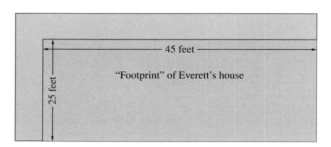

45 feet

25 feet

"Footprint" of Everett's house

Solution   This problem can be approached in many ways. We start by finding the total area that the 1500 bricks will cover. Because each brick (including mortar) takes a space 3 in. × 8 in., it covers 24 in.^{2}. Therefore, 1500 bricks cover an area

$$1500 \text{ bricks} * \frac{24 \text{ in.}^2}{\text{brick}}$$
$$= 36{,}000 \text{ in.}^2$$

Then, because the dimensions of the house are in feet, we convert this area to square feet.

$$36{,}000 \text{ in.}^2 * \frac{1 \text{ ft}}{12 \text{ in.}} * \frac{1 \text{ ft}}{12 \text{ in.}}$$
$$= 250 \text{ ft}^2$$

There are enough bricks to cover 250 square feet; therefore, the area of the sidewalk can be at most 250 ft^2.

Next, we need to deal with the sidewalk and its dimensions relative to the house. Because the width of the sidewalk is unknown, we let w represent its width. We can write an expression for the area of the sidewalk, which we know must be 250, and then solve the resulting equation.

We can write an expression for the area of the sidewalk in many ways, but all simplify to the same expression. Here is a list of three common expression for the area. Try to see where each expression comes from.

- $(45 + w)(25 + w) - 25 * 45$
- $(45 + w)w + 25w$
- $25w + 45w + w^2$

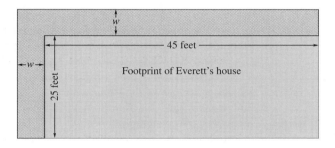

All of these expressions simplify to $w^2 + 70w$. Because the area must be 250, we obtain the equation

$$w^2 + 70w = 250$$

which we can solve using the quadratic formula. First, we rewrite the equation in standard form.

$$w^2 + 70w - 250 = 0$$

Then, $a = 1$, $b = 70$, and $c = {}^-250$, and the quadratic formula yields the following solutions.

$$w = \frac{-b \pm \sqrt{b^2 - 4ac}}{2a}$$

$$w = \frac{{}^-70 \pm \sqrt{70^2 - 4 * 1 * {}^-250}}{2 * 1}$$

$$w \approx 3.4 \qquad \text{or} \qquad w \approx {}^-73.4$$

In the context of this problem, the width cannot be negative. A width of 3.4 feet is reasonable. However, several other factors need to be considered before Everett can determine the exact width of the sidewalk. The bricks can be laid in many different patterns, and each pattern produces a different amount of waste. What we do know is that we have enough bricks to construct a sidewalk that has a reasonable width.

Solving Quadratic Equations Graphically

Although the emphasis in this section is on solving quadratic equations algebraically, we can also solve them graphically just as we did with linear equations.

Example 7

Solve the quadratic equation $0.5x^2 = x + 20$ graphically. Check your solution.

Solution   We can begin this problem in two different ways. We can rewrite the equation in standard form and graph the corresponding function. In this case we look for the horizontal intercepts. Alternatively, we can graph each side of the original equation and find the intersection points of the two graphs. This second method avoids the risk of making a simple algebra mistake. Therefore, we solve this problem by graphing each side of the original equation, $y = 0.5x^2$ and $y = x + 20$.

The equation $y = 0.5x^2$ graphs as a parabola that opens up. The vertex is at $(0, 0)$. The equation $y = x + 20$ graphs as a line with positive slope and a vertical intercept at $(0, 20)$. Using this information, we can find a window that clearly shows both graphs as well as the intersection points of the two graphs.

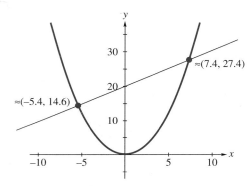

The intersection points are at $\approx(^-5.4, 14.6)$ and $\approx(7.4, 27.4)$. Because the equation we are trying to solve only contains the variable x, the solution to the equation $0.5x^2 = x + 20$ is $x \approx {}^-5.4$ or $x \approx 7.4$.

CHECK

$$x \approx {}^-5.4 \qquad\qquad\qquad x \approx 7.4$$
$$0.5 * ({}^-5.4)^2 \stackrel{?}{\approx} {}^-5.4 + 20 \qquad 0.5 * 7.4^2 \stackrel{?}{\approx} 7.4 + 20$$
$$14.58 \approx 14.6 \quad ✔ \qquad\qquad 27.38 \approx 27.4 \quad ✔$$

The emphasis in this section has been on solving quadratic equations algebraically. The algebraic techniques that we discussed are the **factoring method, the square root method,** and the **quadratic formula.** All of these techniques are good methods for solving quadratic equations. The factoring method and the square root method work only in special cases. The quadratic formula always works, but is not always the most efficient technique.

Typically, the factoring method is used when all of the terms in the equation contain a common variable factor. However, we encountered the equation $(2x + 10)(x - 12) = 0$ where the problem was already written in factored form. This allowed us to apply the zero-product principle to the factors to determine the solution.

The square root method is most often used when the variable *only* occurs inside of a square. For example, we used this technique on the equations $4x^2 - 100 = 0$ and $(m - 7)^2 = 64$.

We can also solve quadratic equations graphically by graphing both sides of the original equation and identifying the intersection points. When we solve a problem algebraically and find no solution, then solving the problem graphically is an ideal means to verify our lack of solutions.

Because we are continually looking at information algebraically and graphically, it is important to understand the connections between the two as well as the differences. In particular, in Sections 7.1 and 7.2 we looked at the two equations $0 = ax^2 + bx + c$ and $y = ax^2 + bx + c$. On the surface, the difference is that one equation is set equal to zero and the other is set equal to y. More important is the difference between the solutions and the connection between the equations. The quadratic equation $0 = ax^2 + bx + c$ has at most two solutions and these solutions correspond the horizontal intercepts of the graph of the function $y = ax^2 + bx + c$. The function $y = ax^2 + bx + c$ has an infinite number of ordered pairs that are solutions, and the graph of these solutions is a parabola.

Based on our examples, we saw quadratic equations with two solutions and one with none. Do you think a quadratic equation can have only one solution?

Problem
Set
7.2

1. Solve each of the following equations algebraically. Try to select the most efficient algebraic method. Round approximate solutions to the thousandths place. Check your solutions.

   a. $-5x^2 + 80 = 0$ **c.** $0.5x^2 - 3x = 0$ **e.** $4x^2 = 60$

   b. $2x^2 + 90x + 1000 = 0$ **d.** $8 + 2x - 3x^2 = 0$ **f.** $(x + 2)^2 - x^2 = 0$

2. Solve each of the following equations algebraically. Try to select the most efficient algebraic method. Round approximate solutions to the thousandths place. Check your solutions.

   a. $x^2 - 5x - 150 = 0$ **e.** $-x^2 = 10$

   b. $x^2 - 1 = 2x$ **f.** $0 = 35x(1 - 0.001x) + 1500$

   c. $16p^2 = 40p$ **g.** $35.4 + 0.025k^2 = 12.7k$

   d. $2 - 3x^2 = 3x(4 - x)$ **h.** $5(x - 1)^2 = 20$

3. **a.** Draw a complete graph of $y = x^2 - 8x + 16$.

   b. Based on your graph from part a, what do you expect the solution of $0 = x^2 - 8x + 16$ to be?

   c. Solve the equation $0 = x^2 - 8x + 16$ using the quadratic formula. Did you get the solution you expected?

4. **a.** Solve the equation $0 = 2x^2 - 3x + 6$ algebraically.

   b. Based on your results from part a, roughly predict how the graph of $y = 2x^2 - 3x + 6$ will look. Graph $y = 2x^2 - 3x + 6$ to verify your prediction.

5. Solve each of the following equations algebraically. Try to select the most efficient algebraic method. Round approximate solutions to the thousandths place. Check your solutions.

   a. $42n - 12n^2 = 0$ **f.** $9B^2 - 12B + 4 = 0$

   b. $-10t^2 + 105t + 30 = 75$ **g.** $(2w - 5)^2 = 15$

   c. $\dfrac{9m^2}{4} - 4 = \dfrac{9m}{2}$ **h.** $(2w - 5)^2 = 15w$

   d. $5x - 15 + 30 = 0$ **i.** $\dfrac{2x^2}{3} = \dfrac{x + 1}{5}$

   e. $x^2 = x - 2$ **j.** $0 = -1.2 * 10^{-4}x^2 + 2.4 * 10^{-2}x - 0.05$

6. Solve each equation graphically and algebraically.

   a. $R^2 - 21R = 46$

   b. $k^2 + 10 = 30 - 2k$

   c. $12 - 8x - 2x^2 = 25$

   d. $(x + 10)^2 = \frac{9}{4}$

7. Identify each equation as linear, quadratic, or other.

   a. $5x^3 = 4.8x$

   b. $4R^2 - 15R = 6$

   c. $6 - 7x = 55$

   d. $\dfrac{5}{m + 3} = 2$

   e. $\dfrac{5}{x + 3} = 2x$

   f. $2p(p - 5) = 7 + p$

   g. $12(x + 5) - 8x = 23$

   h. $\sqrt{x + 4} = 9$

8. Identify each equation as linear, quadratic, or other.

   a. $2x^2 - 15 = 0$

   b. $24 - 3k = 10$

   c. $\dfrac{4}{x} + 15 = 7$

   d. $20x - 4x^2 = 0$

   e. $20x + 4x^2 = 35$

   f. $2.5x^3 = 25.6$

   g. $3m(4m - 1) = 0$

   h. $2(4 - x) + 7x = 0$

   i. $\sqrt[3]{x - 8} = 5$

   j. $\dfrac{6}{R + 8} = 14$

9. Solve each equation from the Problem 8. Round approximate solutions to the thousandths place. Check your solutions.

10. Given the diagram shown here, set up an equation to find the value of *x*. Solve the equation and find both the perimeter and the area of the rectangle.

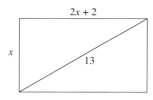

11. How Many Board Feet?   The *National Forest Log Scaling Handbook* used by several northwest bureaus gives the following rule for estimating log volume. To determine the number of board feet in a 16-foot log, use the equation

   $$B = 0.8d^2 - 2.4d$$

   where d = diameter of the log in inches

   B = number of board feet

   In a 16-foot log, what is the diameter necessary to get 105 board feet?

12. A Walk Around the Swimming Pool.   Cody wants to put a sidewalk around his 8.0 foot by 12.0 foot rectangular swimming pool. A friend has concrete materials left from another job that will cover an area of 210 square feet to the depth that Cody's sidewalk would require. If the sidewalk is of uniform width, what is the maximum width it can have?

13. A Walk Around the Fish Pond.   Cody also has a circular fish pond with a diameter of 6.4 feet around which he wants to put a sidewalk. Cody did not use all of the concrete for the walkway around the swimming pool. He has enough concrete left over to cover an area of 65 square feet. Assuming the sidewalk is of uniform width, how wide can the sidewalk around the pond be?

14. Storage Containers.   Maria has a cylindrical storage tank that is 1.5 meters in diameter and 2.1 meters tall. If she is considering replacing it with a spherical storage container, what does the diameter of the new container have to be to hold as much as her current storage tank?

15. The Arrow Revisited.   A 6-foot-tall archer shot an arrow vertically into the air. The height of the arrow after t seconds is given by $h = -16t^2 + 112t + 6$. How long after the arrow is shot does the archer's friend, who is lying on the ground next to the archer, have to get out of the arrow's path as it comes back to earth?

16. Barn Pen.   Chris is building a rectangular pen along the side of a hay barn with the barn serving as one side of the pen. Chris has 50 feet of fencing and wants the pen to be 255 square feet. What possible dimensions can Chris make the pen?

17. *Use the given graph* to approximate the solution to each equation or inequality.
  a. $0 = -0.25x^2 + 7.5x + 126$
  b. $0 < -0.25x^2 + 7.5x + 126$

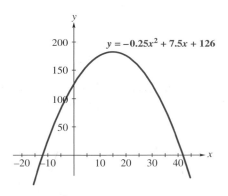

18. *Use the given graph* to solve each equation or inequality.
  a. $0.2x^2 - 4x + 20 = 0$
  b. $0.2x^2 - 4x + 20 \geq 0$
  c. $0.2x^2 - 4x + 20 \leq 0$

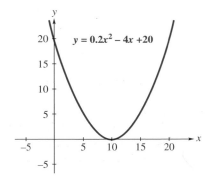

19. *Use the given graph* to approximate the solution to the following equation

$$x(4 - x) = \frac{x - 19}{4}$$

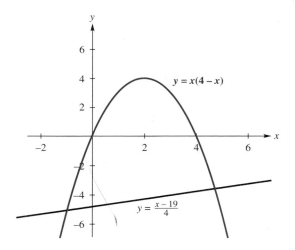

20. The Pipe in the Corner.   A cylindrical tank is to be placed in a corner of a room as shown (top view). The tank has a diameter of 34 inches. You also want to run a pipe up the wall in the corner as shown.

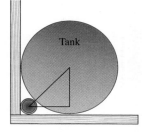

   a. If the pipe has radius r, write expressions, in terms of r, for each side of the triangle shown in the diagram.

   b. Determine the radius of the largest pipe that can be placed in the corner.

21. Designer Windsocks Revisited.   Windless Surfing Company's marketing manager observes that if the price of designer windsocks is $45, then on average 75 windsocks are sold per month. Experience has shown that if the price is raised from $45, then for every $2 increase in price, three fewer windsocks are sold each month.

   a. In Section 4.1 you wrote an equation for the number of windsocks sold per month in terms of the price. Now, write an equation for the monthly revenue in terms of the price of the windsocks.

   b. Determine the price that generates the largest monthly revenue. Show or explain clearly the method used to determine this value.

   c. At the price you determined in part b, how many windsocks are sold?

   d. What prices result in at least $2500 in monthly revenue?

7.3 Factors, Solutions, and the Discriminant

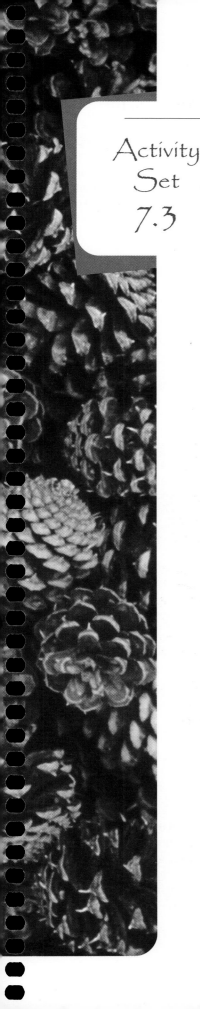

Activity
Set
7.3

1. a. Use the zero-product principle to solve the quadratic equation $(x - 4)(x + 9) = 0$.

   b. Based on your results from part a, how do you expect the graph of $y = (x - 4)(x + 9)$ to look? Graph this function to verify your conjecture.

2. a. Use the zero-product principle to solve the quadratic equation $(x - 2)(x - 15) = 0$.

   b. Based on your results from part a, how do you expect the graph of $y = (x - 2)(x - 15)$ to look? Graph this function to verify your conjecture.

3. a. Use the zero-product principle to solve the quadratic equation $^{-}3(x - 2)(x - 15) = 0$.

   b. Based on your results from part a, how do you expect the graph of $y = {}^{-}3(x - 2)(x - 15)$ to look? Graph this function to verify your conjecture.

   c. How does the graph of $y = (x - 2)(x - 15)$ compare with the graph of $y = {}^{-}3(x - 2)(x - 15)$?

4. a. Use the zero-product principle to solve the quadratic equation $(x + 5)(x + 5) = 0$.

   b. Based on your results from part a, how do you expect the graph of $y = (x + 5)(x + 5)$ to look? Graph this function to verify your conjecture.

5. a. Write a quadratic equation whose solution is $x = 2$ or $x = {}^{-}6$. How many variables are in your equation?

   b. Write the equation of a parabola whose horizontal intercepts are $(2, 0)$ and $(^{-}6, 0)$. How many variables are in your equation?

6. Write equations of three different parabolas, all of which have horizontal intercepts at $(3, 0)$ and $(18, 0)$. Graph each of your equations to verify your results.

7. For each graph, write the equation for the parabola and verify your equation.

a.

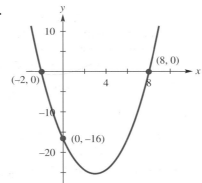

b.

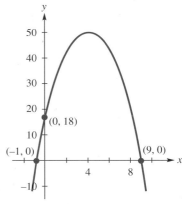

c.

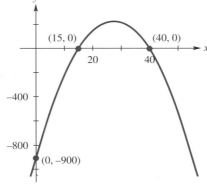

Discussion 7.3

The Discriminant

In the previous two sections, we looked at quadratic functions, $y = ax^2 + bx + c$, and quadratic equations, $0 = ax^2 + bx + c$. We know that the solutions to the equation $0 = ax^2 + bx + c$ correspond to the horizontal intercepts of the graph of $y = ax^2 + bx + c$. Based on our experience graphing quadratic functions, we also know that a parabola might cross the horizontal axis in two places, like the graph of $y = x^2 - 4$. It might touch the horizontal axis just once, like the graph of $y = {}^-x^2 + 10x - 25$, or it might not touch the horizontal axis at all, like the graph of $y = x^2 + 1$.

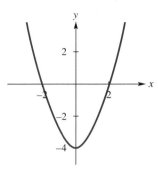

The graph of $y = x^2 - 4$ has two horizontal intercepts: $(-2, 0)$ and $(2, 0)$.

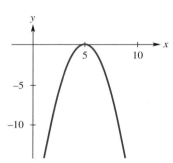

The graph of $y = -x^2 + 10x - 25$ has one horizontal intercept: $(5, 0)$.

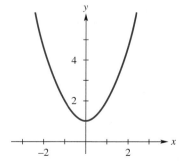

The graph of $y = x^2 + 1$ has no horizontal intercepts.

Therefore, the corresponding equations $0 = x^2 - 4$, $0 = {}^-x^2 + 10x - 25$, and $0 = x^2 + 1$ have two, one, and no solutions, respectively. Our next task is to find a way to predict the number of solutions to a quadratic equation without actually finding the solutions algebraically or graphically. It turns out that we can do this if we look at the quadratic formula closely. It appears from the formula that we always obtain two solutions: one as a result of adding the terms in the numerator of the quadratic formula and one as a result of subtracting the terms. However, as we will see, this is not always the case.

Example 1

For each quadratic equation, *use the quadratic formula* to determine the solution. Work out the formula in steps for the purposes of this example.

$$\text{a. } 0 = {}^-2x^2 + 8x + 10 \qquad \text{b. } 0 = x^2 + 6x + 9 \qquad \text{c. } 0 = x^2 + 6x + 12$$

Solution

a. For this equation, $a = {}^-2$, $b = 8$, and $c = 10$. Using the quadratic formula, we obtain the following solution.

$$x = \frac{-b \pm \sqrt{b^2 - 4ac}}{2a}$$

$$x = \frac{-8 \pm \sqrt{8^2 - 4 * {}^-2 * 10}}{2 * {}^-2}$$

$$x = \frac{-8 \pm \sqrt{64 + 80}}{-4}$$

$$x = \frac{-8 \pm \sqrt{144}}{-4}$$

$$x = \frac{-8 \pm 12}{-4}$$

$$x = {}^-1 \qquad \text{or} \qquad x = 5$$

CHECK

$x = {}^-1$
$0 \stackrel{?}{=} {}^-2 * ({}^-1)^2 + 8 * {}^-1 + 10$
$0 = 0$ ✔

$x = 5$
$0 \stackrel{?}{=} {}^-2 * 5^2 + 8 * 5 + 10$
$0 = 0$ ✔

The solution to $0 = {}^-2x^2 + 8x + 10$ is $x = {}^-1$ or $x = 5$. In this case we obtained two real solutions.

b. In the equation $0 = x^2 + 6x + 9$, $a = 1$, $b = 6$, and $c = 9$. The quadratic formula produces the following.

$$x = \frac{-b \pm \sqrt{b^2 - 4ac}}{2a}$$

$$x = \frac{-6 \pm \sqrt{6^2 - 4 * 1 * 9}}{2 * 1}$$

$$x = \frac{-6 \pm \sqrt{36 - 36}}{2}$$

$$x = \frac{-6 \pm \sqrt{0}}{2}$$

$$x = \frac{-6}{2}$$

$$x = {}^-3$$

CHECK

$x = {}^-3$
$0 \stackrel{?}{=} ({}^-3)^2 + 6 * {}^-3 + 9$
$0 = 0$ ✔

The solution to $0 = x^2 + 6x + 9$ is $x = {}^-3$. Notice, in this case we obtained only one real solution. This is because the expression under the radical, called the radicand, is equal to zero. Whether we add zero or subtract zero we obtain the same amount. Therefore, there is only one real solution.

c. For the equation $0 = x^2 + 6x + 12$, $a = 1$, $b = 6$, and $c = 12$. The quadratic formula produces the following.

$$x = \frac{-b \pm \sqrt{b^2 - 4ac}}{2a}$$

$$x = \frac{-6 \pm \sqrt{6^2 - 4 * 1 * 12}}{2 * 1}$$

$$x = \frac{-6 \pm \sqrt{36 - 48}}{2}$$

$$x = \frac{-6 \pm \sqrt{-12}}{2}$$

There are no real solutions because the radicand is negative.

From the results of Example 1, we can see that the sign of the radicand in the quadratic formula determines the number of solutions to the quadratic equation. We summarize these observations in a list.

- If the expression $b^2 - 4ac$ is a positive number, then $\sqrt{b^2 - 4ac}$ is also a positive number. In this case, there are two real solutions to the equation $ax^2 + bx + c = 0$.
- If the expression $b^2 - 4ac$ is zero, then $\sqrt{b^2 - 4ac}$ is also zero. In this case, there is one real solution to the equation $ax^2 + bx + c = 0$.
- If the expression $b^2 - 4ac$ is a negative number, then $\sqrt{b^2 - 4ac}$ is not a real number. In this case, there are no real solutions to the equation $ax^2 + bx + c = 0$.

Because the sign of the expression $b^2 - 4ac$ determines the number of solutions to the quadratic equation $0 = ax^2 + bx + c$, $b^2 - 4ac$ is called the **discriminant.** Using the discriminant to predict the number of solutions is called the discriminant test.

Discriminant Test
- If $b^2 - 4ac > 0$, then $ax^2 + bx + c = 0$ has two real solutions.
- If $b^2 - 4ac = 0$, then $ax^2 + bx + c = 0$ has one real solution.
- If $b^2 - 4ac < 0$, then $ax^2 + bx + c = 0$ has no real solutions.

Factors of Quadratic Equations and Functions

Quadratic equations are sometimes written in factored form. In the activities, we saw a relationship between the factors of a quadratic equation and the solutions to that equation. For example, in Activity 1a, we solved the equation $(x - 4)(x + 9) = 0$. Using the zero-product principle we obtained the following.

$$x - 4 = 0 \quad \text{or} \quad x + 9 = 0$$
$$x = 4 \quad \text{or} \quad x = {}^-9$$

So, the quadratic equation with factors of $x - 4$ and $x + 9$ has solution $x = 4$ or $x = {}^-9$.

Until now, we usually started with a quadratic equation or function and algebraically or graphically found the solutions or horizontal intercepts. With this new connection between the factors of a quadratic equation and the solutions we can now go in the reverse order. The next two examples illustrate some uses of this connection.

Example 2

Write a quadratic equation whose solution is $x = 3$ or $x = {}^-8$. Is there only one equation with this solution? Explain.

Solution To help us understand the process, we can work our way backwards. The solution $x = 3$ may have come from solving the equation $x - 3 = 0$. Similarly, $x = {}^-8$ was a result of solving the equation $x + 8 = 0$. Therefore, $x - 3$ and $x + 8$ are factors of the quadratic equation. So, a quadratic equation that has this solution is

$$0 = (x - 3)(x + 8)$$

Notice that this equation is set equal to zero and not y. This is because we are writing a quadratic equation that only has two solutions rather than a quadratic function with an infinite number of solutions.

Let's check to see that this equation really has the indicated solutions.

CHECK

$x = 3$	$x = {}^-8$
$0 = (3 - 3)(3 - 8)$	$0 = ({}^-8 - 3)({}^-8 + 8)$
$0 = 0$ ✔	$0 = 0$ ✔

However, many equations have these solutions. We can include any numerical factor as long as we maintain the factors of $x - 3$ and $x + 8$. For example,

$$0 = {}^-4(x - 3)(x + 8).$$

is a quadratic equation with solution $x = 3$ or $x = {}^-8$. We can easily check this. If we substitute $x = 3$, the second factor is zero, producing a zero on the right-hand side. If $x = {}^-8$, the third factor is zero, producing a zero on the right-hand side.

To conclude, an infinite number of quadratic equations have the solution $x = 3$ or $x = {}^-8$. We have listed two possible equations in this example.

Example 3

Write the equation for the following parabola and verify your equation.

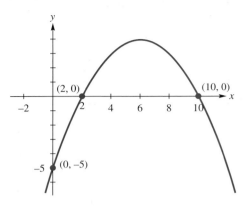

Solution   We can start by identifying information from the figure. The parabola has horizontal intercepts $(2, 0)$ and $(10, 0)$. The vertical intercept is $(0, {}^-5)$. Because the parabola crosses the *x*-axis at 2 and 10, the quadratic function has factors $x - 2$ and $x - 10$. With this we might try the function $y = (x - 2)(x - 10)$; however, this quadratic function graphs as a parabola that opens up, whereas the one we are given opens down. Therefore, we must somehow modify this equation.

Just as in Example 2, we can include any numerical factor in addition to the algebraic factors. Because we are not sure what this numerical factor needs to be, let's call it *k*. The quadratic function

$$y = k(x - 2)(x - 10)$$

has the correct horizontal intercepts. Next, we need to determine the value of *k* so that the function graphs as the desired parabola. In addition to the horizontal intercepts, we also know that the parabola needs to pass through the point $(0, {}^-5)$. Therefore, if we substitute $x = 0$ and $y = {}^-5$ into the equation, we can determine the value of *k*.

$$^-5 = k(0 - 2)(0 - 10)$$
$$^-5 = 20k$$
$$^-0.25 = k$$

With this, we can write the equation of the parabola as $y = {}^-0.25(x - 2)(x - 10)$. We can check this function numerically by substituting the three known points into the equation, or we can do it graphically by graphing the function on our calculator using a window similar to the one shown.

Notice that in the previous example the equation is set equal to *y* instead of zero. Can you explain why?

In this section we looked at two different ideas related to quadratic equations and functions. The first was the **discriminant.** The discriminant is the radicand, $b^2 - 4ac$, in the quadratic formula. It can be used to determine the number of real solutions to a quadratic equation $ax^2 + bx + c = 0$ or the number of horizontal intercepts to the quadratic function $y = ax^2 + bx + c$.

The second idea involved the relationship between the factors of a quadratic equation and the solutions to that equation. Then, because the solutions to a quadratic equation $ax^2 + bx + c = 0$ correspond the horizontal intercepts of the quadratic function $y = ax^2 + bx + c$, we can use this connection to write the equation of a quadratic function when we have three distinct intercepts.

Problem
Set
7.3

1. Solve each equation algebraically and check your solutions. Express approximate solutions to three significant digits.

    a. $\sqrt{m + 21} = m + 1$

    b. $\dfrac{2t}{t - 1} - 5 = 7$

    c. $\dfrac{2x^2}{x - 3} = 1 + \dfrac{18}{x - 3}$

    d. $\dfrac{3P}{2P - 1} = \dfrac{5P + 3}{P}$

2. Solve each equation algebraically and check your solutions. Round approximate solutions to the thousandths place.

    a. $\dfrac{3x^2}{x + 5} - 1 = \dfrac{7}{x + 5}$

    b. $\sqrt{4x^2 - 1} = x + 3$

    c. $\dfrac{5T}{8T - 5} = \dfrac{8T + 5}{3T}$

    d. $(2x + 11)^3 = 64$

3. Use the discriminant to predict the number of real solutions to each of the following quadratic equations. Do *not* solve the equations.

    a. $t^2 + 2t + 2 = 0$

    b. $0 = \frac{1}{4}x^2 + 5x + 25$

    c. $4x(x - 3) = 7$

    d. $7.8k^2 = 5.6 - 2.5k$

4. Use the discriminant to predict the number of horizontal intercepts for the graph of each of the following quadratic functions. Verify your prediction graphically.

    a. $y = -3x^2 + 5x + 14$

    b. $y = 0.2x^2 - 4x + 20$

    c. $y = x^2 - 14.8x + 153.7$

    d. $y = -1.1x^2 - 2.2x + 5$

    e. $y = -x^2 + 2.2x - 5$

    f. $y = 0.025x^2 + 35.4$

5. Consider the function $y = ax^2 + bx + c$. Draw one possible graph of this function if we know that $a < 0$, $c > 0$, and $b^2 - 4ac > 0$.

6. **a.** Write a quadratic equation that has solution $x = -3$ or $x = 7$.

    b. Write a quadratic equation that has solution $x = 5$ or $x = 20$.

    c. Write a quadratic equation that has solution $x = \frac{1}{2}$ or $x = -6$.

    d. Write a quadratic equation that has the solution $x = 8$ only.

7. **a.** Write two different quadratic equations that have the solution $x = -7$ or $x = 24$.

    b. Write two different quadratic equations that have the solution $x = -42$ or $x = -35$.

    c. Write two different quadratic equations that have the solution $x = -5$ only.

8. For each graph, write the equation and verify your equation.

    a.

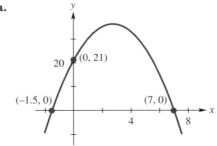

    b.

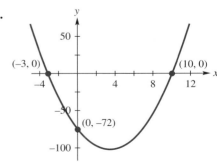

9. For each graph, write the equation and verify your equation.

a.

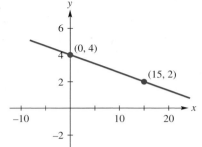

b.

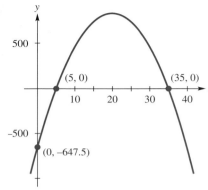

c.

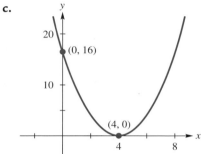

d.

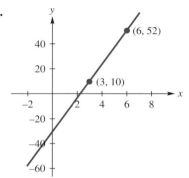

10. Fountain.   An empirically established formula states that the vertical reach of a water stream coming from a 1-inch-diameter nozzle is given by the formula

$$R = \frac{(195 - P)P}{100}$$

where R is the vertical reach of the water stream in feet, and P is the nozzle pressure in pounds per square inch (psi).

a. Find the vertical reach of the water stream when the nozzle pressure is 85 psi.

b. What nozzle pressure produces the maximum vertical reach for the water stream?

c. The atrium of the new city library building includes plans for a fountain that shoots streams of water vertically from 1-inch-diameter pipes. The ceiling of the atrium is 35 feet high. What do you recommend for the maximum nozzle pressure for this fountain? The fountain calls for water streams at three different heights. What heights do you recommend, and what are the nozzle pressures needed for each?

11. Population Growth Rate.   The following table shows the average annual growth rate for each decade in the United States.

    a. Plot the data given in the chart.

    b. Which two decades do not seem to fit the pattern? Can you think of reasons why?

    c. Eliminate the two points that do not fit and then draw an eyeball fit line through the remaining data.

    d. Write an equation for the line you drew.

    e. Assuming the trend continues, when will the annual growth rate for the United States reach zero?

Decade Beginning in Year	Average Annual Growth Rate (%)
1900	2.10
1910	1.50
1920	1.62
1930	0.72
1940	1.45
1950	1.85
1960	1.34
1970	1.14
1980	1.02
1990	0.7 (estimated)

Activity
Set
7.4

7.4 Literal Quadratics

1. Consider the formula $TW^2 - 3WT + T + 5 = 0$.

   a. Substitute $T = 5$ into the formula. What is the coefficient of W^2? What is the coefficient of W? What is the constant term? Solve the formula for W when $T = 5$.

   b. Substitute $T = {}^-3$ into the formula. What is the coefficient of W^2? What is the coefficient of W? What is the constant term? Solve the formula for W when $T = {}^-3$.

   c. Using your results from parts a and b complete the first two rows of the following table.

   d. Looking at the literal formula, what is the coefficient of W^2 in terms of T? What is the coefficient of W in terms of T? What is the constant term in terms of T? Solve the formula for W in terms of T.

T	Coefficient of W^2	Coefficient of W	Constant	Solution
5				
$^-3$				
T				

Discussion 7.4

In Section 7.2, we used three different algebraic techniques to solve quadratic equations: the factoring method, the square root method, and the quadratic formula. In this section, we will use these same ideas to solve literal quadratics. However, because we will be looking at equations that contain more than one variable, we must be able to determine whether an equation is quadratic in a specific variable. To help us in this transition, we will begin by looking at equations that contain only one variable.

Example 1

Identify each equation as linear or quadratic.

   a. $4(x - 1) = 15$

   b. $2.5(x^2 - 1) = 0$

   c. $14x + 3 = 5x^2$

   d. $2x - 10x - 7 = 0$

   e. $7(x - 1)(2x + 5) = 9$

Solution   To begin, let's review the different forms of linear and quadratic equations. A linear equation in the variable x can be written in the form $mx + b = 0$, where $m \neq 0$. The standard form of a quadratic equation in the variable x is $ax^2 + bx + c = 0$, where $a \neq 0$. Notice that in the linear equation, the power of x is one, whereas in the quadratic equation, x *must* occur to the second power and may also occur to the first power.

a. Because x only occurs to the first power, the equation $4(x - 1) = 15$ is *linear*.

b. In the equation $2.5(x^2 - 1) = 0$, x occurs to the second power; therefore, this is a *quadratic* equation.

c. Because x occurs to the first power and the second power, $14x + 3 = 5x^2$ is a *quadratic* equation.

d. The equation $2x - 10x - 7 = 0$ can be rewritten as $^-8x - 7 = 0$, which is a *linear* equation.

e. If the left-hand side of the equation $7(x - 1)(2x + 5) = 9$ is simplified, then x occurs to the second power. Therefore, this equation is *quadratic*.

Example 2

Identify each equation as linear or quadratic in the indicated variable.

a. $10(x + y^2) = 13$ for x

b. $10(x + y^2) = 13$ for y

c. $4pm^2 - 7m = 0$ for m

d. $4pm^2 - 7m = 0$ for p

e. $(R + 4)(R + N) = 20$ for R

Solution

a. In the equation $10(x + y^2) = 13$, x only occurs to the first power. Therefore, this equation is *linear in the variable x*.

b. Because y occurs to the second power, $10(x + y^2) = 13$ is *quadratic in y*.

c. In the equation $4pm^2 - 7m = 0$, m occurs to the first power in one of the terms and to the second power in the other, so this equation is *quadratic in m*.

d. The variable p only occurs to the first power. Therefore, $4pm^2 - 7m = 0$ is *linear in the variable p*.

e. If we simplify the left-hand side by performing the multiplication, then we obtain a term containing R^2 and terms containing R. Therefore, $(R + 4)(R + N) = 20$ is *quadratic in R*.

Remember that we can use three different algebraic techniques to solve quadratic equations: the factoring method, the square root method, and the quadratic formula. To solve linear equations, we typically isolate the variable by performing inverse operations. Occasionally, we need to factor a common variable when the variable we are solving for occurs in multiple terms that cannot be combined. Therefore, it is important to identify the type of equation we are solving because the techniques for solving them differ.

Example 3

Solve each equation for the indicated variable.

a. $10(x + y^2) = 13$ for x

b. $10(x + y^2) = 13$ for y

c. $4pm^2 - 7m = 0$ for m

d. $4pm^2 - 7m = 0$ for p

e. $(R + 4)(R + N) = 20$ for R

Solution   a. Because the equation $10(x + y^2) = 13$ is linear in the variable x, we need to isolate x.

$$10(x + y^2) = 13$$
$$10x + 10y^2 = 13 \qquad \text{Simplify the left-hand side.}$$
$$10x = 13 - 10y^2 \qquad \text{Subtract } 10y^2 \text{ from both sides.}$$
$$x = \frac{13 - 10y^2}{10} \qquad \text{Divide both sides by 10.}$$
$$x = 1.3 - y^2 \qquad \text{Divide each term in the numerator by 10.}$$

b. The equation $10(x + y^2) = 13$ is quadratic in the variable y. Because y only occurs to the second power, we can use the square root method to solve the equation.

$$10(x + y^2) = 13$$
$$10x + 10y^2 = 13 \qquad \text{Simplify the left-hand side.}$$
$$10y^2 = 13 - 10x \qquad \text{Subtract } 10x \text{ to isolate the term containing } y^2.$$
$$y^2 = \frac{13 - 10x}{10} \qquad \text{Divide both sides by 10.}$$
$$y^2 = 1.3 - x \qquad \text{Divide each term in the numerator by 10.}$$
$$(y^2)^{1/2} = (1.3 - x)^{1/2} \qquad \text{Raise both sides to the } \tfrac{1}{2} \text{ power.}$$
$$|y| = \sqrt{1.3 - x} \qquad \text{An even power followed by an even root produces an absolute value. The } \tfrac{1}{2} \text{ power on the right can be written in radical form (if desired).}$$
$$y = \pm\sqrt{1.3 - x}$$

c. The equation $4pm^2 - 7m = 0$ is quadratic in the variable m. The variable m occurs to the first power in one term and the second power in the other. Because these are the only two terms in the equation, we can factor this equation and use the zero-product principle.

$$4pm^2 - 7m = 0$$
$$m(4pm - 7) = 0 \qquad \text{Factor } m \text{ out of the right-hand side.}$$
$$m = 0 \quad \text{or} \quad 4pm - 7 = 0 \qquad \text{Set each factor equal to zero.}$$
$$m = 0 \quad \text{or} \quad m = \frac{7}{4p} \qquad \text{Solve each equation for } m.$$

d. In terms of the variable p, this equation is linear. Therefore, to solve $4pm^2 - 7m = 0$ for p, we perform inverse operations to isolate p.

$$4pm^2 - 7m = 0$$
$$4pm^2 = 7m \qquad \text{Add } 7m \text{ to both sides.}$$
$$p = \frac{7m}{4m^2} \qquad \text{Divide both sides by } 4m^2.$$
$$p = \frac{7}{4m} \qquad \text{Cancel one factor of } m.$$

e. The equation $(R + 4)(R + N) = 20$ is quadratic in the variable R. The left-hand side is written in factored form. However, the right-hand side is not zero; therefore, we are *not* able to use the zero-product principle. Let's begin by simplifying the left-hand side.

$$(R + 4)(R + N) = 20$$
$$R^2 + RN + 4R + 4N = 20$$

Because this equation has terms that contain R, R^2, and some that do not contain the factor R, we need to use the quadratic formula. To begin let's rewrite this equation in standard form.

$$R^2 + RN + 4R + 4N - 20 = 0$$

This does not look quite like the standard form we are used to. There are two different terms containing the factor R. To make this look more like standard form and to help us identify the coefficients necessary to use the quadratic formula, we factor the two terms containing the variable R.

$$R^2 + (N + 4)R + 4N - 20 = 0$$

Next, we need to identify

a: the coefficient of R^2

b: the coefficient of R

c: the constant term

From the equation $R^2 + (N + 4)R + 4N - 20 = 0$, we can see that $a = 1$, $b = N + 4$ and $c = 4N - 20$. When we substitute these expressions into the quadratic formula, we obtain the following solution for R.

$$R = \frac{-(N + 4) \pm \sqrt{(N + 4)^2 - 4 * 1 * (4N - 20)}}{2 * 1}$$

Although it is not necessary to simplify this expression, it can be rewritten as

$$R = \frac{-N - 4 \pm \sqrt{N^2 - 8N + 96}}{2}$$

You may have noticed that we did not verify any of our solutions in Example 3. The verification process for these literals is the same as it has always been. However, because the solutions to quadratic equations often contain two possibilities, the verification process becomes quite cumbersome. In addition, when you pick values for the variable you need to be sure that the expression is defined for those values. For these reasons, we do not insist that you verify solutions to literal quadratic equations.

In Section 1.2 we briefly discussed the equation $y^2 = x$. We decided that this equation did not represent a function because an input of $x = 1$, produced two different outputs $y = 1$ or $y = {}^-1$. Our calculator only graphs functions. Therefore, we are not able to graph this equation on our calculator in its current form. However, if we solve this equation for y, we can graph it on our calculator.

$$y^2 = x$$
$$(y^2)^{1/2} = x^{1/2}$$
$$|y| = \sqrt{x}$$
$$y = \pm\sqrt{x}$$

Notice that y is equal to two different expressions, $y = \sqrt{x}$ or $y = {}^-\sqrt{x}$. Therefore, we can graph this on our calculator by entering both of these expressions simultaneously.

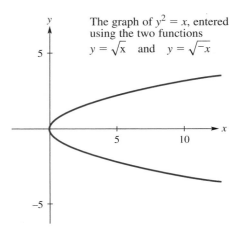

The graph of $y^2 = x$, entered using the two functions $y = \sqrt{x}$ and $y = \sqrt{{}^-x}$

Notice that the graph does not pass the vertical line test. Both of the equations $y = \sqrt{x}$ and $y = -\sqrt{x}$ represent functions. Together these equations represent the relation $y^2 = x$, which is not a function. The graph of $y^2 = x$ appears to be a "sideways" parabola. Does this seem reasonable?

In this section, we looked at solving literal quadratic equations. Before we can solve these types of equations we must recognize that they are quadratic in the variable we are solving for. Then we are able to use the three algebraic methods from Section 7.2 to solve these equations. At times the original equation may not be quadratic, but can be rewritten so it is quadratic. Therefore, to solve the problems in this section you need to use all of the techniques you learned for solving equations.

Here we list the general processes for solving the equations we encountered so far.

- Solve linear equations by isolating the variable.
- Solve power and radical equations by applying reciprocal powers.
- Solve rational equations by clearing fractions.
- Solve quadratic equations by factoring, using a square root, or using the quadratic formula.

Problem Set 7.4

1. Identify each equation as linear or quadratic for the indicated variable.
   - **a.** $2pt^2 = 10$ for p
   - **b.** $2pt^2 = 10$ for t
   - **c.** $mp^2 + 3p = 0$ for p
   - **d.** $kx - 3kx^2 + 10 = 0$ for k
   - **e.** $2xy + 5x - y^2 = 15$ for y
   - **f.** $2xy + 5x - y^2 = 15$ for x

2. Identify each equation as linear or quadratic for the indicated variable.
   - **a.** $5pm^2 = 35$ for m
   - **b.** $0 = 4t - pt^2$ for t
   - **c.** $12R - S^2T = 0$ for R
   - **d.** $\dfrac{5(x - 7)}{4} = 9y$ for x
   - **e.** $ARM^2 - 3AM + 4 = 0$ for M
   - **f.** $ARM^2 - 3AM + 4 = 0$ for R
   - **g.** $ARM^2 - 3AM + 4 = 0$ for A

3. In Problem 2 you identified each equation as linear or quadratic in a specific variable. Now solve each of these equations for the indicated variable.
   - **a.** $5pm^2 = 35$ for m
   - **b.** $0 = 4t - pt^2$ for t
   - **c.** $12R - S^2T = 0$ for R
   - **d.** $\dfrac{5(x - 7)}{4} = 9y$ for x
   - **e.** $ARM^2 - 3AM + 4 = 0$ for M
   - **f.** $ARM^2 - 3AM + 4 = 0$ for R
   - **g.** $ARM^2 - 3AM + 4 = 0$ for A

4. Solve each of the following equations for the indicated variable.
   - **a.** $8r^2m = H$ for r
   - **b.** $Rx^2 - x - Rp = 0$ for x
   - **c.** $(y + p)(y - q) = 0$ for y
   - **d.** $(2m + 5)(m + k) = 9$ for m

5. Solve for the indicated variable.

   a. From electronics $0 = VI - I^2R$ for R

   b. From electronics $0 = VI - I^2R$ for I

   c. From electronics $LCM^2 + RCM + 1 = 0$ for M

   d. From business $R = \dfrac{C - S}{t}$ for S

   e. General $M = \dfrac{ab}{a + b}$ for b

   f. General $\dfrac{D}{g + 1} = \dfrac{g + 1}{g}$ for g

6. The volume of a right circular cone has the formula $V = \frac{1}{3}\pi r^2h$, where r is the radius of the base and h is the height of the cone. Solve this formula for the radius.

7. Machine Technology. From machine technology the diameter of the circle can be determined using the following formula.

$$d = \frac{\left(\dfrac{w}{2}\right)^2 + h^2}{h}$$

   where h = depth of cut

           w = width of cut

           d = diameter of circle

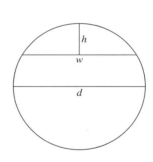

   as shown in the figure.

   a. Solve this formula for h.

   b. Find the depth of the cut (h) needed to produce a 0.125-inch flat (w) on the cross section of a cylindrical shaft 0.75 inches in diameter (d).

8. Solve the following equations. Round all approximate solutions to five significant digits. Check your solutions.

   a. $\dfrac{x}{3} = \dfrac{12x}{5x - 1}$ **d.** $\dfrac{1}{y - 4} + 6 = \dfrac{3}{y}$

   b. $\dfrac{4p - 3}{5} + 7 = \dfrac{20}{p}$ **e.** $\sqrt{r + 6} = 2r$

   c. $\sqrt[3]{5k + 1} = 4$

9. Solve the following equations. Round all approximate solutions to five significant digits. Check your solutions.

   a. $\dfrac{x + 5}{2} = \dfrac{4}{x}$ **d.** $\sqrt{B + 2} + 2B - 1 = 5$

   b. $2\sqrt{2Q} + 5 = Q$ **e.** $(5x^2 - 178)^3 = 8$

   c. $\dfrac{3m^2 + 12}{m - 3} = 3 + 2m + \dfrac{13m}{m - 3}$

10. The Trucker.   A trucker maintained a constant speed with an empty truck and traveled 480 miles. On the return trip, he reduced his speed by 20 mph, and the trip took 4 hours longer. What was his speed with the empty truck?

11. Wall Art.   A museum director wishes to order a wall hanging to display in the new lobby. The wall selected is 32 feet long by 20 feet high. The director wants the hanging to be 2 feet from both the floor and the ceiling of the lobby. The director wishes the dimensions of the hanging to match those of the Golden Rectangle.

The **golden rectangle** is important in art and architecture and has dimensions of width W and height H that satisfy the following formula.

$$\frac{W + H}{H} = \frac{H}{W}$$

Determine the dimensions of the wall hanging that satisfy all of the director's wishes.

12. a. Solve the equation $x^2 + y^2 = 25$ for y. Enter both solutions into your calculator, and view the graph in a square window. Describe your graph.

   b. Solve the equation $x^2 + y^2 = 49$ for y. Enter both solutions into your graphing calculator, and view the graph in a square window. Describe your graph.

   c. Solve the equation $x^2 + y^2 = 7$ for y. Enter both solutions into your graphing calculator, and view the graph in a square window. Describe your graph.

13. Solve the equation $(x - 3)^2 + y^2 = 25$ for y. Enter both solutions into your graphing calculator, and view the graph in a square window. Describe your graph.

7.5 Nonlinear Systems and Inequalities

Activity Set 7.5

1. In your team, solve the following system. Have half of your team solve the system graphically and the other half solve the system algebraically. Share your processes and results when you are done.

$$\begin{cases} y = 0.5x^2 - 32 \\ y = x - 8 \end{cases}$$

2. In your team, use your graphs from Problem 1 to determine the solution to the following inequality. Write your solution using inequality notation, and graph your solution on a number line.

$$0.5x^2 - 32 \le x - 8$$

3. In your team, use your graphs from Problem 1 to determine the solution to the following equation.

$$0.5x^2 - 32 = x - 8$$

4. Solve the following system as you did in Problem 1, but switch roles. The team members who solved the first system algebraically should solve this system graphically and vice versa.

$$\begin{cases} y = 2x - 12 \\ y = {}^{-}x^2 + 10x + 4 \end{cases}$$

5. In your team, use your graphs from Problem 4 to determine the solution to the following inequality. Write your solution using inequality notation, and graph your solution on a number line.

$$2x - 12 > {}^{-}x^2 + 10x + 4$$

6. Solve the following system graphically.

$$\begin{cases} x^2 + y^2 = 36 \\ x + y = 3 \end{cases}$$

7. Solve each inequality graphically. Write your solution using inequality notation, and graph your solution on a number line.
   a. $(x - 8)(x + 12) \ge 0$
   b. $x^2 - 16x + 73 < {}^{-}3x + 103$
   c. $9 - x^2 \le 10 + x$

Discussion 7.5

Nonlinear Systems and Inequalities

The content of this section is a natural extension of the material we have studied. In Section 2.4 we solved linear systems algebraically and graphically. The algebraic methods that we used were the elimination method and the substitution method. In this section, we will solve systems of nonlinear equations graphically and algebraically using methods we learned previously. However, the method of elimination will not always work for nonlinear systems.

In Section 4.1, we solved linear inequalities algebraically and graphically. The algebraic process for solving linear inequalities was similar to solving linear equations except that when we multiply or divide by a negative number we must reverse the direction of the inequality. Because the algebraic process for solving nonlinear equations involves much more than isolating the variable or undoing operations, it is unclear when the direction of an inequality might need to be reversed. Therefore, we will solve nonlinear inequalities *only* graphically.

Example 1

Solve the following system algebraically and graphically.

$$\begin{cases} y = 20 - 0.2x^2 \\ y = {}^-1.8x + 6 \end{cases}$$

Solution   To solve this system algebraically, we can choose either the method of elimination or the substitution method. Because both equations are already solved for y, let's use the method of substitution. We can substitute the expression $20 - 0.2x^2$ for y into the second equation. This results in a quadratic equation in one variable, which we can solve using the quadratic formula.

$$20 - 0.2x^2 = {}^-1.8x + 6$$
$$^-0.2x^2 + 1.8x + 14 = 0$$

For this equation $a = {}^-0.2$, $b = 1.8$, and $c = 14$. The quadratic formula produces the following solution.

$$x = \frac{^-1.8 \pm \sqrt{1.8^2 - 4 * {}^-0.2 * 14}}{2 * {}^-0.2}$$
$$x = {}^-5 \qquad \text{or} \qquad x = 14$$

Next, because the original problem was a system of equations in two variables, we must determine the corresponding values of y. We can substitute the values of x into either of the original equations. Let's use the second one.

When $x = {}^-5$,

$$y = {}^-1.8 * {}^-5 + 6$$
$$y = 15$$

When $x = 14$,

$$y = {}^-1.8 * 14 + 6$$
$$y = {}^-19.2$$

From our algebraic process, the solution to the system is $({}^-5, 15)$ or $(14, {}^-19.2)$. Because we are also going to solve this system graphically, that serves as our check.

To solve this system graphically we must graph both of the original equations, $y = 20 - 0.2x^2$ and $y = {}^-1.8x + 6$. Let's list what we know about the graph of $y = 20 - 0.2x^2$.

- The graph of $y = 20 - 0.2x^2$ is a parabola that opens down.
- The line of symmetry is $x = 0$. When the term bx is missing, we know the graph is not shifted horizontally and, therefore, the line of symmetry remains the y-axis.
- The vertex and the vertical intercept are both $(0, 20)$.

Similarly, let's list what we know about the graph of $y = {}^-1.8x + 6$.

- The graph of $y = {}^-1.8x + 6$ is a line.
- The slope of the line is $^-1.8$.
- The vertical intercept of the line is $(0, 6)$.

With this information we can find a window that clearly shows both graphs. It may take more than one try to obtain a window that also shows both points of intersection. The following graph shows the important information.

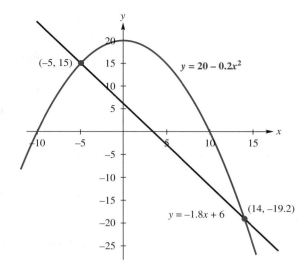

The points of intersection correspond to the solution. Therefore, we can see that the solution to the system is $(^-5, 15)$ or $(14, ^-19.2)$, which agrees with our previous result.

Example 2

Use the graph from Example 1 to determine the solution to the inequality $20 - 0.2x^2 \geq {}^-1.8x + 6$. Write your solution using inequality notation, and graph your solution on a number line.

Solution   In this example, we are only solving for x. From the graph in Example 1, we know that the two sides of the inequality are equal when $x = {}^-5$ or when $x = 14$. These are the border points of our solution region. Because our inequality includes equals $(=)$, these points are included in our solution. Next, we must check the three regions created by these two points.

From the inequality statement, $20 - 0.2x^2 \geq {}^-1.8x + 6$, we want the output values of the parabola to be greater than or equal to the output values of the line. If x is less than $^-5$, then the graph of the parabola is *below* the graph of the line, which means the inequality is false for those values. Similarly, if x is greater than 14, the graph of the parabola is *below* the graph of the line and the inequality is false. If x is between $^-5$ and 14, the graph of the parabola is *above* the graph of the line. Therefore, the inequality is true when x is between $^-5$ and 14.

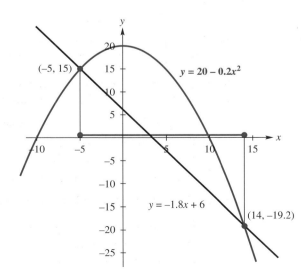

The solution to the inequality $20 - 0.2x^2 \geq {}^-1.8x + 6$ is all the values of x between $^-5$ and 14, including $^-5$ and 14. This can be written either as $x \geq {}^-5$ and $x \leq 14$ or as $^-5 \leq x \leq 14$. The graph of the solution is shown on the following number line.

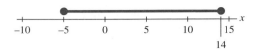

Example 3

The Flight of the Golf Ball.   The height of a golf ball in terms of the horizontal distance from the tee is given by

$$h = {}^-0.00340d^2 + 0.476d$$

where h is the vertical height in yards, and d is the horizontal distance in yards.

The golf ball is hit toward a hill that lies 100 yards away and is at a 20.0% grade. Determine the point on the hill where the golf ball lands. How far does the person who hit the golf ball have to walk to reach the ball?

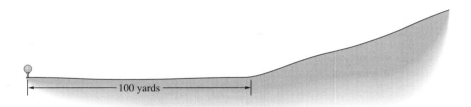

100 yards

Solution   First let's sketch the path of the golf ball. We know that $h = {}^-0.00340d^2 + 0.476d$ is a quadratic equation in terms of d. The graph is a downward parabola. The line of symmetry is

$$d = \frac{^-0.476}{2 * {}^-0.00340} = 70.0$$

Therefore, the d-coordinate of the vertex is 70. We can find the h-coordinate by substituting $d = 70.0$ into the equation and solving for d.

$$h = {}^-0.00340 * 70.0^2 + 0.476 * 70.0$$

$$h \approx 16.7$$

Therefore, the vertex is the point $\approx(70.0, 16.7)$. The horizontal intercepts can be found by setting $h = 0$ and solving for d.

$$0 = {}^-0.00340d^2 + 0.476d$$

$$0 = d({}^-0.00340d + 0.476)$$

$$d = 0 \quad \text{or} \quad {}^-0.00340d + 0.476 = 0$$

$$d = 0 \quad \text{or} \quad d = \frac{^-0.476}{^-0.00340}$$

$$d = 0 \quad \text{or} \quad d = 140$$

With this information we can sketch a graph of the path of the golf ball.

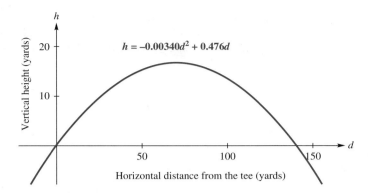

To determine where the golf ball lands on the hill we need to write an equation that represents the side of the hill. We know the hill has a 20.0% grade and begins 100 yards from the tee. This means that the equation is a line that passes through the point $(100, 0)$ and has a slope of $20.0\% = \frac{20.0}{100} = 0.200$. We can use this information to write the equation.

Using the same variables from the golf ball model and the slope–intercept form of the equation of a line, we obtain the following equation. Remember, in this example, we are using h instead of y for the dependent variable and d instead of x for the independent variable.

$h = m * d + b$

$h = 0.200d + b$ Substitute 0.200 into the equation for m.

$0 = 0.200 * 100 + b$ Substitute the point $(100, 0)$ for d and h to determine the value of b.

$^-20.0 = b$ Solve for b.

The equation of the line that represents the height of the hill at a point that is a horizontal distance of d from the tee is

$h = 0.20d - 20.0$

Next, we need to determine where the golf ball hits the hill. This is the intersection of the path of the golf ball and the hill. Therefore, this corresponds to solving the following system.

$$\begin{cases} h = {}^-0.00340d^2 + 0.476d \\ h = 0.20d - 20 \end{cases}$$

We can solve this system graphically or algebraically. Because we already have the graph of the first equation. Let's solve the system graphically. From the graph, the intersection point is $\approx(127, 5.5)$. This means that the golf ball lands on the hill at a point that is a horizontal distance of about 127 yards from the tee and a vertical height of about 5.5 yards. Does this seem reasonable?

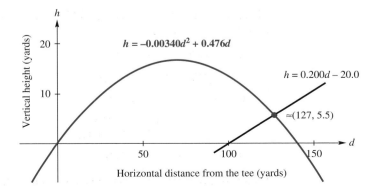

Finally, we need to determine the distance the person must walk to reach the ball. We know that he or she must first walk 100 yards to the base of the hill. Then, we need to determine the distance from the base of the hill to the ball. We can find this distance using the Pythagorean theorem.

$$(\text{distance up the hill})^2 = 27^2 + 5.5^2$$
$$\text{distance up the hill} = \sqrt{27^2 + 5.5^2}$$
$$\text{distance up the hill} \approx 28$$

Therefore, the person must walk a total of about 128 yards to reach the golf ball.

The Equation of a Circle Centered at the Origin

If you did Problem 12 from the problem set in Section 7.4, you may have noticed that an equation of the form $x^2 + y^2 = K$ graphed as a circle. The radius of this circle was $\sqrt{K}$. Let's see if we can understand why this type of equation graphs as circle. Suppose a circle is centered at the origin with radius r.

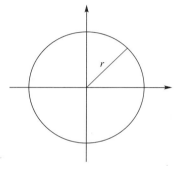

Let's label the point where the radius in the figure intersects the circle (x, y). With this point and the radius we can form a right triangle whose legs have lengths x and y.

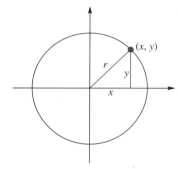

If we then apply the Pythagorean theorem to this right triangle, we obtain the following equation.

$$x^2 + y^2 = r^2$$

> **Equation of a Circle Centered at the Origin**
> Any equation of the form $x^2 + y^2 = r^2$ graphs as a circle centered at the origin with radius r.

Does the equation of a circle represent a function?

Example 4

Solve the following system graphically and algebraically. Round your solution to three significant digits.

$$\begin{cases} x^2 + y^2 = 36 \\ y = x^2 - 3 \end{cases}$$

Solution   We start by solving the system graphically. We know that $y = x^2 - 3$ is the equation of a parabola that opens up. The line of symmetry is $x = 0$. The vertex and the vertical intercept are both the point $(0, {}^-3)$. With this information we can draw a rough graph.

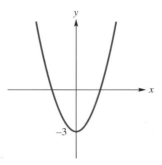

Next, we know that $x^2 + y^2 = 36$ is the equation of a circle centered at the origin with radius 6.

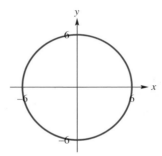

Now that we have an idea of how each graph will look, we can graph them on our calculator. However, before we can do this, we must first solve $x^2 + y^2 = 36$ for y.

$$x^2 + y^2 = 36$$

$$y^2 = 36 - x^2$$

$$(y^2)^{1/2} = (36 - x^2)^{1/2}$$

$$|y| = \sqrt{36 - x^2}$$

$$y = \pm\sqrt{36 - x^2}$$

Therefore, to graph this system we must graph the three equations

$$y = x^2 - 3$$

$$y = \sqrt{36 - x^2}$$

$$y = {}^-\sqrt{36 - x^2}$$

When the three equations are graphed simultaneously, we obtain a circle and a parabola.

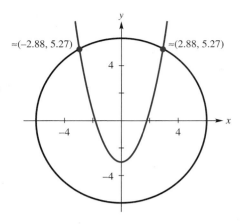

From the graph, the solution appears to be $\approx(^-2.88, 5.27)$ or $\approx(2.88, 5.27)$.

Next, let's solve the system algebraically.

$$\begin{cases} x^2 + y^2 = 36 \\ \quad\;\; y = x^2 - 3 \end{cases}$$

Because the second equation is already solved for y, we might consider substituting $x^2 - 3$ into the first equation for y. However, if we do this we obtain an equation containing x^4. At this point, we do not know how to solve such an equation.

Alternatively, we can solve the first equation for x^2 and substitute the resulting expression into the second equation for x^2.

$x^2 + y^2 = 36$	We start with the first equation.
$x^2 = 36 - y^2$	Subtract y^2 from both sides.

Now we can substitute the expression $36 - y^2$ into the second equation for x^2.

$y = x^2 - 3$	Next, we use the second equation.
$y = (36 - y^2) - 3$	The expression for x^2 substituted into the second equation.
$y^2 + y - 33 = 0$	Rewrite the equation in standard form.

$$y = \frac{-1 \pm \sqrt{1^2 - 4 * 1 * {}^-33}}{2 * 1}$$

$$y \approx 5.27 \quad \text{or} \quad y \approx {}^-6.27$$

Next, we must find the corresponding values for the variable x. Using the second equation we know that $x^2 = y + 3$.

When $y \approx 5.27$, we obtain the following.   When $y \approx {}^-6.27$, we obtain the following.

$$x^2 = y + 3 \qquad\qquad\qquad\qquad x^2 = y + 3$$
$$x^2 \approx 5.27 + 3 \qquad\qquad\qquad x^2 \approx {}^-6.27 + 3$$
$$x^2 \approx 8.27 \qquad\qquad\qquad\quad x^2 \approx {}^-3.27$$
$$(x^2)^{1/2} \approx 8.27^{1/2} \qquad\qquad\quad \text{No real solution}$$
$$|x| \approx 2.88$$
$$x \approx \pm 2.88$$

Therefore, the solution to the system is $\approx(2.88, 5.27)$ or $\approx(^-2.88, 5.27)$.

Because we solved this system two different ways and obtained the same solution, it is not necessary to check our solution numerically.

Early in this text we learned how to solve linear systems and linear inequalities graphically and algebraically. In this section, we used those same techniques to solve nonlinear systems. However, nonlinear inequalities were solved only graphically.

We also added to our collection of equations the equation of a circle centered at the origin, $x^2 + y^2 = r^2$. To graph this equation on our calculator, we must solve for y. The result is two equations that must both be entered and graphed to obtain the complete circle.

Problem Set 7.5

1. Solve each system graphically and algebraically. Round approximate solutions to the thousandths place. Be sure to sketch the graphs you use because you will need them again in Problem 2.

   a. $\begin{cases} y = 10 - x^2 \\ y = x + 3 \end{cases}$

   b. $\begin{cases} y = 2x^2 - 6x - 140 \\ 2y = x - 24 \end{cases}$

   c. $\begin{cases} y = 2.0x^2 + 6.2x - 4 \\ y - 5.1x = {}^{-}12 \end{cases}$

   d. $\begin{cases} 5x - y = 48 \\ 3x - 7y = 16 \end{cases}$

2. Solve each inequality graphically.

   a. $10 - x^2 > x + 3$

   b. $\frac{1}{2}x - 12 \le 2x^2 - 6x - 140$

   c. $2.0x^2 + 6.2x - 4 > 5.1x - 12$

   d. $10x^2 + 25 < 0$

   e. $10 - x^2 < 0$

   f. $10.2x + 24 \ge 4.0x^2 + 12.4x - 8$

3. Solve each nonlinear system graphically. Round approximate solutions to the thousandths place. Check your solutions. Be sure to sketch the graphs you use because you will need them again in Problem 4.

   a. $\begin{cases} x^2 + y^2 = 25 \\ y = 2x - 3 \end{cases}$

   b. $\begin{cases} y = 3(x^2 - 1.2) \\ x^2 + y^2 = 7 \end{cases}$

   c. $\begin{cases} y = |3.5x - 7| \\ y = 5 - 0.5x \end{cases}$

   d. $\begin{cases} y = |4x - 60| \\ y = x(20 - x) \end{cases}$

4. Solve each inequality graphically.

   a. $|3.5x - 7| > 5 - 0.5x$

   b. $x(20 - x) \ge |4x - 60|$

5. Use the graph to answer the following questions.

   a. Solve the system

   $$\begin{cases} y = ax^2 + bx + c \\ y = mx + k \end{cases}$$

   b. Solve the inequality

   $$ax^2 + bx + c > 0$$

   c. Solve the inequality

   $$ax^2 + bx + c \le mx + k$$

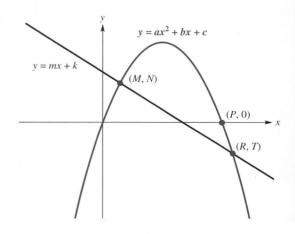

6. Use the graph to answer the following questions .

   a. Solve the system

   $$\begin{cases} y = ax^2 + bx + c \\ y = dx^2 + ex + f \end{cases}$$

   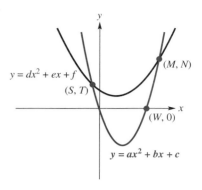

   b. Solve the inequality

   $$ax^2 + bx + c \le 0$$

   c. Solve the inequality

   $$ax^2 + bx + c > dx^2 + ex + f$$

7. The Tip Top Tennis Balls manufacturing company makes and sells tennis balls. The revenue (in dollars) that the company receives in selling x cases (12 cans per case, three balls per can) of tennis balls in one week is given by $R = {}^-0.51x^2 + 180x$. The cost (in dollars) is given by $C = 78x + 1500$. All of the following questions refer to a one-week period, and only full cases of balls are sold.

   a. Graph both equations on the same set of axes.

   b. Determine the break-even points graphically and algebraically.

   c. Write an equation for the profit P the company makes in producing and selling x cases. (Remember that *profit = revenue − cost*.)

   d. Graph the profit equation on the same set of axes.

   e. Determine the horizontal intercepts of your profit graph algebraically.

   f. What is the relationship between the break-even points you found in part b and the horizontal intercepts you found in part e? How is this relationship illustrated on your graph?

8. Write the equation for the line tangent to the graph of $x^2 + y^2 = 169$ at the point $(^-12, 5)$. (*Hint:* A line tangent to a circle is perpendicular to the radius at the point of tangency.) Graph both the equation $x^2 + y^2 = 169$ and your line to verify your equation.

9. For each part, write a 2×2 system that satisfies the conditions.

   a. The system consists of two lines and has one solution.

   b. The system consists of two lines and has no solution.

10. For each part, write a 2×2 system that satisfies the conditions.

    a. The system consists of a circle and a line and has two solutions.

    b. The system consists of a circle and a line and has one solution.

    c. The system consists of a circle and a line and has no solutions.

11. For each part, write a 2×2 system that satisfies the conditions.

    a. The system consists of a parabola and a line and has two solutions.

    b. The system consists of a parabola and a line and has one solution.

    c. The system consists of a parabola and a line and has no solutions.

12. For each part, write a 2×2 system that satisfies the conditions.

  a. The system consists of a parabola and a circle and has two solutions.

  b. The system consists of a parabola and a circle and has four solutions.

  c. The system consists of a parabola and a circle and has one solution.

  d. The system consists of a parabola and a circle and has no solutions.

  e. The system consists of a parabola and a circle and has three solutions.

13. For each part, write a 2×2 system that satisfies the conditions.

  a. The system consists of two parabolas and has two solutions.

  b. The system consists of two parabolas and has one solution.

  c. The system consists of two parabolas and has no solutions.

  d. The system consists of two upward parabolas and has no solutions.

7.6 Solving Quadratic Equations by Factoring

Activity
Set
7.6

1. Solve the quadratic equations.

   a. $(x + 4)(x + 8) = 0$ **c.** $3(w - 2)(w + 1) = 0$ **e.** $5x(x - 10) = 0$

   b. $(R - 2)(R + 7) = 0$ **d.** $P(2P + 11) = 0$ **f.** $2m^2 + 6m = 0$

2. Complete the following table by finding integers a and b such that their sum and product are the values given in the table.

a	b	$a * b$	$a + b$
		12	7
		8	6
		8	9
		10	$^-7$
		25	$^-10$
		$^-21$	$^-4$
		$^-18$	7
		20	$^-12$
		$^-36$	0

3. Multiply each of the following expressions.

   a. $(x + 3)(x + 4)$ **d.** $(x - 5)(x - 5)$

   b. $(x + 1)(x + 8)$ **e.** $(x - 2)(x + 9)$

   c. $(x - 5)(x - 2)$

4. Consider the expression $x^2 + 9x + 18$. List all of the pairs of integer factors whose product is 18. Which pair of factors from your list sum to 9? Use the pair of factors you found to factor the expression $x^2 + 9x + 18$.

5. Use the same process from Activity 4 to factor the following expressions.

   a. $x^2 + 6x + 8$ **c.** $x^2 - 4x - 21$ **e.** $x^2 - 9x - 10$

   b. $x^2 - 10x + 25$ **d.** $x^2 - 8x + 12$ **f.** $x^2 - 8x - 9$

6. Solve the following quadratic equations by first factoring the expression and then using the zero-product principle.

   a. $x^2 - 11x + 18 = 0$ **d.** $x^2 - 8x = 0$

   b. $x^2 + 13x + 12 = 0$ **e.** $x^2 + 6x + 9 = 0$

   c. $x^2 - 5x - 24 = 0$

Discussion
7.6

Earlier we saw that three methods can be used to solve quadratic equations: the factoring method, the square root method, and the quadratic formula. We found that if a quadratic equation in standard form was written as $0 = ax^2 + bx$, then factoring was the most efficient method for solving the equation. We know that we can always factor expressions written as $ax^2 + bx$ because x is common to both terms. In the activities, we found that we can also factor some expressions written in the form $x^2 + bx + c$. Let's see how this can be done. One way to remember how to multiply expressions like $(x + 5)(x + 6)$ is to use the mnemonic FOIL. FOIL stands for **f**irst, **o**uter, **i**nner, and **l**ast.

$$(x + 5)(x + 6) = x^2 + 5x + 6x + 30$$
$$= x^2 + 11x + 30$$

Then to factor an expression like $x^2 - 6x + 8$ we need to work backwards.

$$x^2 - 6x + 8 = (?\ \pm\ ?)(?\ \pm\ ?)$$

Because the product of the first terms must be x^2, each of the first terms most likely is x.

$$x^2 - 6x + 8 = (x\ \pm\ ?)(x\ \pm\ ?)$$

Next, let's look at the constant term, 8. The 8 comes from the product of the last terms in the factored form. What two integers multiply to produce 8? The possibilities are the following.

    1 and   8
    $^-$1 and $^-$8
    2 and   4
    $^-$2 and $^-$4

To choose the pair of integers to use in our factored form, we must also consider the sums of the outer and inner terms. In this case the sum must be ^-6x. Which pair of integers has a product of 8 and a sum of $^-$6? We see from our list that the pair $^-$2 and $^-$4 does. We use these numbers to factor the expression as shown.

$$x^2 - 6x + 8 = (x - 2)(x - 4)$$

When $^-$2 and $^-$4 are substituted into the previous expression, you may have noticed that we wrote it as $(x - 2)(x - 4)$, not as $(x + {}^-2)(x + {}^-4)$. Although either is correct, using subtraction is simpler.

We can verify that $x^2 - 6x + 8$ factors to $(x - 2)(x - 4)$ by numerical substitution. Instead, we check our results by multiplying out the factored expression because this is a fairly easy task and quick to do.

$(x - 2)(x - 4)$ Start with the factored form.

$= x^2 - 2x - 4x + 8$ Multiply out the expression.

$= x^2 - 6x + 8$

We conclude that $x^2 - 6x + 8$ factors to $(x - 2)(x - 4)$.

NOTE: Multiplication can be performed in any order, so we can also say that $x^2 - 6x + 8$ factors to $(x - 4)(x - 2)$ because $(x - 2)(x - 4) = (x - 4)(x - 2)$.

Example 1

Rewrite each of the following expressions in factored form.

a. $x^2 + 8x + 15$ c. $m^2 - 8m + 12$

b. $x^2 + 6x - 7$ d. $w^2 - 25$

Solution

a. We want to factor the expression $x^2 + 8x + 15$ into a product for which each factor has two terms.

$$x^2 + 8x + 15 = (x \pm \;?)(x \pm \;?)$$

We need to find a pair of integers whose product is 15 and sum is 8. Let's start by listing the pairs of integers whose product is 15.

    1 and   15
   $^-$1 and $^-$15
    3 and   5
   $^-$3 and $^-$5

The pair 3 and 5 also sums to 8, so we use this pair to write the factors.

$$x^2 + 8x + 15 = (x + 3)(x + 5)$$

To verify our result, we can multiply the factored expression.

CHECK

$\quad (x + 3)(x + 5)$
$= x^2 + 5x + 3x + 15$ Multiply out the factored expression.
$= x^2 + 8x + 15 \quad$ ✔   This matches our original expression, so the factored form is equivalent.

We conclude that $x^2 + 8x + 15$ factors into $(x + 3)(x + 5)$.

b. We want to factor the expression $x^2 + 6x - 7$ into a product in which each factor has two terms.

$$x^2 + 6x - 7 = (x \pm \;?)(x \pm \;?)$$

The pairs of integer factors that multiply to $^-7$ are the following.

    1 and $^-$7
   $^-$1 and   7

Because the pair $^-$1 and 7 sums to 6, we use this pair to write the factors.

$$x^2 + 6x - 7 = (x - 1)(x + 7)$$

CHECK

$\quad (x - 1)(x + 7)$
$= x^2 + 7x - x - 7$ Multiply out the factored expression.
$= x^2 + 6x - 7 \quad$ ✔   This matches our original expression, so the factored form is equivalent.

We conclude that $x^2 + 6x - 7$ factors into $(x - 1)(x + 7)$.

c. We want to factor the expression $m^2 - 8m + 12$ into a product for which each factor has two terms.

$$m^2 - 8m + 12 = (m \pm ?)(m \pm ?)$$

The pairs of integer factors that multiply to 12 are the following.

$$\begin{array}{rl} 1 \text{ and } & 12 \\ {}^-1 \text{ and } & {}^-12 \\ 2 \text{ and } & 6 \\ {}^-2 \text{ and } & {}^-6 \\ 3 \text{ and } & 4 \\ {}^-3 \text{ and } & {}^-4 \end{array}$$

Because the pair $^-2$ and $^-6$ sums to $^-8$, we use this pair to write the factors.

$$m^2 - 8m + 12 = (m - 2)(m - 6)$$

CHECK

$$\begin{aligned} &(m - 2)(m - 6) \\ &= m^2 - 6m - 2m + 12 \\ &= m^2 - 8m + 12 \quad \checkmark \end{aligned}$$

We conclude that $m^2 - 8m + 12$ factors into $(m - 2)(m - 6)$.

d. This expression looks different from the others. It has only two terms. We try to factor the expression $w^2 - 25$ using the same process.

$$w^2 - 25 = (w \pm ?)(w \pm ?)$$

The pairs of integer factors that multiply to $^-25$ are the following.

$$\begin{array}{rl} 1 \text{ and } & {}^-25 \\ {}^-1 \text{ and } & 25 \\ 5 \text{ and } & {}^-5 \end{array}$$

Because the sum of the outer and inner terms must be zero, we use 5 and $^-5$.

$$w^2 - 25 = (w + 5)(w - 5)$$

CHECK

$$\begin{aligned} &(w + 5)(w - 5) \\ &= w^2 + 5w - 5w - 25 \\ &= w^2 - 25 \quad \checkmark \end{aligned}$$

We conclude that $w^2 - 25$ factors into $(w + 5)(w - 5)$.

In part d of Example 1, we see a special factoring case called the difference of two squares. The general pattern for factoring the difference of two squares is

$$a^2 - b^2 = (a + b)(a - b).$$

We can use factoring to solve some quadratic equations.

Example 2

Solve the following quadratic equations using the factoring method.

$$\text{a. } x^2 - 2x - 15 = 0 \qquad \text{b. } x^2 = {}^-10x$$

Solution a. Because this quadratic equation is already equal to zero, we will begin by factoring the left side of the equation.

$$x^2 - 2x - 15 = 0$$

$$(x + 3)(x - 5) = 0 \qquad \text{Factor the left side of the equation.}$$

$$x + 3 = 0 \quad \text{or} \quad x - 5 = 0 \qquad \text{Using the zero-product principle, set each factor equal to 0.}$$

$$x = {}^-3 \quad \text{or} \quad x = 5$$

CHECK

$$x = {}^-3 \qquad\qquad\qquad\qquad x = 5$$

$$x^2 - 2x - 15 = 0 \qquad\qquad x^2 - 2x - 15 = 0$$

$$({}^-3)^2 - 2 * {}^-3 - 15 \overset{?}{=} 0 \qquad\qquad 5^2 - 2 * 5 - 15 \overset{?}{=} 0$$

$$0 = 0 \quad\checkmark \qquad\qquad\qquad 0 = 0 \quad\checkmark$$

We conclude that the solution to the equation $x^2 - 2x - 15 = 0$ is $x = {}^-3$ or $x = 5$.

b.
$$x^2 = {}^-10x$$

$$x^2 + 10x = 0 \qquad \text{First set one side equal to zero.}$$

$$x(x + 10) = 0 \qquad \text{Factor the left side of the equation.}$$

$$x = 0 \quad \text{or} \quad x + 10 = 0 \qquad \text{Using the zero-product principle, set each factor to 0.}$$

$$x = 0 \quad \text{or} \quad x = {}^-10$$

CHECK

$$x = 0 \qquad\qquad\qquad x = {}^-10$$

$$x^2 = {}^-10x \qquad\qquad x^2 = {}^-10x$$

$$0^2 = {}^-10 * 0 \qquad\qquad ({}^-10)^2 = {}^-10 * {}^-10$$

$$0 = 0 \quad\checkmark \qquad\qquad 100 = 100 \quad\checkmark$$

We conclude that the solution to the equation $x^2 = {}^-10x$ is $x = 0$ or $x = {}^-10$.

You may be wondering why we did not use the factoring method earlier in this chapter. There are two very good reasons for not introducing this method earlier. The first reason is that not all expressions factor into expressions containing only integers. For example, let's try to factor $x^2 + 12x + 16$.

$$x^2 + 12x + 16 = (x \pm ?)(x \pm ?)$$

The integer factors of the last term are

$$1 \text{ and } 16$$
$${}^-1 \text{ and } {}^-16$$
$$2 \text{ and } 8$$
$${}^-2 \text{ and } {}^-8$$
$$4 \text{ and } 4$$
$${}^-4 \text{ and } {}^-4$$

But the sum of the outer and inner terms must be $12x$. None of the pairs of integer factors of 16 result in a sum of 12. We conclude that $x^2 + 12x + 16$ does not factor over the set of integers.

The second reason for not introducing this technique earlier is that unless the expression has very simple numbers, factoring is not an efficient choice. Do you want to try to factor the expression $x^2 + 78x + 1512$? The numbers in the equation $x^2 + 78x + 1512 = 0$ are too large to factor quickly. Therefore, it is more efficient to use the quadratic formula to solve this equation.

Recall that when we find the solutions to a quadratic equation, we also know the factors. For example, using the quadratic formula we find that the solution to the equation $x^2 + 78x + 1512 = 0$ is $x = {}^-36$ or $x = {}^-42$. Because $x = {}^-36$ is a solution, $x + 36$ is a factor. Similarly, $x + 42$ is a factor. Verify for yourself that $x^2 + 78x + 1512$ factors into $(x + 36)(x + 42)$.

Most applications involving quadratic expressions cannot be factored or are too difficult to factor easily. Because of this, be prepared to use the quadratic formula or other methods discussed in this chapter to solve these equations.

In this section, we saw that some equations in the form $x^2 + bx + c = 0$ can be easily factored. Notice that we restrict this discussion to quadratic equations when the coefficient of x^2 is 1. If the factors are easy to find, then we can use the factoring method together with the zero-product principle to solve the equation. Knowing that the solutions to the quadratic equation $x^2 + bx + c = 0$ correspond to the horizontal intercepts of the quadratic function $y = x^2 + bx + c$, we can also use this method to find horizontal intercepts. If a quadratic equation in standard form includes all three terms and is difficult to factor, we continue to use the quadratic formula as the most efficient method.

Problem Set 7.6

1. Factor the following expressions, if possible.
    a. $x^2 + 10x + 21$
    b. $x^2 - 3x$
    c. $x^2 - 6x - 7$
    d. $x^2 + 6x - 55$

2. Factor the following expressions, if possible.
    a. $x^2 - 5x + 4$
    b. $x^2 + 13x + 36$
    c. $2x^2 + 5x$
    d. $x^2 - 6x + 8$
    e. $x^2 - 8x + 9$
    f. $x^2 - 36$

3. Solve the following quadratic equations using the factor method.
    a. $0 = x^2 - x - 30$
    b. $0 = x^2 + 12x + 35$
    c. $16 = x(x - 6)$
    d. $x^2 + 21 = 10x$
    e. $(x + 12)(x + 3) = 36$
    f. $x^2 + 2x - 16 = 2x$

4. Solve the following equations using any method.
    a. $0 = x^2 - 12x - 13$
    b. $w^2 + 12 = 7w$
    c. $0 = x^2 - x - 12$
    d. $(x - 7)(x - 8) = 6$
    e. $0 = x^2 - 5x + 12$
    f. $^-24 = m^2 + 10m$
    g. $12 = 6x^2 + x$
    h. $x(x + 2) = 224$
    i. $9x^2 = 25$

5. a. Write a quadratic equation that has the solution $x = 2$ or $x = {}^-7$.
    b. Write two different quadratic equations that have the solution $R = 15$ or $R = 10$.

6. For each graph, write the equation and verify your equation.

a.

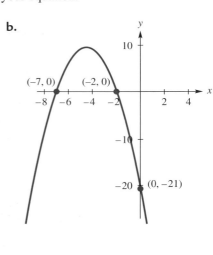

b.

7. The L-Shaped Patio.   Weikko is planning an L-shaped patio around his new home. The length along the east side of the house is to be 10 yards, and the length along the north side is to be 8 yards as shown in the following figure. How wide should the patio be if Weikko has enough concrete to cover 45 square yards?

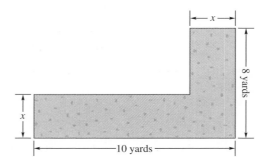

Chapter Seven Summary

In this chapter, we studied quadratic functions in two variables, $y = ax^2 + bx + c$, and quadratic equations in one variable, $0 = ax^2 + bx + c$.

- The **quadratic function $y = ax^2 + bx + c$,** with $a \neq 0$, has an infinite number of solutions. The solutions are ordered pairs of numbers. The graph of these solutions is a parabola.
- The **quadratic equation $0 = ax^2 + bx + c$,** with $a \neq 0$, has at most two solutions. These solutions correspond to the horizontal intercepts of the graph of the function $y = ax^2 + bx + c$.

The Quadratic Function $y = ax^2 + bx + c$, with $a \neq 0$

We learned the relationship between the parameters of the function $y = ax^2 + bx + c$, with $a \neq 0$, and its graph. These are summarized in the following box.

Graphs of Quadratic Functions

The graph of the quadratic function $y = ax^2 + bx + c$, where $a \neq 0$, is a parabola.

- If $a > 0$ then the parabola opens up.
- If $a < 0$ then the parabola opens down.
- The vertical intercept is the point $(0, c)$.
- The equation of the line of symmetry is $x = \frac{-b}{2a}$.
- The x-coordinate of the vertex is $x = \frac{-b}{2a}$.

The Quadratic Equation $0 = ax^2 + bx + c$, with $a \neq 0$

Three algebraic techniques can be used to solve quadratic equations in one variable: the factoring method, the square root method, and the quadratic formula. All of these techniques are good methods for solving quadratic equations.

- The **factoring method** is generally used when all of the terms in the equation contain a common variable factor. We factor out the common factor and then apply the zero-product principle.

  We encountered other equations that we could factor or were already in factored form like $(x + 10)(x - 12) = 0$. In these cases, we also applied the zero-product principle to the factors to determine the solution.

- The **square root method** is most often used when the variable *only* occurs inside of a square. For example, we used this technique on the equations $4x^2 - 100 = 0$ and $(m - 7)^2 = 64$.

- The **quadratic formula** can be used to solve all quadratic equations, but is not always the most efficient technique.

We can solve **literal quadratic equations** using these same algebraic techniques. To do this, we must first be able to recognize that an equation is quadratic in the variable that we are solving for.

At times the original equation may not be quadratic, but can be rewritten so that it is quadratic. Therefore, to solve equations algebraically we may need to use all of the techniques we learned for solving various kinds of equations.

- Solve linear equations by isolating the variable.
- Solve power or radical equations by applying reciprocal powers.
- Solve rational equations by clearing fractions.
- Solve quadratic equations by factoring, applying a square root, or using the quadratic formula.

We can also **solve quadratic equations graphically,** by graphing both sides of the original equation and identifying the intersection points. The solution is the horizontal coordinates of these intersection points. When we solve a problem algebraically and find no solution, then solving the problem graphically is an ideal way to verify our lack of solutions.

A quadratic equation $ax^2 + bx + c = 0$ may have zero, one, or two real solutions. The discriminant can be used to determine the number of solutions without actually knowing what the solutions are. The **discriminant** is the radicand $b^2 - 4ac$ in the quadratic formula.

Combining much of the information we learned in this course, we can now solve nonlinear systems and nonlinear inequalities. Nonlinear systems can be solved algebraically and graphically. However, we only solved nonlinear inequalities graphically.

We also added to our collection of equations the equation $x^2 + y^2 = r^2$. This graphs as a circle, centered at the origin, with radius r. To graph this equation on our calculator, we must solve for y. The result is two equations that must both be entered and graphed to obtain the complete circle.

Chapter Eight

Geometry and Trigonometry

We will begin Chapter 8 with a review of similar triangles and scale drawings. We need to be prepared to solve problems by drawing and measuring. These can be very powerful problem-solving tools and will help us in understanding more abstract mathematics. Right triangle trigonometry is a natural extension of these concepts. We will see how right triangle trigonometry can be used to solve many of the same types of problems more efficiently.

8.1 Similar Triangles

1. a. Using a ruler and protractor, draw a triangle that is twice as big as the one shown here.

  b. Using a ruler and protractor, draw a triangle that is three-fifths the size of the one pictured in the figure.

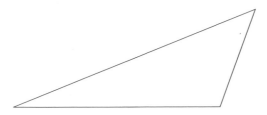

2. a. Calculate the ratios of corresponding sides between the first triangle you drew and the original triangle. Is the ratio what you expected?

  b. What do you predict the ratios of corresponding sides to be between the second triangle you drew and the original? Check your prediction.

In Activity 1, you drew triangles that were similar to the original triangle. Intuitively, being similar means that the new triangles have the same shape but can be a different size.

We can talk about similar figures of any shape. Informally, a similar figure is an enlargement or reduction of the original figure. We focus on similar triangles in this section. Some properties of similar triangles are not true for other shapes.

> Definition ——————————————————————————
>
> A pair of triangles are called **similar triangles** if and only if the measures of each of the angles of one triangle are equal to the measures of each of the angles in the other triangle. The angles whose measures are equal are called **corresponding angles.**

> Definition ——————————————————————————
>
> In a pair of similar triangles, the sides opposite corresponding angles are called **corresponding sides.**

From the definition, if two triangles are similar, then their corresponding angles are equal, and if corresponding angles are equal, then the triangles are similar.

Some notation makes it easier to talk about similar triangles. It is a common convention to label the vertices of a triangle with capital letters and the sides opposite the angles with the corresponding lowercase letters. This is shown in the following triangles.

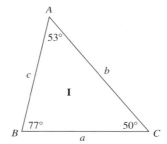
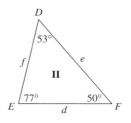

In talking about triangles I and II, we can say that $\triangle ABC \sim \triangle DEF$. The symbol "$\sim$" is read as "is similar to." The order in which the vertices of the triangles are given indicates corresponding angles in the two triangles. Therefore, by saying $\triangle ABC \sim \triangle DEF$ we mean that $\angle A$ and $\angle D$ are corresponding angles, $\angle B$ and $\angle E$ are corresponding, and $\angle C$ and $\angle F$ are corresponding. Once we know corresponding angles, we automatically know corresponding sides.

Because the measure of angle A, which can be written symbolically as $m\angle A$, is equal to $m\angle D$, vertex A corresponds to vertex D and side a corresponds to side d. Likewise, because $m\angle B = m\angle E$, vertex B corresponds to vertex E and side b corresponds to side e. Because $m\angle C = m\angle F$, vertex C corresponds to vertex F and side c corresponds to side f. For the triangles shown, we can also say that $\triangle BAC \sim \triangle EDF$. Can we say $\triangle ACB \sim \triangle DEF$? Is $\triangle CAB \sim \triangle FDE$?

You should have found that the ratios of the sides of the first triangle you drew in Activity 1 to the given triangle were all approximately 2. This ratio is sometimes called a **scale factor.** In this case, it indicates that each side in the triangle that you drew is approximately twice as long as the corresponding side in the original triangle. What was the scale factor you identified for the second triangle you drew and the original triangle?

The following box summarizes the properties of similar triangles.

> *Properties of Similar Triangles*
> 1. Similar triangles have equal angles, and, conversely, triangles with equal angles are similar.
> 2. The ratio of corresponding sides in similar triangles is a constant. Conversely, if the ratios of the corresponding sides are constant, the triangles are similar.

Example 1

In the following figures $\triangle GHJ \sim \triangle KPN$. Find the lengths of the missing sides.

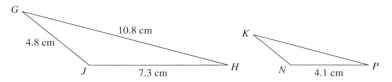

Solution   Because the two triangles are similar, we know that the ratios of corresponding sides are equal. The following figures show the corresponding sides visually differentiated by the way the lines are drawn. You can do the same by using different colors to draw the lines.

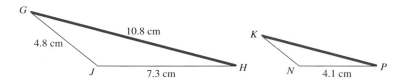

Side *NP* corresponds to side *JH*, and side *KN* corresponds to *GJ*. Therefore,

$$\frac{NP}{JH} = \frac{KN}{GJ}$$ Ratios must be set up so that the relationships on each side are consistent. This ratio is

$$\frac{\text{middle side of } \triangle KPN}{\text{middle side of } \triangle GHJ} = \frac{\text{shortest side of } \triangle KPN}{\text{shortest side of } \triangle GHJ}$$

$$\frac{4.1 \text{ cm}}{7.3 \text{ cm}} = \frac{KN}{4.8 \text{ cm}}$$ Substitute the values that we know.

$$\frac{4.1}{7.3} = \frac{KN}{4.8 \text{ cm}}$$ The units on the left cancel.

$$4.8 \text{ cm} * \frac{4.1}{7.3} = KN$$ Multiply both sides by 4.8 cm.

$$2.7 \text{ cm} \approx KN$$

In the fourth step, we also could have cross multiplied and then solved for *KN*. When the variable we are solving for is in the numerator, it is usually easier to just undo the division by multiplying.

Similarly,

$$\frac{NP}{JH} = \frac{KP}{GH}$$ This ratio is

$$\frac{\text{middle side of } \triangle KPN}{\text{middle side of } \triangle GHJ} = \frac{\text{longest side of } \triangle KPN}{\text{longest side of } \triangle GHJ}$$

$$\frac{4.1 \text{ cm}}{7.3 \text{ cm}} = \frac{KP}{10.8 \text{ cm}}$$ Substitute the values that we know.

$$\frac{4.1}{7.3} = \frac{KP}{10.8 \text{ cm}}$$ The units on the left cancel.

$$10.8 \text{ cm} * \frac{4.1}{7.3} = KP$$ Multiply both sides by 10.8 cm.

$$6.1 \text{ cm} \approx KP$$

We conclude that *KN* is approximately 2.7 cm, and *KP* is approximately 6.1 cm.

Example 2

a. In the following figure, segment *DE* is parallel to segment *CB*. Show that triangle *ABC* is similar to triangle *AED*, that is, △*ABC* ~ △*AED*.

b. Find the length of segment *CD* if *DE* = 4 ft, *CB* = 6 ft, and *AD* = 5 ft.

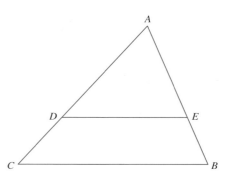

Solution

a. If we can show that the measures of corresponding angles in the two triangles are equal, then we know that the triangles are similar.

We can see that ∠*A* in △*ABC* corresponds to ∠*A* in △*AED*. Clearly, $m\angle A = m\angle A$. Because *DE* is parallel to *CB*, ∠*ACB* and ∠*ADE* are corresponding angles. Therefore, $m\angle ACB = m\angle ADE$. Similarly, because *DE* is parallel to *CB*, ∠*ABC* and ∠*AED* are corresponding angles. Therefore, $m\angle ABC = m\angle AED$. Because the measures of each of the angles of one triangle are equal to the measures of each of the corresponding angles of the other triangle, we conclude that △*ABC* ~ △*AED;* the two triangles are similar. Notice that we talk about corresponding angles when a transversal cuts parallel lines, and we also talk about corresponding angles in similar triangles. The context should make it clear which we mean.

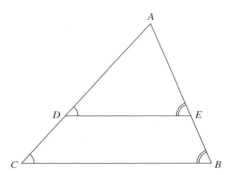

b. The figure following shows the corresponding angles marked and the given lengths labeled. Angle *A* corresponds to itself.

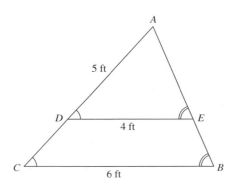

Separating the two similar triangles can make it easier to see corresponding sides. Again corresponding sides are indicated with the same style line.

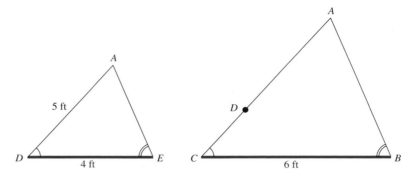

Because the two triangles are similar, we know that ratios of corresponding sides are equal.

$$\frac{CB}{DE} = \frac{AC}{AD}$$

$$\frac{6 \text{ ft}}{4 \text{ ft}} = \frac{AC}{5 \text{ ft}}$$

$$5 \text{ ft} * \frac{6}{4} = AC$$

$$7.5 \text{ ft} = AC$$

The length we want is *CD*. Because *CD* = *AC* − *AD*,

$$CD = 7.5 \text{ ft} - 5 \text{ ft} = 2.5 \text{ ft}$$

We conclude that the length of *CD* is 2.5 ft.

In this section, we were introduced to **similar triangles.** We saw how to label similar figures to make it easier to identify the **corresponding angles and sides.** Similar triangles were used to solve problems.

We used previous knowledge of angle relationships in showing that pairs of angles in triangles are equal. This in turn can help us show that the triangles are similar. If two triangles are similar, we can form proportions using pairs of corresponding sides. As long as we know three of the lengths in a proportion, we can solve for the unknown length.

Problem Set 8.1

1. In the following figure, $\triangle ABC \sim \triangle TUV$. Find the lengths of the missing sides.

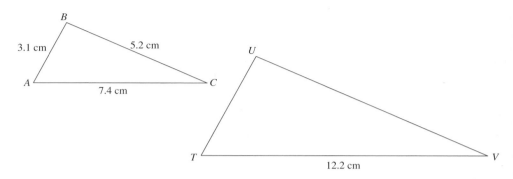

2. In the following figure, $\triangle DEF \sim \triangle NPL$. Find the lengths of the missing sides.

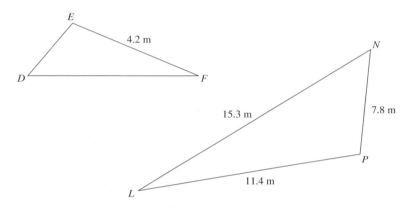

3. In the following figure, $\triangle ABC \sim \triangle DEF$. Find the missing sides.

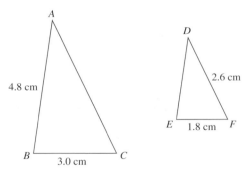

4. Given that segment *AE* is parallel to segment *DC*, identify the similar triangles in the following figure. Indicate corresponding sides by the way you name the triangles. Explain how you know the triangles are similar.

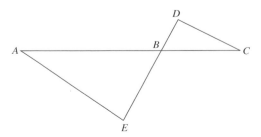

5. a. Which of the following triangles are similar? Explain your reasoning.

   b. Determine the length of side *GI*.

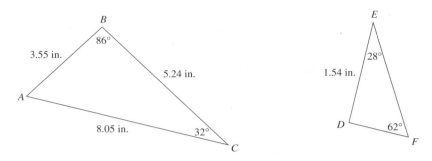

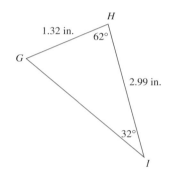

6. If *CD* is parallel to *AB*, find the length of *CD*. When using similar triangles, first explain how you know the triangles are similar.

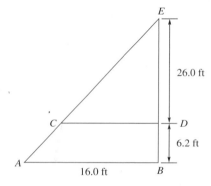

7. The Flagpole. A 5-foot-8-inch person standing beside a flagpole casts a shadow of 12 feet. If the flagpole casts a shadow of 27 feet, use similar triangles to determine the height of the flagpole.

8. a. In the following figure, $m\angle W = m\angle Z = 85°$. Determine if the two triangles are similar. Explain how you decided.

   b. If *VW* = 0.94 inch, *WX* = 1.25 inch, *XY* = 0.95 inch, and *YZ* = 0.69 inch, find the lengths of the missing sides.

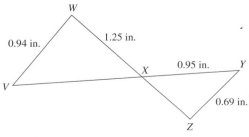

9. Identify the three similar triangles in the following drawing. Explain how you know these triangles are similar. Indicate by the way you name them which sides and angles are corresponding.

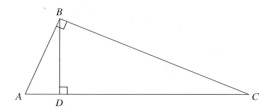

10. a. Which of the following triangles are similar? Explain your reasoning.

    b. Determine the length of sides *WX* and *NM*.

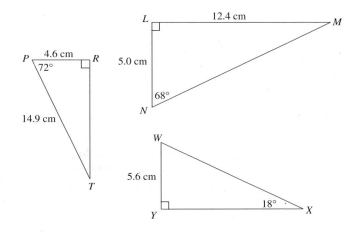

11. In the following figures, $\triangle GHJ \sim \triangle MNK$. Find the lengths of the missing sides.

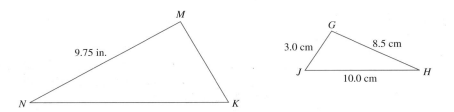

12. In the following figure, *QR* is parallel to *MP*. If you know that *QR* = 9.5 meters, *NR* = 8.2 meters, and *RP* = 3.8 meters, find the length of *MP*. When using similar triangles, first explain how you know the triangles are similar.

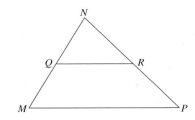

13. Width of a River.   Chris and Sam are trying to determine the width of a river that runs between their two properties. They have devised a plan to accomplish this. Chris is on one side of the river and measures lengths *AB* and *BC*. Sam is on the other side of the river and measures length *DE*. Explain how they can use these measurements to determine the width of the river.

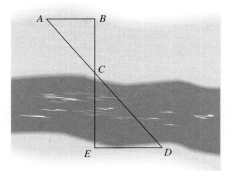

14. Width of a River, Revisited.   Suppose that Chris needs to figure the width of the river without Sam's help. He is only able to make measurements on one side of the river. After giving this some thought, Chris devises a second plan. Chris set markers at points *A*, *B*, *C*, and *D* as shown in the diagram. Side *BD* is parallel to *CE*. Chris measures and finds *AB* = 4.3 m, *BC* = 15.2 m, and *BD* = 2.6 m.

a. Chris's plan involves similar triangles. Explain which triangles are similar and why.

b. Determine the width of the river.

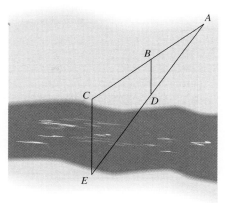

15. The Downgrade.   There is a sign on highway 26 returning from Mt. Hood that says "6% downgrade for the next 5 miles."

a. What is the slope of the road?

b. What is the elevation change in that 5 miles?

5 miles

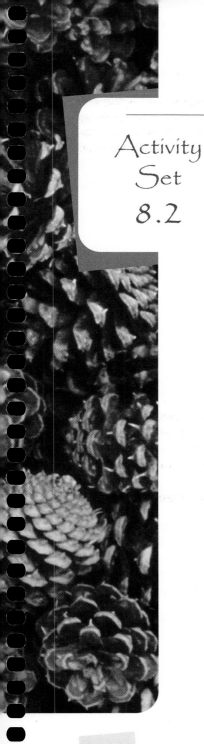

8.2 Scale Drawings

1. Attempting to find the height of the flagpole in front of the college, a team of students measured a distance of 16 meters from the base of the flagpole and determined the angle from their baseline to the tip of the pole to be approximately 36°. The following drawing is not to scale. Using a scale of 1 meter = 0.5 centimeter, create a scale drawing, and use it to approximate the actual height of the pole.

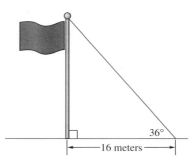

2. To find the height of the flagpole by the football stadium, a second team of students measure a distance of 32.5 feet from the base of the flagpole and determine the angle from their base-line to the tip of the pole to be approximately 54°. Additionally, this flagpole is located on a hill, so they measure the angle between the base of the pole and the hill to be 85° as seen in the following figure. Using a scale of 2.5 feet = 1 centimeter, create a scale drawing, and use it to approximate the actual height of the pole.

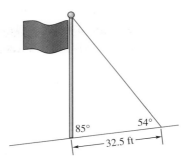

3. To determine the height of a tree, a person stands at a distance 35 feet from the base of the tree and measures the angle of elevation to the top of the tree. The ground between the person and the tree is flat, and the angle of elevation measures 62°. Make a scale drawing to determine the height of the tree.

When we make a scale drawing, our drawing and the original object are similar, that is, they have the same shape but are not the same size. Scale drawings are often used to show objects on paper that would be either too large or too small to be shown their actual size. Blueprints, maps, and detailed pictures of microorganisms are examples of scale drawings.

As we saw in the activities, scale drawings can be used to solve problems when making direct measurements would be difficult. When making a scale drawing, the angle measurements in the drawing are equal to the original angles, and the lengths of the sides are proportional to the original. The ratio of the drawing to the original figure is called the scale ratio or scale factor.

Example 1

Transco Tower. Architects often create a scale model of a building before the actual building is constructed. A model of the Transco Tower in Houston, Texas, was created with a 1-inch to 8-meter scale. If the height of the model was $34\frac{3}{8}$ inches, how tall is the actual tower?

Solution   We can use unit fractions to convert between the scale model measurements and the actual measurements. We use the scale factor like a unit fraction. So in this problem, 1-inch to 8-meter can be written as

$$\frac{1 \text{ in.}}{8 \text{ m}} \quad \text{or} \quad \frac{8 \text{ m}}{1 \text{ in.}}$$

Start with the given information.

$$34\frac{3}{8} \text{ in.}$$
$$= 34\frac{3}{8} \text{ in.} * \frac{8 \text{ m}}{1 \text{ in.}}$$
$$= 275 \text{ m}$$

The height of the Transco Tower is about 275 meters.

Example 2

Welton Family Reunion. The Welton family is having a family reunion at a campsite in the Bitterroot National Forest. They invested in walkie–talkies to keep in contact with some of the younger members of the group who plan to leave the base camp to do some exploring. Yesterday several of the youth were practicing their orienteering skills. Leaving from the base camp they discovered a small fishing lake by going 520 feet along a heading that is 32° to the east of north and then 930 feet due east. While still at the lake, can they hear the dinner call on their walkie–talkies, which have a quarter-mile range?

Solution   First we do a rough sketch of the situation to get an idea of the layout. Then we make a scale drawing to determine the direct distance from the base camp to the lake.

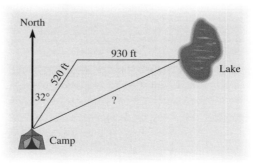

To do the scale drawing, we need to decide on a scale to use. If we choose to let 1 cm represent 100 feet, then 520 feet corresponds to 5.2 cm and 930 feet is represented by 9.3 cm. Using a protractor and ruler, we can draw the following figure.

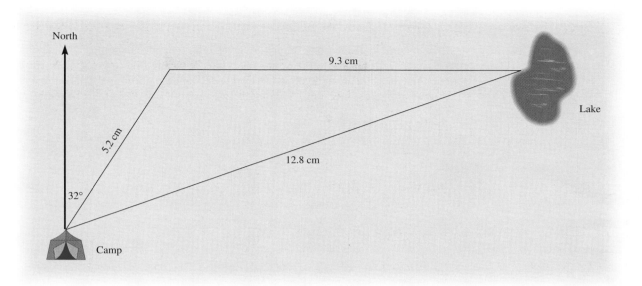

Measuring from the camp to the lake on the map, we find the length to be about 12.8 cm. This corresponds to

$$12.8 \text{ cm} * \frac{100 \text{ ft}}{1 \text{ cm}} = 1280 \text{ feet}.$$

So the exploring group was approximately 1280 feet from the base camp. Because a quarter mile is $\frac{5280 \text{ ft}}{4}$, or 1320 feet, they were close enough to hear the dinner call if the walkie–talkies worked as promised.

Angles in Context

In many problem situations, angles are given with reference to compass directions. The angles are measured with respect to the north–south line. If we are told to go N38°W, we first locate north and then measure an angle of 38° toward west from the north line. This direction is shown in the following figure. Directions given in this way are often referred to as **bearings** or **headings.**

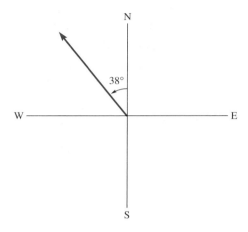

The following figure shows a bearing of S64°E.

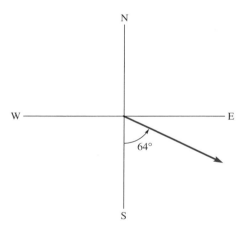

Notice that in both examples, north or south is the direction given first. We start at the north–south line and indicate at what angle toward east or west to head. Bearing angles are often read starting with the angle measurement. For example N38°W can be read as 38° west of north.

Example 3

Outdoor School.   A sixth-grade outdoor school class just learned how to read a compass. They are going to try their new skills on a course that was set up by one of their counselors. Here are the directions that they are to follow.

- Walk 15 feet due east.
- Head N20°E for 24 feet.
- Head N85°E for 10 feet.
- Walk 35 feet due south.
- Head S30°W for 12 feet.
- At this point you should be able to find a marker. Sign your name on the marker, and then head back to where you started for lunch.

Assume that the sixth-graders make it to the marker successfully and return to the starting point in a straight course, how far do they need to walk back to the starting point?

Solution   We start by making a rough sketch of the situation.

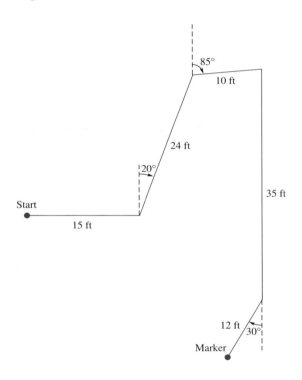

Next, we need to choose a scale. We have some fairly large lengths. The largest length is 35 ft. If we let 1 cm = 2 ft, then 35 feet is 17.5 cm. It is not clear if this will all fit on a page with this scale. Therefore, let's choose our scale to be 1 cm = 3 ft. Then we can convert all of our lengths into centimeters.

$$15 \text{ ft} * \frac{1 \text{ cm}}{3 \text{ ft}} = 5 \text{ cm} \qquad 35 \text{ ft} * \frac{1 \text{ cm}}{3 \text{ ft}} \approx 11.7 \text{ cm}$$

$$24 \text{ ft} * \frac{1 \text{ cm}}{3 \text{ ft}} = 8 \text{ cm} \qquad 12 \text{ ft} * \frac{1 \text{ cm}}{3 \text{ ft}} = 4 \text{ cm}$$

$$10 \text{ ft} * \frac{1 \text{ cm}}{3 \text{ ft}} \approx 3.3 \text{ cm}$$

With these lengths we can now make our scale drawing. Notice that we must indicate our scale with our drawing.

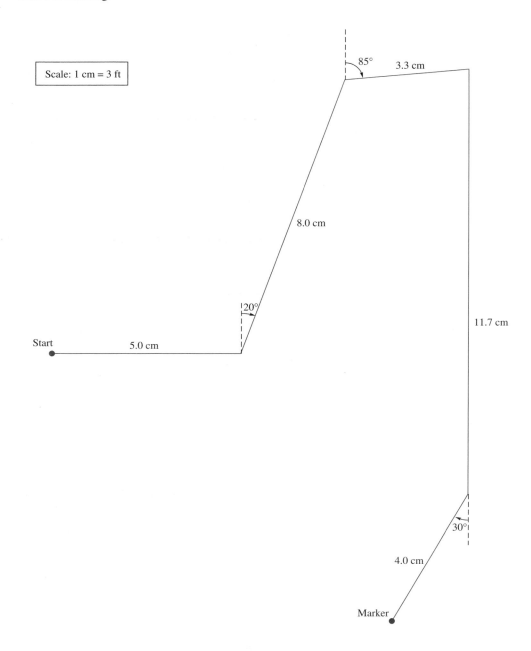

Scale: 1 cm = 3 ft

Finally, we can measure the distance between the marker and the start. Then, convert this distance to feet.

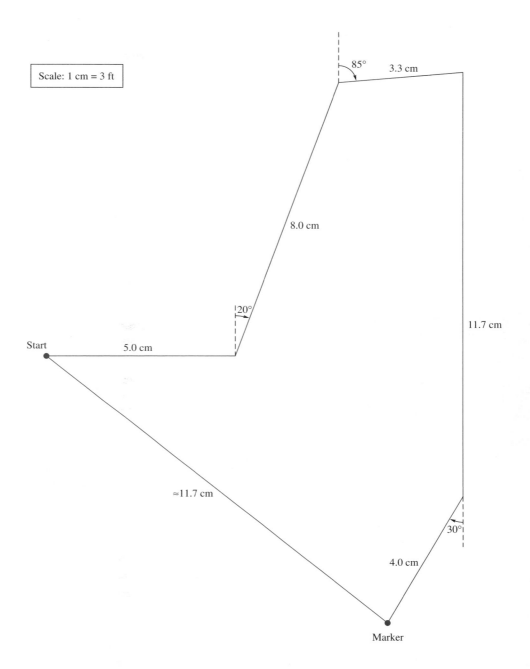

The distance measures approximately 11.6 cm.

$$11.7 \text{ cm} * \frac{3 \text{ ft}}{1 \text{ cm}} = 35.1 \text{ ft}$$

We conclude that they must walk about 35 feet back to the start.

Angles can also be measured from a horizontal reference line. An angle measured upward from the horizontal is called an **angle of elevation** or **angle of inclination**. An angle measured downward from the horizontal is called an **angle of depression** or an **angle of declination**.

In this section, we explored ways of using scale drawings to solve problems. As you worked on some of these problems with others in your class, you may have noticed that answers did not always match exactly. Any time we rely on our abilities to measure and draw accurately, we must expect some error due to lack of precision.

Problem Set 8.2

1. A right triangle has legs of lengths 62 cm and 45 cm.

   a. Determine the length of the hypotenuse.

   b. Using a scale drawing, approximate the measures of the angles of the triangle.

2. **The Bowl.**   An ad in the Sunday paper shows a bowl for sale. The bowl is not shown actual size. What is the scale factor? How wide is the actual bowl across the top?

Shown smaller than actual size of $6\frac{1}{2}$ in. high.

3. The following blueprint is drawn with a scale of $\frac{1}{8}$ inch to 1 foot. In the blueprint, the bold lines indicate windows and the two line segments and arcs represent doors. Determine the actual area and perimeter of the room. Compute the entire perimeter of the room, including the doors.

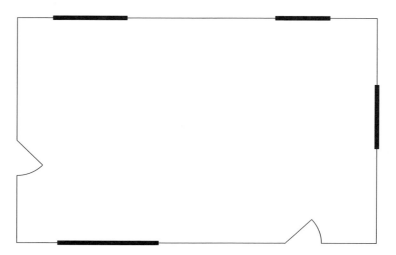

4. The Photograph. Clara has a 4″ × 6″ photograph that she wants to enlarge to fit a 8″ × 11″ frame. Explain why a simple enlargement does not fit. Describe how you might enlarge the picture and cut the edges so that it will fit the frame.

5. The Copy Machine.   The photocopier in your office enlarges or reduces copies. What are the dimensions of the image of a figure 3 inches by 5 inches if the copier instructions are set at 127%? At 69%?

6. Use careful drawings to determine the measures of the missing angles and sides in triangles a–f.

a.

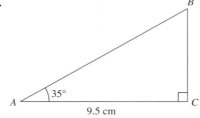

b.

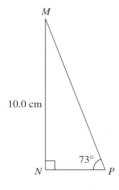

c.

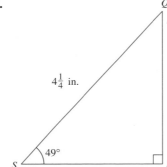

d.

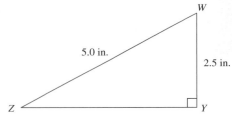

e.

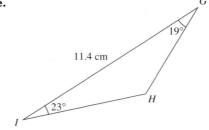

f.

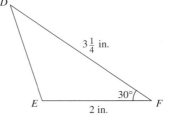

7. Use scale drawings to determine the lengths of the missing sides in the figures.

a.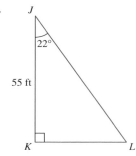

b.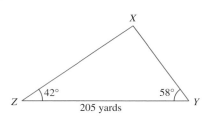

8. The scale of a map of the United States is 1 inch: 200 miles. Explain why the ratio between the distances on the map and the actual distances is *not* $\frac{1}{200}$.

9. Camera Distance.   From the top of a building a photographer is focusing on a point on the street below. The angle of depression between the camera and the point is 36°. The height of the building is about 43 feet. Use a scale drawing to determine the distance between the point and the camera. Ignore the height of the photographer.

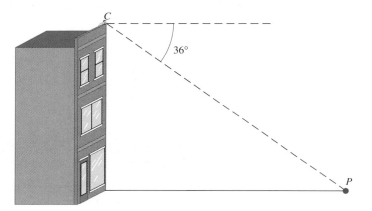

10. Distance Across Another River.   To determine the distance across a river, a group of people sited a rock directly across the river. Then, they hiked along the river approximately 12 meters to a second point and measured the angle between the path they hiked and the rock as seen in the following figure. This angle measured 40°. Use a scale drawing to determine the width of the river.

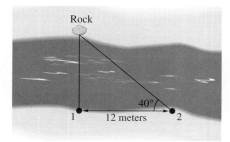

11. Rocket Launch.   Doug plans to launch a model rocket in a field. He is sitting a safe distance of 20 yards from the rocket when it takes off straight upward. He estimates that the angle of elevation from his eye level to the highest point the rocket reaches is about 75°. How high did the rocket fly? (*Note:* The rocket flies fairly high and the angle of elevation is just a quick estimate; therefore, we can safely ignore the height of Doug's eye level in computing the height of the rocket.)

12. Second River.   Along the shore of a river, from a cliff at a height of 310 feet above the river, the angle of depression to the closest point on the opposite shore is 22°. The eye level of the person finding the angle of depression is 5 feet. What is the width of the river?

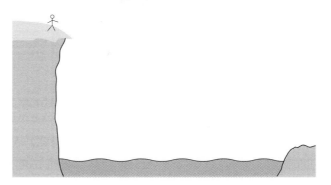

13. The Glider.   If a glider is soaring at an altitude of 7500 meters and the angle of depression from the glider to the control tower is 15°, approximate the horizontal distance, in kilometers, from the glider to a point directly over the tower. (*Note:* The height of the tower is negligible given the size of the other measurements.)

14. Road Construction.   In road construction the following terminology is used.

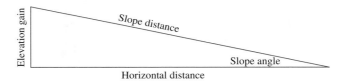

A slope distance of 625 ft. is measured between two points that differ in elevation by 23.0 ft. Approximate the slope angle and the horizontal distance between the two points.

15. How Steep?   A road rises 375 feet in elevation in a mile as measured on the road itself. What is the approximate measure of the angle between the road and the horizontal? (*Note:* This is called the slope angle.)

16. Laying Drainage Pipe.   Find the approximate length of drainage pipe needed to follow the path *ABCD* shown in the following figure, given that *AB* = 15.1 meters, *MN* = 43.0 meters, and *CD* = 16.2 meters.

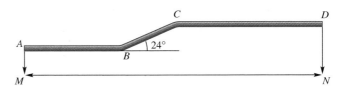

17. Area of a Piece of Land.   A surveyor stands at a compass station at the SW corner of a piece of land that is known to be a true rectangle. The west side of the land measures 58 chains and runs true north from the compass station. The surveyor sights on a flagger at the NE corner and gets a reading of N34°E. What is the approximate area of the piece of land in acres?

18. Orienteering.   A sixth-grade outdoor school class just learned how to read a compass, and they are going to try their skills on a course set up by one of the counselors. Their directions are as follows.

- Walk 10 yards due west.
- Turn to N32°E and proceed for 18 yards.
- Go in a direction of N68°E for 30 yards.
- Walk due west again for 48 yards.
- Go in a direction of N22°W for 18 yards.
- At this point you should be able to find the marker. Sign your name on the marker, and then head back to where you started for lunch.

Assuming they can return to the start by a straight course, how far do they need to hike to get back for lunch? What compass bearing should they take as they go back for lunch?

19. Captain Sam.   Captain Sam is heading into a new port in a fog. Right now she is sitting at the lighthouse, which marks the entrance to the harbor. She cannot find her map of the narrow rocky harbor entrance, so she radios for directions. Her directions are to turn due east into the harbor just after passing the lighthouse. Continue for 1.3 nautical miles, then turn S22°E for 1.4 nautical miles. Next head N50°E for 0.8 nautical miles, turn again to S60°E for 2.6 nautical miles, and then head due north for 2.8 nautical miles. This should bring Captain Sam into the dock. Make a scale drawing of these directions. How far does Captain Sam need to travel in the fog to reach the dock? How far did the radio signal need to travel, assuming it was sent from the dock?

8.3 Right Triangle Trigonometry

Activity Set 8.3

1. In Section 8.2, you solved the following problem by making a scale drawing.

   Height of a Tree.   To determine the height of a tree, a person stands at a distance 35 feet from the base of the tree and measures the angle of elevation to the top of the tree. The angle measures 62°. Make a scale drawing to determine the height of the tree.

   In this section, we see how this problem can be solved in a different way.

   a. Each team member needs to draw a triangle whose angles measure 90°, 62°, and 28°. Have one person in your team draw a small triangle, one a medium-size triangle, and one a large triangle.

   b. Measure the sides of your triangle, and compute the following ratios.

   $$\frac{\text{leg opposite the } 62° \text{ angle}}{\text{hypotenuse}}$$

   $$\frac{\text{leg adjacent } 62° \text{ angle}}{\text{hypotenuse}}$$

   $$\frac{\text{leg opposite } 62° \text{ angle}}{\text{leg adjacent } 62° \text{ angle}}$$

   c. Compare your ratios with those of your teammates. Are the ratios the same? Are they close? Should they be exactly the same? Why or why not?

2. Can any of the ratios that you wrote in Activity 1 be used to solve the original problem, "Height of a Tree"?

Discussion 8.3

In this section, we will be looking at three trigonometric ratios: the sine ratio, the cosine ratio, and the tangent ratio. Before we begin, we must first discuss some function notation.

One common formula is for the area of a circle, $A = \pi r^2$. In this formula, the area of a circle depends on the circle's radius. Another way to say this is that the area of a circle is a function of its radius. A notation used to indicate this dependence is called **function notation.** Instead of simply writing $A = \pi r^2$, we write $A(r) = \pi r^2$. We read $A(r)$ as "A of r."

The notation $A(r)$ is used to indicate that r is the independent variable and A is dependent on r (or is a function of r). In other words, r is the input variable into the function. For example, $A(4)$ represents the area of a circle when the radius is 4. Therefore, to find the value of $A(4)$ we need to substitute $r = 4$ into the function.

$$A(4) = \pi * 4^2$$
$$\approx 50$$

Similarly, $A(10)$ represents the area of a circle when the radius is 10. That is,

$$A(10) = \pi * 10^2$$
$$\approx 314$$

Therefore, $A(10) \approx 314$.

NOTE: This notation can be confusing. In some contexts, $A(r)$ represents the product of A and r. However, in this context when A is the dependent variable of the function, the notation $A(r)$ indicates that the area is a function of the radius rather than indicating the product of A and r.

In the activities, we looked at ratios of sides in similar triangles. Because corresponding sides of similar triangles are proportional, we know that the ratios of the sides of similar triangles are constant. To make this more concrete, consider angle A, which measures 62°, in the following triangle. Then, the ratio $\frac{\text{leg opposite 62° angle}}{\text{hypotenuse}}$ is constant for any right triangle containing a 62° angle because any right triangle containing this angle is a similar triangle. Can you explain why?

Likewise, the ratio $\frac{\text{leg adjacent to angle } A}{\text{hypotenuse}}$ is constant for any right triangle containing angle A and the ratio $\frac{\text{leg opposite angle } A}{\text{leg adjacent to angle } A}$ is constant.

For this reason, given any right triangle, these three ratios are given the names sine, cosine, and tangent, respectively. All three of these ratios depend on a given acute angle; therefore, we use function notation to refer to these ratios. The ratios are abbreviated as follows.

sine of angle A	is abbreviated	$\sin(A)$
cosine of angle A	is abbreviated	$\cos(A)$
tangent of angle A	is abbreviated	$\tan(A)$

The definitions of these ratios are summarized in the following box.

Trigonometric Ratios

Angle A is an acute angle in a right triangle.

$$\sin(A) = \frac{\text{length of leg opposite angle } A}{\text{length of hypotenuse}} = \frac{\text{opp}}{\text{hyp}}$$

$$\cos(A) = \frac{\text{length of leg adjacent to angle } A}{\text{length of hypotenuse}} = \frac{\text{adj}}{\text{hyp}}$$

$$\tan(A) = \frac{\text{length of leg opposite angle } A}{\text{length of leg adjacent to angle } A} = \frac{\text{opp}}{\text{adj}}$$

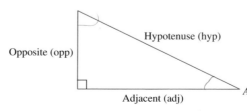

Notice that the *input* for sine, cosine, and tangent is the *measure of an angle* and that the *output* is a *ratio*. Because the output is a ratio of lengths whose units must be the same, the output has no units.

By drawing a right triangle with an acute angle of 62° in the activities, we approximated the ratio $\frac{\text{opp}}{\text{adj}}$ to be about 1.8. Because this ratio, by definition, is tan(62°), we find that tan(62°) ≈ 1.8. However, our scale drawings lack precision. We can use our calculator to obtain a more accurate value. First, be sure that your calculator is in degree mode. Then, rounded to the nearest ten-thousandth, tan(62°) ≈ 1.8807.

Example 1

Use the following figure to determine each ratio.

a. sin(A)

b. sin(B)

c. cos(B)

d. tan(A)

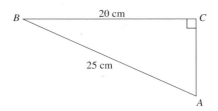

Solution   a. The sine ratio is $\frac{\text{opp}}{\text{hyp}}$. The side opposite angle A measures 20 cm. The hypotenuse measures 25 cm. Therefore,

$$\sin(A) = \frac{20 \text{ cm}}{25 \text{ cm}} = \frac{4}{5}$$

b. In part b, we are looking for sin(B). From the figure, we are not given the length of the side opposite angle B. So, first we can use the Pythagorean theorem to find the length of side AC.

$$(20 \text{ cm})^2 + AC^2 = (25 \text{ cm})^2$$

$$AC^2 = (25 \text{ cm})^2 - (20 \text{ cm})^2$$

$$AC = \sqrt{(25 \text{ cm})^2 - (20 \text{ cm})^2}$$

$$AC = 15 \text{ cm}$$

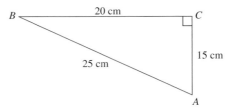

With this side length, we can now find the sine of angle B.

$$\sin(B) = \frac{15 \text{ cm}}{25 \text{ cm}} = \frac{3}{5}$$

c. $\cos(B) = \dfrac{\text{adj}}{\text{hyp}} = \dfrac{20 \text{ cm}}{25 \text{ cm}} = \dfrac{4}{5}$

d. $\tan(A) = \dfrac{\text{opp}}{\text{adj}} = \dfrac{20 \text{ cm}}{15 \text{ cm}} = \dfrac{4}{3}$

Example 2

Determine the lengths of the missing sides in the following triangles.

a.

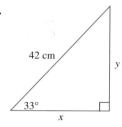

b.

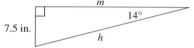

Solution   a. Because we are given the length of only one side in the triangle, we cannot use the Pythagorean theorem. However, we also know one of the acute angles, so we can use a trigonometric ratio. We want to choose a ratio that involves only one of the unknown lengths because this allows us to solve for that length.

Because the hypotenuse is the known side, we can use either the sine or the cosine ratio. If we start with the 33° angle, then y is opposite the angle and x is adjacent to the angle. Therefore, we can write

$$\sin(33°) = \frac{y}{42 \text{ cm}} \qquad \text{or} \qquad \cos(33°) = \frac{x}{42 \text{ cm}}$$

Let's start with the left equation and solve for y.

$$\sin(33°) = \frac{y}{42 \text{ cm}}$$

$42 \text{ cm} * \sin(33°) = y$ Multiply both sides by 42 cm.

$23 \text{ cm} \approx y$ Evaluate using calculator.

At this point, we can find x either by using the other equation we wrote or by using the Pythagorean theorem. Let's use our previous equation.

$$\cos(33°) = \frac{x}{42 \text{ cm}}$$

$42 \text{ cm} * \cos(33°) = x$ Multiply both sides by 42 cm.

$35 \text{ cm} \approx x$ Evaluate using a calculator.

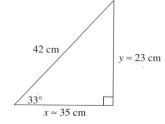

Are these values for x and y reasonable? From the drawing they do not appear reasonable; however, the triangle may not be drawn to scale. Therefore, we should analyze this further. We know that the side opposite a smaller angle should be shorter than the side opposite the larger angle. In addition, because one acute angle measures 33°, the other must be 57°. Putting these two pieces of information together, the leg opposite the 33° angle should be shorter than the other leg, and it is. Therefore, the values for x and y are reasonable.

b. We can find the missing lengths in the second triangle in a similar manner. However, to add an alternative, we do not need to use the given acute angle. In a right triangle, if we know one acute angle we can find the other. The sum of the two acute angles in a right triangle must be 90°. Therefore, the angle opposite the side labeled m measures $90° - 14° = 76°$.

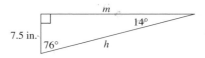

If we use the 14° angle, then side m is the leg adjacent to the angle. If we use the 76° angle, then side m is opposite the angle. Let's use the 76° angle. Because m is opposite the 76° angle and the 7.5 inches is adjacent to this angle, we need to use the tangent ratio.

$$\tan(76°) = \frac{m}{7.5 \text{ in.}}$$

$$7.5 \text{ in.} * \tan(76°) = m \qquad \text{Multiply both sides by 7.5 in.}$$

$$30.1 \text{ in.} \approx m \qquad \text{Evaluate using calculator.}$$

At this point, we again have two choices. To find h we can use the Pythagorean theorem or we can use a different ratio. Because h is the hypotenuse, we can use either the cosine or the sine ratio.

$$\cos(76°) = \frac{7.5 \text{ in.}}{h}$$

$$h * \cos(76°) = 7.5 \text{ in.} \qquad \text{Multiply both sides by } h.$$

$$h = \frac{7.5 \text{ in.}}{\cos(76°)} \qquad \text{Divide both sides by cos (76°).}$$

$$h \approx 31.0 \text{ in.}$$

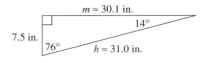

Are these values for h and m reasonable? Because a 14° angle is very small, the length of the side opposite the 14° angle should be small, relative to the other sides, and it is. The hypotenuse is always the longest side. And finally, the length of the leg adjacent to the 14° angle and the hypotenuse are somewhat close to the same because the 14° angle is so small. We can also check these values in the Pythagorean theorem.

$$(7.5 \text{ in.})^2 + (30.1 \text{ in.})^2 \overset{?}{\approx} (31.0 \text{ in.})^2$$

$$962 \text{ in.}^2 \overset{?}{\approx} 961 \text{ in.}^2 \qquad ✔$$

Example 3

In the activities in Section 8.2, you encountered the following problem. Use a trigonometric ratio to find the height of the pole.

Attempting to find the height of the flagpole in front of the college, a team of students measured a distance of 16 meters from the base of the flagpole and determined the angle from their baseline to the tip of the pole to be approximately 36°.

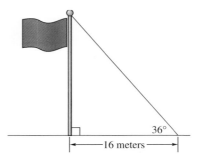

Solution   Let *h* represent the height of the flagpole. Because *h* is opposite the 36° angle and the 16 meters is adjacent to the angle, we can use the tangent ratio.

$$\tan(36°) = \frac{h}{16 \text{ m}}$$

$$16 \text{ m} * \tan(36°) = h$$

$$12 \text{ m} \approx h$$

The flagpole is about 12 meters tall.

Does the solution to Example 3 agree with what you found when you drew the situation to scale? In general, you can verify your solution to these kinds of problems by creating a scale drawing.

Example 4

A Bullet in the Wall.   A bullet is found embedded in the wall of a room. The bullet entered the wall 0.55 meters above the floor, going downward at an angle of 8°, that is, at an angle of depression of 8°. If the gun was held at a height of 4 ft. 3 in., how far from the wall was the gun fired?

Solution   Let's draw a picture of the situation.

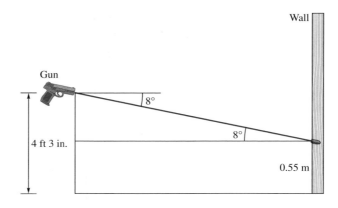

Next, 4 ft. 3 in. is the same as 4.25 ft and 0.55 m is approximately 1.80 ft. Therefore, the length of the side opposite the 8° angle is the difference between 4.25 ft and 1.80 ft, that is, 2.45 ft. Because we know the length of the side opposite the 8° angle and we are looking for the length of the side adjacent to the angle, we can use the tangent ratio to solve the problem. Let D represent the distance from the wall to the gun. Then,

$$\tan(8°) = \frac{2.45 \text{ ft}}{D}$$

$D * \tan(8°) = 2.45 \text{ ft}$ Multiply both sides by D.

$$D = \frac{2.45 \text{ ft.}}{\tan(8°)}$$ Divide both sides by $\tan(8°)$.

$D \approx 17.4 \text{ ft}$

We conclude that the gun was approximately 17.4 feet from the wall when it was fired.

In the discussion in Section 8.2, we solved the following problem using a scale drawing. Let's revisit the problem and see how we can use trigonometric ratios to answer the question.

Example 5

Can You Hear Me, Revisited.   The Welton family is having a family reunion at a campsite in Bitterroot National Forest. They invested in walkie–talkies to keep in contact with some of the younger members of the group who plan to leave the base camp to do some exploring. Yesterday several of the youth were practicing their orienteering skills. From camp, they discovered a small fishing lake by going 520 feet along a heading that is 32° to the east of north and then 930 feet due east. While at the lake, could they hear the dinner call on their walkie–talkies that have a quarter-mile range?

Solution   If we can determine the vertical and horizontal distances that the lake is from camp, we can use the Pythagorean theorem to determine the distance between the camp and the lake.

We start by drawing extra lines in our figure to create right triangles. We know that line AC is at a bearing of N32°E. We can use this to determine the measure of $\angle ACB$.

$32° + m\angle ACB = 90°$

$m\angle ACB = 58°$

Next, we can use the sine function to determine the length of AB.

$$\sin(58°) = \frac{AB}{520 \text{ ft}}$$

$520 \text{ ft} * \sin(58°) = AB$

$441 \text{ ft} \approx AB$

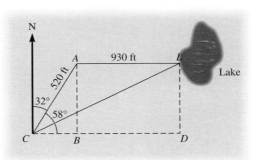

Similarly, we can use the cosine function to determine the length of *BC*.

$$\cos(58°) = \frac{BC}{520 \text{ ft}}$$

$$520 \text{ ft} * \cos(58°) = BC$$

$$276 \text{ ft} \approx BC$$

With these distances and those given in the problem, we can determine the length of *CD*. We know that the length of *CD* is the sum of the length of *CB* and the length of *BD*.

$$CD = CB + BD$$

$$CD \approx 276 \text{ ft} + 930 \text{ ft}$$

$$CD \approx 1206 \text{ ft}$$

The length of *DL* is the same as *AB*. Therefore, the length of *DL* is approximately 441 ft. These distances are labeled in the following figure.

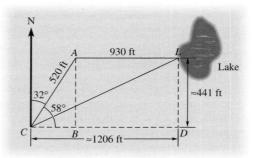

Now we can use the Pythagorean theorem to determine the distance between the camp and the lake.

$$(CL)^2 \approx (1206 \text{ ft})^2 + (441 \text{ ft})^2$$

$$CL \approx \sqrt{(1206 \text{ ft})^2 + (441 \text{ ft})^2}$$

$$CL \approx 1280 \text{ ft}$$

To the nearest 10 feet, the distance between the camp and the lake is 1280 feet. Because 1 mile is 5280 ft, one-quarter mile is 1320 ft. Therefore, the youths were able to hear their dinner call from the lake.

Slope is used to express the rise and fall in roadbeds, drainage areas, and so on.

Definition

Percent slope is defined to be the slope expressed as a percent. For example, a slope of $\frac{3}{10}$ can be expressed as a percent slope of 30%.

Example 6

Determine the percent slope of a bank whose angle of inclination is 40°.

Solution   The slope of the bank is equal to the ratio of the vertical change to the horizontal change. The vertical change corresponds to the length of the side opposite the 40° angle and the horizontal change corresponds to the side adjacent to the 40° angle. This ratio is the same as the tangent ratio. Therefore, we can use the tangent ratio to determine the slope of the bank.

$$\tan(40°) = \frac{\text{opposite}}{\text{adjacent}}$$

$$\text{slope of bank} = \tan(40°) \approx 0.839$$

The slope of the bank is approximately 0.839, which means that the percent of slope is approximately 83.9%.

In Section 8.1, we reviewed the ideas and properties of similar triangles. These properties can be used to solve many kinds of problems. In Section 8.2, we solved problems by creating scale drawings. A scale drawing of an object or situation is a figure similar to the original. Scale drawings are excellent problem-solving tools but can be tedious and lack precision. In this section, we found that in any right triangle, three ratios are associated with that triangle. Furthermore, any other right triangle that is similar has the exact same ratios due to the properties of similar triangles. These three ratios are given names: the **sine** function, the **cosine** function, and the **tangent** function. The inputs into these functions are acute angles, and the outputs are ratios. These functions allow us to solve the same types of problems without creating scale drawings.

Problem Set 8.3

1. Use the triangle to determine these trigonometric values.

   a. $\sin(X)$
   b. $\cos(X)$
   c. $\tan(X)$
   d. $\sin(Y)$

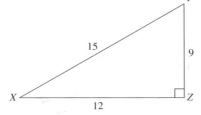

2. **a.** Draw a right triangle and label it ABC with $m\angle A = 90°$. Carefully measure the sides and angles of your triangle.

   b. Use your measurements of the sides to determine $\sin(B)$ and $\cos(B)$ to three significant digits.

   c. Use your angle measurements and a calculator to determine $\sin(B)$ and $\cos(B)$.

   d. Compare your results from parts b and c. Discuss why the results may be slightly different.

3. Use the triangle to determine these trigonometric values. The measure of ∠ACB is 90°.

   a. sin(A)

   b. cos(A)

   c. tan(A)

   d. cos(B)

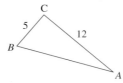

4. In the problem set in Section 8.2, you found the lengths of the missing sides by making a scale drawing. Now determine the value of x using trigonometric ratios. Compare this value to your previous results.

   a.

   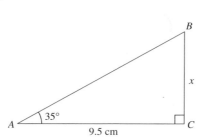

   b.

   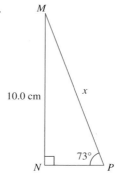

5. In the problem set in Section 8.2, you found the lengths of the missing sides by making a scale drawing. Now determine the lengths of the missing sides using trigonometric ratios. Compare these values to your previous results.

   a.

   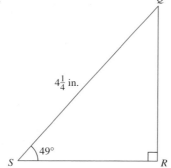

   b.

   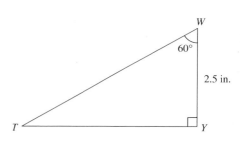

6. Camera Distance, Revisited.   From the top of a building a photographer is focusing on a point on the street below. The angle of depression between the camera and the point is 36°. The height of the building is about 43 feet. Use trigonometric ratios to determine the distance between the point and the camera. Ignore the height of the photographer.

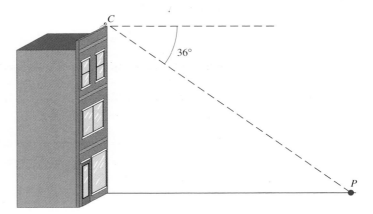

7. Distance Across Another River, Revisited.   To determine the distance across a river, a group of people sited a rock directly across the river. Then they hiked along the river approximately 12 meters to a second point and measured the angle between the path they hiked and the rock as seen in the following figure. This angle measured 40°. Use trigonometric ratios to determine the width of the river.

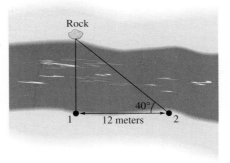

8. Second River, Revisited.   Along the shore of a river, from a cliff at a height of 310 feet above the river, the angle of depression to the closest point on the opposite shore is 22°. The eye level of the person finding the angle of depression is 5 feet. What is the width of the river?

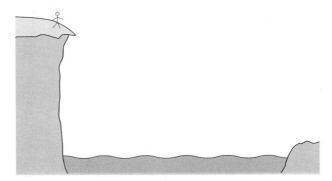

9. How Steep, Revisited.   A road rises 375 feet in elevation in a mile as measured on the road itself. What is the percent slope of the road?

10. The Glider, Revisited.   If a glider is soaring at an altitude of 7500 meters and the angle of depression from the glider to the control tower is 15°, approximate the horizontal distance, in kilometers, from the glider to a point directly over the tower. (*Note:* The height of the tower is negligible given the size of the other measurements.)

11. Determine the measure of an angle whose cosine is $\frac{2}{5}$.

12. Rocket Launch, Revisited.   Doug plans to launch a model rocket in a field. He is sitting a safe distance of 20 yards from the rocket when it takes off straight upward. He estimates that the angle of elevation to the highest point the rocket reaches is about 75°. How high did the rocket fly? (*Note:* The rocket flies fairly high and the angle of elevation is just a quick estimate; therefore, we can safely ignore the height of Doug's eye level in computing the height of the rocket.)

13. Height of a Tree.   To determine the height of a tree, Terry paced 15 feet back from the base. She then used a clinometer to determine that the angle of elevation to the top of the tree was 31.5°. Terry is 5 feet 8 inches tall. How tall is the tree?

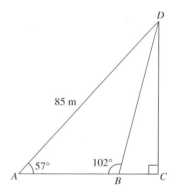

14. a. Determine the lengths of the sides of △*ABD* in the following figure.
   b. Find the area of △*ABD*.

15. The figure is the cross section of a roof. The angle the rafter makes with the joist is 22.5° and the span is 32.0 feet. What is the length of the rafter? What is the rise?

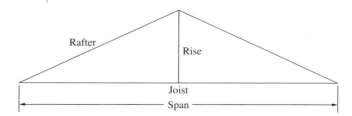

16. Laying Drainage Pipe, Revisited. Find the approximate length of drainage pipe needed to follow the path $ABCD$ shown in the figure, given that $AB = 15.1$ meters, $MN = 43.0$ meters, and $CD = 16.2$ meters.

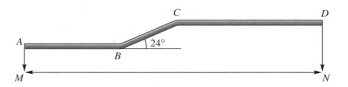

17. a. Determine the area of the following triangle. (*Hint:* Draw a line from R to the opposite side to form two right triangles.)

   b. Determine the perimeter of triangle *TRI*.

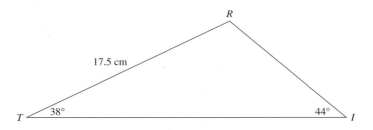

18. Rain Gutter. A rain gutter can be made by taking a 12-inch-wide strip of metal and folding it up at 60° angles 4 inches from the edges into a trapezoidal shape. See the figure.

   a. Determine the cross-sectional area of the rain gutter.

   b. If the gutter is 20 feet long, how many gallons of water can it hold?

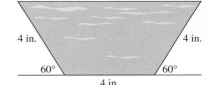

19. The Plane. A plane at an altitude of 10,000 feet is approaching an island. At a given moment the angle of depression to the near side of the island is 55°, and the angle of depression to the far side of the island is 47°. Determine the length of the island, in miles, as seen in the figure.

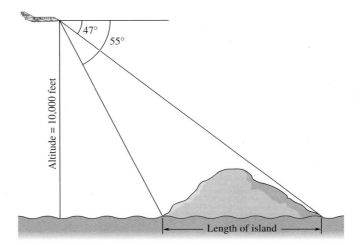

20. Hex-Head Cap Screw.   The distance across the flat of a certain hex-head cap screw is 1.875 inches as shown in the following figure.

    a. What is the width of the head of the screw at its widest point?

    b. The head of the screw is to be recessed into a cylindrical hole. What is the minimum diameter needed for the hole if a socket with a 0.062-inch-thick wall is to be used to tighten the screw?

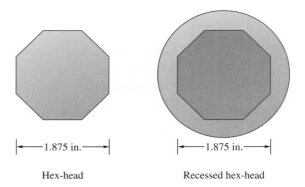

          Hex-head   Recessed hex-head

21. **a.** Draw a line with a positive slope, on the following coordinate axis, that passes through the origin and forms an angle of 65° with the positive horizontal axis.

    b. Determine $\sin(65°)$, $\cos(65°)$, and $\tan(65°)$ to the nearest hundredth.

    c. Carefully estimate the slope of the line that you drew in part a. Use your protractor to determine the perpendicular lines needed to determine the vertical and horizontal change.

    d. Determine which trigonometric ratio corresponds to the slope of the line. Explain why this trigonometric ratio always outputs the slope of a line.

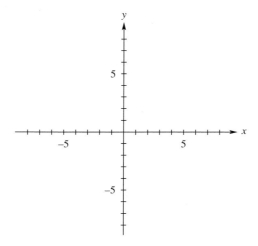

22. Use the partial trigonometric table to answer the following questions. The table is precise to the nearest ten-thousandth.

a. Determine sin(30°), cos(30°), and tan(30°).

b. Determine sin(45°), cos(45°), and tan(45°).

c. Determine sin(75°), cos(75°), and tan(75°).

d. What observations can you make about the three trigonometric ratios as the angle increases from 5° to 85°?

e. Can the sine ratio ever output a number greater than 1? Why or why not?

f. For what angle is $\cos(x) = \frac{1}{2}$? $\tan(x) = 1$?

Angle x (in degrees)	sin(x)	cos(x)	tan(x)
5	0.0872	0.9962	0.0875
10	0.1736	0.9848	0.1763
15	0.2588	0.9659	0.2679
20	0.3420	0.9397	0.3640
25	0.4226	0.9063	0.4663
30	0.5000	0.8660	0.5774
35	0.5736	0.8192	0.7002
40	0.6428	0.7660	0.8391
45	0.7071	0.7071	1.0000
50	0.7660	0.6428	1.1918
55	0.8192	0.5736	1.4281
60	0.8660	0.5000	1.7321
65	0.9063	0.4226	2.1445
70	0.9397	0.3420	2.7475
75	0.9659	0.2588	3.7321
80	0.9848	0.1736	5.6713
85	0.9962	0.0872	11.4301

8.4 Inverse Trigonometry

1. **a.** Individually draw a right triangle in which the ratio of the legs is 2:1.

   b. Determine the measure of the angle opposite the longer leg.

   c. Compare your angle measure with those of your teammates.

2. **a.** In Activity 1 you determined the approximate measure of an angle. Is that measure the same as the measure of any of the angles in the following triangle? Explain why or why not.

   b. What trigonometric ratio of angle *A* is equivalent to the ratio 2:1 in this triangle?

   c. Create a table using the trigonometric ratio you chose for part b. Use this table to find the measure of angle *A* to the nearest tenth of a degree.

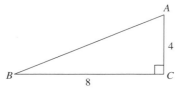

3. Identify the trigonometric ratio that can be used to find the measure of angle *R* in each figure.

a. 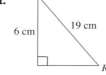   **b.**    **c.**

In Section 8.3 we learned that for a given angle in a right triangle, the sine function outputs a ratio of two sides of the triangle and that this ratio is the same regardless of the size of the triangle. The same holds true for the cosine ratio and tangent ratio. These ratios can be found using a scientific calculator. We then used these ratios to find lengths of sides in right triangles when we knew the measures of the angles and at least one side.

What if we don't know the measures of the acute angles in a right triangle? Is it possible to determine their measures if we know the lengths of the sides? As we saw in Activity 1, we can determine the measure of an angle by drawing the triangle to scale. However, drawing and measuring to determine the measure of angles is time-consuming and not very precise.

Consider the following right triangle whose side lengths are 3 cm, 4 cm, and 5 cm.

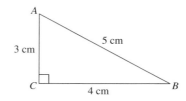

Is there a way to determine the measure of angle B without drawing the triangle to scale? Let's start by writing down a trigonometric ratio using angle B, like $\cos(B) = \frac{4}{5}$. Although at this point we have no technique to solve for B, we can certainly guess and check using our calculator. If we guess that $m\angle B = 45°$, then $\cos(45°) \approx .707$, but $\frac{4}{5} = 0.8$, so we are not close enough. If $m\angle B = 50°$, then $\cos(50°) \approx 0.643$. Because this angle produces a ratio farther away from 0.8, we need to choose an angle smaller than 45°. If $m\angle B = 40°$, then $\cos(40°) \approx 0.766$. We can continue this process until we are satisfied with the precision of our answer. In this case, $\cos(36.8°) \approx 0.801$. Therefore, $m\angle B \approx 36.8°$.

In each of the trigonometric functions, the input is an angle and the output is a ratio. Because we are trying to determine an angle, we want a function whose input is the ratio and whose output is an angle. These functions are the inverses of the trigonometric functions.

For each of the three trigonometric functions, there is a corresponding function that is its inverse. For example, the sine ratio takes an angle as its input and outputs the ratio $\frac{\text{opp}}{\text{hyp}}$. The inverse of the sine ratio is called the inverse sine, or **arcsine**. The arcsine takes the ratio $\frac{\text{opp}}{\text{hyp}}$ as its input and outputs the measure of the angle. Similarly, the inverse of the cosine ratio is the **arccosine** and the inverse of the tangent ratio is the **arctangent**.

The Inverses of the Trigonometric Ratios

$$\arcsin\left(\frac{\text{opp}}{\text{hyp}}\right) = \text{measure of angle } A$$

$$\arccos\left(\frac{\text{adj}}{\text{hyp}}\right) = \text{measure of angle } A$$

$$\arctan\left(\frac{\text{opp}}{\text{adj}}\right) = \text{measure of angle } A$$

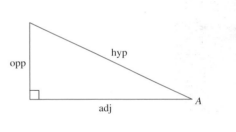

In words,

  arctan(x) produces the measure of the angle whose tangent is x.

  arcsin(x) produces the measure of the angle whose sine is x.

  arccos(x) produces the measure of the angle whose cosine is x.

To reiterate, the *input* for arctangent, arcsine, and arccosine is a *ratio,* and the *output* for these functions is the *measure of an angle.* Because the output for arctangent, arcsine, and arccosine is the measure of an angle, be sure that your calculator is set in degree mode.

On most scientific calculators, the arcsine is denoted by $\sin^{-1}$, the arccosine by $\cos^{-1}$, and the arctangent by $\tan^{-1}$. It is important that we understand that $\tan^{-1}(x)$ is the measure of the angle whose tangent is x and not $\frac{1}{\tan(x)}$. Because this notation can be confusing, we use arctan(x), arcsin(x), and arccos(x) in this text.

$$\tan^{-1}(x) = \arctan(x)$$

$$\tan^{-1}(x) \neq \frac{1}{\tan(x)}$$

Example 1

For each triangle, determine the measures of the angles.

a.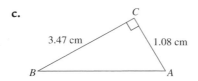

b.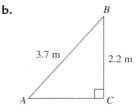

c.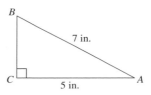

Solution   For each of the triangles, we know that the measure of angle C is 90°. If we can determine the measure of either angle A or B, we can then use the sum of the angles to determine the measure of the third angle.

a. We can take a couple different approaches to solve this problem. We can start by finding either the measure of angle A or B. If we start by finding the measure of angle A, we need to determine the relationship between the known sides and angle A.

The side whose length is 5 inches is adjacent to angle A, and the hypotenuse is 7 inches long. Because the ratio of these sides is the cosine of angle A, we know that $\cos(A) = \frac{5}{7}$. Because we want the measure of the angle, we need to use the arccosine function.

$$\arccos\left(\tfrac{5}{7}\right) = m\angle A$$
$$44.4° \approx m\angle A$$

Next, because the sum of the measures of the angles in a triangle is 180° and $m\angle C = 90°$,

$$m\angle A + m\angle B = 90°$$
$$44.4° + m\angle B \approx 90°$$
$$m\angle B \approx 45.6°$$

As an alternative, we can find the measure of angle B using an inverse trigonometric ratio. In this case, we use the arcsine.

$$\arcsin\left(\tfrac{5}{7}\right) = m\angle B$$
$$45.6° \approx m\angle B$$

Therefore, $m\angle C = 90°$, $m\angle A \approx 44.4°$, and $m\angle B \approx 45.6°$.

b. Again, we can start with either angle A or B. Let's again start with angle A. We know the side opposite angle A, and we know the hypotenuse. Therefore, $\sin(A) = \frac{2.2}{3.7}$. Because we want to find the measure of the angle, we need to use the arcsine function.

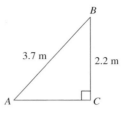

$$\arcsin\left(\frac{2.2}{3.7}\right) = m\angle A$$
$$36.5° \approx m\angle A$$

Then using the sum of the measures of angles in a triangle,

$$m\angle B \approx 90° - 36.5°$$
$$m\angle B \approx 53.5°$$

Therefore, $m\angle C = 90°$, $m\angle A \approx 36.5°$, and $m\angle B \approx 53.5°$.

c. This time let's start with angle B. The known sides are the legs of the triangle. We know the side opposite angle B and the side adjacent to angle B. Therefore,

$$\tan(B) = \frac{1.08}{3.47}$$

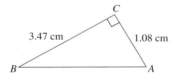

Again, because we want to find the measure of the angle, we need to use the arctangent function.

$$\arctan\left(\frac{1.08}{3.47}\right) = m\angle B$$
$$17.3° \approx m\angle B$$

Then using the sum of the measures of angles in a triangle,

$$m\angle A \approx 90° - 17.3°$$
$$m\angle A \approx 72.7°$$

Therefore, $m\angle C = 90°$, $m\angle A \approx 72.7°$, and $m\angle B \approx 17.3°$.

Example 2

A Bird's Eye View. The following sketch shows the position of a camera that is mounted on a tripod 3.7 feet above the roof it rests on. A reporter is attempting to film a press conference that is taking place 890 feet from the base of the building. Determine the angle of depression, that is, the angle that the camera must be tilted from the horizontal to film the press conference below.

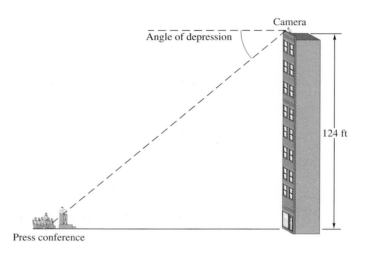

Solution   We assume that the ground is flat and parallel to the line of the horizon. The angle labeled P and the angle of depression are equal because they are alternate interior angles. Therefore, if we can determine the measure of angle P in the triangle, we will know the angle that the camera must be tilted.

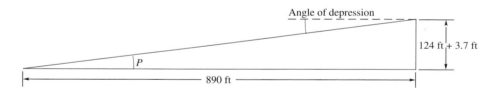

The height from the ground to the camera is 127.7 feet and is opposite $\angle P$. The distance from the press conference to the building is 890 feet and is adjacent to $\angle P$. Therefore, $\tan(P) = \frac{127.7}{890}$. Because we want to find the measure of angle P, we need to use arctangent.

$$\arctan\left(\frac{127.7}{890}\right) = m\angle P$$
$$8.2° \approx m\angle P$$

Therefore, the angle of depression should be approximately 8.2°.

In the previous two examples, you may have noticed that we did not include units in the ratios. In each of these examples, the units on a given problem were the same; therefore, they canceled in the ratio. If the units had been different, we would have needed to convert one or both of the measurements so that the units were the same.

In all the problems we encountered so far, the measures of the angles were recorded to the nearest degree, tenth of a degree, or hundredth of a degree. When measures of angles are written in this way, we say the angles are written in **decimal degrees,** abbreviated **DD.** In some

applications the measures of angles are given in **degrees-minutes-seconds,** abbreviated **DMS.**
Similar to time, one degree is equal to 60 minutes and one minute is equal to 60 seconds.

> *Conversions Between Decimal Degrees to Degrees-Minutes-Seconds*
>
> 1 degree = 60 minutes   denoted $1° = 60'$
>
> 1 minute = 60 seconds   denoted $1' = 60''$

We can convert 0.25° to minutes using these conversion facts and unit fractions.

$$0.25°$$
$$= 0.25° * \frac{60'}{1°}$$
$$= 15'$$

Therefore, 0.25° is the same as 15′.

We can visualize this conversion on the following clock.

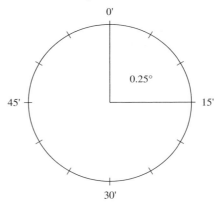

One degree = 60 minutes

Example 3

a. Rewrite 4.609° in degrees-minutes-seconds.

b. Rewrite 45° 25′ 30″ in decimal degrees.

Solution   a. Because we want to convert to degree-minutes-seconds and we already know the number of
degrees, we just need to convert the fraction of a degree, 0.609, to minutes and seconds.

$$0.609° * \frac{60'}{1°} = 36.54'$$

This means that 0.609° is 36 minutes plus part of a minute. So, now we need to convert the
fraction of a minute, 0.54′, to seconds.

$$0.54' * \frac{60''}{1'} = 32.4''$$

Therefore, 4.609° = 4° 36′ 32.4″.

b. To go from degrees-minutes-seconds to decimal degrees, we need to take each part and convert to degrees. Then, we can add the decimals to determine the measure of the angle in decimal degrees.

Convert 25′ to decimal degrees.

$$25' * \frac{1°}{60'} \approx 0.41667°$$

Convert 30″ to decimal degrees.

$$30' * \frac{1'}{60''} * \frac{1°}{60'} \approx 0.00833° \; \cdot$$

Therefore,

$$45° \, 25' \, 30''$$
$$\approx 45° + 0.41667° + 0.00833°$$
$$= 45.425°.$$

After having gone through these examples, you will be pleased to know that most calculators convert between decimal degrees and degrees-minutes-seconds.

Example 4

Percent of Slope.   The road design department must construct a road over a small hill. The hill has a slope of approximately 23%. The proposed road is to have a slope of 11%. Determine the measure of the angle of the cut that must be made.

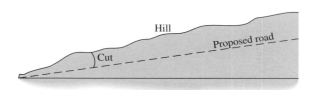

Solution   We begin by simplifying the picture of the hill and labeling the new sketch.

First we look at the slope of the hill and the relationships in triangle *ABC*. We know that the slope of the hill is 23%. Slope is $\frac{\text{rise}}{\text{run}}$, so we know that the tangent of $\angle ACB$ is equal to 23%, or 0.23.

$$\tan(\angle ACB) = 0.23$$

We can find the measure of $\angle ACB$ by using arctangent.

$$\arctan(0.23) = m\angle ACB$$
$$13.0° \approx m\angle ACB$$

Next, we will look at the slope of the proposed road and the relationships in triangle *DBC*. The slope of the road is 11%, or 0.11. So, tan(∠*DCB*) = 0.11. Therefore,

arctan(0.11) = *m*∠*DCB*
6.3° ≈ *m*∠*DCB*

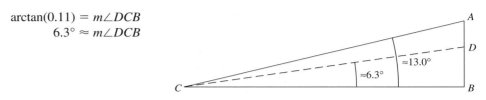

The angle of the cut is ∠*ACD*. The measure of angle *ACD* is the difference of the measures of angle *ACB* and angle *DCB*.

m∠*ACD* = 13.0° − 6.3° = 6.7°

The measure of the angle of the cut for the proposed roadway is approximately 6.7°.

In this section we defined the arcsine, arccosine, and arctangent. These functions are the inverses to the three trigonometric ratios sine, cosine, and tangent, respectively. The input into the arcsine, arccosine, or arctangent is the appropriate ratio in a right triangle. The output is the measure of the corresponding angle. Because the output for these functions is the measure of an angle, the answer can be written in **decimal degrees** or **degrees-minutes-seconds.**

Problem
Set
8.4

1. For each triangle, find the measure of angle *T*.

a.

b.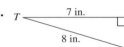

c.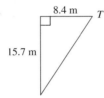

d.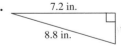

2. For each triangle, find the value of the angle or side indicated by the given variable.

a.

b.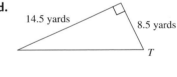

c.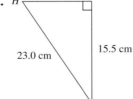

d.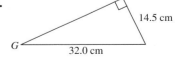

3. Road Construction. Given a slope distance of 512 feet and a horizontal distance of 505 feet, determine the angle of elevation and the difference in elevation.

4. Survey Stations.   Two survey stations differ in elevation by 2 chains and are a horizontal distance apart of 15 chains. Determine the slope distance in feet and the slope angle.

5. Wheelchair Ramp.   Mrs. Peters recently purchased an old office building. To bring the office up to code, she must build a wheelchair ramp. The local building code specifies that a wheelchair ramp cannot exceed a slope of $\frac{1}{12}$.

    a. Determine the maximum angle that the ramp makes with the ground if it is built to code.

    b. Below are two pictures of the situation. The first picture is the finished ramp and the second picture is the ramp before the lumber is cut. The second picture indicates the angles to be cut. Determine the angles at which to cut the lumber if the finished ramp is to have a slope of $\frac{1}{12}$.

The Finished Ramp

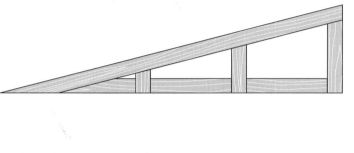

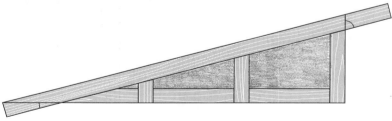

6. The Pendulum.   The total angle that a 30.5-foot-long pendulum swings through is 23.6°. Find the horizontal distance between the extreme ends of one pendulum swing.

7. Grade of the Road.   A road rises 375 feet in elevation in a mile as measured on the road itself. What is the slope angle of the road in degrees?

8. Percent Grade.   The grade of a road is the slope of that road expressed as a percent. A stretch of highway has a uniform grade of +1.2%. Two points on the pavement are measured with a steel tape resting on the pavement and found to be 517 feet apart. Determine the horizontal distance between these two points and the difference in elevation between the two points.

9. Railroad.   A railroad track has an angle of elevation of 2°. What is the percent of slope of the railroad track?

10. Piston.   A piston moves up and down as rod R revolves around the wheel. A stroke is measured as the piston moves from its highest point to its lowest point, or one-half revolution of the wheel. Find the length of rod R. Determine the distance traveled by the piston in one stroke.

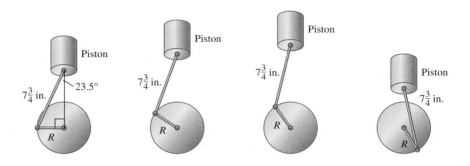

11. The Shed.   The following figure is a front view of the framing for a garden shed. Below the framing is a figure showing the rafter. Determine the angles necessary to cut the rafter if the roof has a $\frac{7}{12}$ pitch.

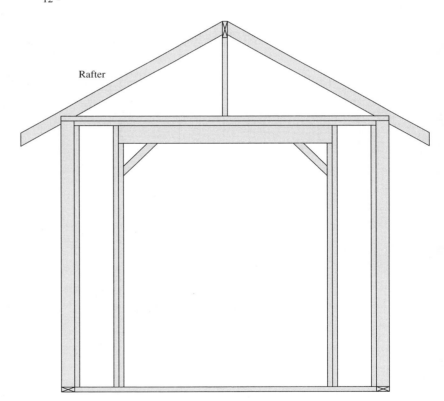

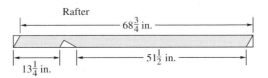

12. Detour from Centerline.   When laying out the centerline for a roadway, an obstacle, consisting of trees and a marsh, was encountered by the survey crew. To get around the obstacle, they had to deflect to the right N34°32′E from a bearing of true north. They chained out a distance of 300.1 ft. At this point they were in a position to return to the centerline.

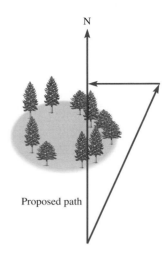

N

Proposed path

a. With respect to their present bearing, what angle must they turn to intersect their center line at a right angle?

b. How far must they travel to reach the centerline?

13. If the rise of a roof is 6 feet and the span is 36 feet, find the angle that the rafter makes with the joist.

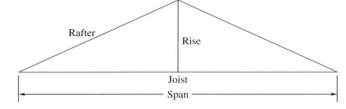

Rafter

Rise

Joist

Span

14. *Use the table* to answer the following questions. The table is precise to the nearest ten-thousandth.

    a. Determine

        sin(20°)

        cos(28°)

        tan(36°)

    b. Determine

        arcsin(0.5299)

        arccos(0.9848)

        arctan(0.4040)

    c. Estimate arccos(0.90).

Angle x (in degrees)	sin(x)	cos(x)	tan(x)
10	0.1736	0.9848	0.1763
12	0.2079	0.9781	0.2126
14	0.2419	0.9703	0.2493
16	0.2756	0.9613	0.2867
18	0.3090	0.9511	0.3249
20	0.3420	0.9397	0.3640
22	0.3746	0.9272	0.4040
24	0.4067	0.9135	0.4452
26	0.4384	0.8988	0.4877
28	0.4695	0.8829	0.5317
30	0.5000	0.8660	0.5774
32	0.5299	0.8480	0.6249
34	0.5592	0.8290	0.6745
36	0.5878	0.8090	0.7265

15. *Use the table* to answer the following questions. The table is precise to the nearest thousandth.

a. Determine

arcsin(0.2)

arccos(0.55)

arctan(1.00)

b. Estimate

sin(23°)

cos(80°)

tan(14.5°)

ratio x	arcsin(x) (in degrees)	arccos(x) (in degrees)	arctan(x) (in degrees)
0.000	0.000	90.000	0.000
0.050	2.866	87.134	2.862
0.100	5.739	84.261	5.711
0.150	8.627	81.373	8.531
0.200	11.537	78.463	11.310
0.250	14.478	75.523	14.036
0.300	17.458	72.542	16.699
0.350	20.487	69.513	19.290
0.400	23.578	66.422	21.801
0.450	26.744	63.256	24.228
0.500	30.000	60.000	26.565
0.550	33.367	56.633	28.811
0.600	36.870	53.130	30.964
0.650	40.542	49.458	33.024
0.700	44.427	45.573	34.992
0.750	48.590	41.410	36.870
0.800	53.130	36.870	38.660
0.850	58.212	31.788	40.365
0.900	64.158	25.842	41.987
0.950	71.805	18.195	43.531
1.000	90.000	0.000	45.000

Chapter Eight Summary

In this chapter we studied several methods for solving problems related to triangles, including properties of similar triangles, right triangle trigonometry, and inverses of trigonometric ratios. We also found that figures in general can be solved using carefully measured scale drawings. However, the results from scale drawings are not always accurate due to measuring errors.

The following box summarizes the properties of similar triangles.

> *Properties of Similar Triangles*
>
> 1. Similar triangles have equal angles and, conversely, triangles with equal angles are similar.
> 2. The ratio of corresponding sides in similar triangles is a constant. Conversely, if the ratios of the corresponding sides are constant, the triangles are similar.

When solving problems using similar triangles, keep in mind the following points.

- In some cases, when we are solving for unknown sides of similar triangles, we are told that the triangles are similar. The order of the vertices in the statement $\triangle ABC \sim \triangle END$ indicates the corresponding angles and sides of the two triangles.
- In other cases we must first determine if triangles are similar. If we can show two angles of one triangle are congruent to two angles of another, we can use the first property in the box to say that the triangles are similar.
- Ratios of corresponding sides must be set up so that the relationships on each side are consistent. For example, a ratio might look like

$$\frac{\text{longest side of } \triangle ABC}{\text{longest side of } \triangle END} = \frac{\text{shortest side of } \triangle ABC}{\text{shortest side of } \triangle END}$$

In right triangles we defined three ratios: sine, cosine, and tangent.

> *Trigonometric Ratios*
>
>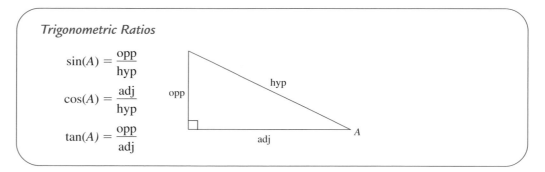
>
> $$\sin(A) = \frac{\text{opp}}{\text{hyp}}$$
>
> $$\cos(A) = \frac{\text{adj}}{\text{hyp}}$$
>
> $$\tan(A) = \frac{\text{opp}}{\text{adj}}$$

Remember that the *input* for sine, cosine, and tangent is the *measure of an acute angle in a right triangle* and that the *output* is a *ratio*. Because the output is a ratio of lengths whose units must be the same, the output has no units.

> *Inputs and Outputs of Trigonometric Ratios*
>
> input of any trigonometric ratio = measure of an angle
>
> output of any trigonometric ratio = ratio of sides

The arcsine, arccosine, and arctangent functions are the inverses of the trigonometric functions.

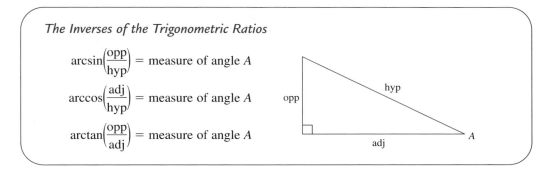

The Inverses of the Trigonometric Ratios

$$\arcsin\left(\frac{\text{opp}}{\text{hyp}}\right) = \text{measure of angle } A$$

$$\arccos\left(\frac{\text{adj}}{\text{hyp}}\right) = \text{measure of angle } A$$

$$\arctan\left(\frac{\text{opp}}{\text{adj}}\right) = \text{measure of angle } A$$

In words,

arctan(x) produces the measure of the angle whose tangent is x.

arcsin(x) produces the measure of the angle whose sine is x.

arccos(x) produces the measure of the angle whose cosine is x.

Again, remember what kinds of measurements the inputs and outputs of the inverse trigonometric functions are.

Inputs and Outputs of Inverse Trigonometric Ratios

input of the *inverse* of any trigonometric ratio = ratio of sides

output of the *inverse* of any trigonometric ratio = measure of an angle

When you use these trigonometric functions or their inverses to solve problems involving right triangles, be sure your calculator is set so angles are calculated in degrees (degree mode).

Chapter Nine

Functions

In Section 1.2, we learned that a specific type of relation is called a function. Now we will learn the formal language and notation associated with functions. Some functions have inverse functions. We will also look at the relationships between functions and their inverses.

9.1 Functions and Notation

Discussion 9.1

In Section 1.2, we defined a function as a rule that assigns a single output for every input. We now introduce formal notation, called function notation, that more clearly indicates the relationship between the variables, that is, which variable is independent and which is dependent.

One common formula is for the area of a circle, $A = \pi r^2$. In this formula, the area of a circle depends on its radius. Another way to say this is that the area of a circle is a function of its radius. A notation used to indicate this dependence is called **function notation.** Instead of simply writing $A = \pi r^2$, we write $A(r) = \pi r^2$. We read $A(r)$ as "A of r."

The notation $A(r)$ is used to indicate that r is the independent variable and A is dependent on r (or is a function of r). In other words, r is the input variable into the function. For example, $A(4)$ represents the area of a circle when the radius is 4. Therefore, to find the value of $A(4)$ we need to substitute $r = 4$ into the function.

$$A(4) = \pi * 4^2$$
$$\approx 50$$

Similarly, $A(10)$ represents the area of a circle when the radius is 10. That is,

$$A(10) = \pi * 10^2$$
$$\approx 314$$

Therefore, $A(10) \approx 314$.

NOTE: This notation can be confusing. In some contexts, $A(r)$ represents the product of A and r. However, in this context, in which A is the dependent variable of the function, the notation $A(r)$ indicates that the area is a function of the radius rather than indicating the product of A and r.

Example 1

Cooking at High Altitudes.   If T represents the temperature in degrees Fahrenheit at which water begins to boil and A represents the altitude above sea level in thousands of feet, then T depends on A. This relationship can be modeled by the function

$$T(A) = {}^-1.71A + 212$$

Determine the value of each expression and interpret the meaning of each in the context of this problem situation.

   a. $T(0)$
   b. $T(6)$
   c. $T(12)$

Solution   a. The notation $T(0)$ is read "T of 0." We are looking for the value of the function when the input is 0.

$$T(0) = {}^-1.71 * 0 + 212$$
$$T(0) = 212$$

At 0 feet above sea level, water boils at 212°F.

b. The notation $T(6)$ is read "T of 6." The value of the function when the input is 6 is

$$T(6) = {}^-1.71 * 6 + 212$$
$$T(6) \approx 202$$

At 6000 feet above sea level, water boils at approximately 202°F.

c. The notation $T(12)$ is read "T of 12." The value of the function when the input is 12 is

$$T(12) = {}^-1.71 * 12 + 212$$
$$T(12) \approx 191$$

At 12,000 feet above sea level, water boils at approximately 191°F.

In application problems it often makes sense to use letters to help us remember what the variables represent. In Example 1, the letter T represented the temperature in degrees Fahrenheit of the boiling point of water and A represented altitude in thousands of feet. In many abstract equations we use x and y as our variables. When writing in function notation, the convention is to use letters such as f, g, h, or k to name the function and $f(x)$, $g(x)$, and so on to indicate the outputs for an input of x. With this notation, a function written in the form

$$y = \text{expression in terms of } x$$

can be rewritten as

$$f(x) = \text{expression in terms of } x$$

Function Notation

The function

$$y = \text{expression in terms of } x$$

can be rewritten using function notation as

$$f(x) = \text{expression in terms of } x$$

In this notation,

- f is the name of the function.
- The independent variable x is called the **argument.**
- $f(x)$ is the output of the function for an input of x.

Example 2

Let $f(x) = \sqrt{x+3}$.

a. Determine $f(6)$.   d. Determine $f(x + 2)$.

b. Determine $f({}^-3)$.   e. Determine $f({}^-5)$.

c. Determine $f(b)$.

Solution a. The expression $f(6)$ represents the value of the function when x is 6.

$$f(6) = \sqrt{6+3}$$
$$= \sqrt{9}$$
$$= 3$$

b. The expression $f(^-3)$ represents the value of the function when x is $^-3$.

$$f(^-3) = \sqrt{^-3 + 3}$$
$$= \sqrt{0}$$
$$= 0$$

c. The expression $f(b)$ represents the value of the function when x is b.

$$f(b) = \sqrt{b + 3}$$

This cannot be further simplified.

d. In the expression $f(x)$, remember that x represents the input into the function f. The expression $f(x + 2)$ represents the value of the function when the input is $x + 2$, that is, $f(\text{input}) = \sqrt{\text{input} + 3}$. The rule for the function f is to add 3 to the input and then take the square root of that result. Everywhere that we see x in $f(x) = \sqrt{x + 3}$, we replace it with the new input of $x + 2$.

$$f(x + 2) = \sqrt{(x + 2) + 3}$$
$$= \sqrt{x + 5}$$

e. The expression $f(^-5)$ represents the value of the function when x is $^-5$.

$$f(^-5) = \sqrt{^-5 + 3}$$
$$= \sqrt{^-2}$$

The square root of $^-2$ is not a real number. This means $f(^-5)$ is not a real number. Therefore, the function f is not defined for an input value of $^-5$.

We looked at functions defined by an algebraic rule. Remember that a function can also be defined by a graph or by a table.

Example 3

The following graph defines a function $G(x)$. Assume that a dot at an endpoint indicates the function stops, and a line without a dot indicates the function continues in that direction.

Determine the value of each of the following expressions.

a. $G(7)$ d. $G(^-1)$

b. $G(1)$ e. $G(15)$

c. $G(4)$

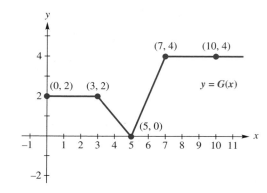

Solution   a. To find $G(7)$, we must find the output value for an input of 7. The point $(7, 4)$ on the graph indicates that for an input of 7, the output is 4. Therefore, $G(7) = 4$.

b. To find $G(1)$, we locate the input value of 1 on the x-axis. Looking up from that point we see the point $(1, 2)$ on the graph. This means that for an input of 1, the output is 2. Therefore, $G(1) = 2$.

c. To determine $G(4)$, we must find the output value for an input of 4. The point $(4, 1)$ appears to be on the graph. Using the slope of the line segment, we can verify that this is true. Therefore, $G(4) = 1$.

d. When we locate $^-1$ on the horizontal axis, we find no corresponding point on the graph. This means that $G(^-1)$ is undefined.

e. The input value of 15 is off of our graph. However, because of the pattern that is established, we see that any input greater than 7 has an output of 4. Therefore, $G(15) = 4$.

Domains and Ranges

In Example 3d, we saw that for $x = {}^-1$, the output was not defined. This means that when we ask for the output of a function for a given input value we must restrict the input to those values for which the output is defined and is a real number.

> **Definition** _____
>
> We call the set of all input values that produce a real number output the **domain** of the function.

> **Definition** _____
>
> The **range** is the set of all output values of the function.

In this chapter we restrict all function input and output values to the real numbers. In later courses you may expand these ideas to include the complex numbers.

Example 4

a. What are the domain and range of $g(x) = x^2 + 3$?

b. What are the domain and range of $f(x) = \sqrt{x + 3}$?

Solution   a. There are no restrictions on the input values for the function $g(x) = x^2 + 3$. Therefore, the domain of the function is all real numbers.

To determine the range of a function, it is often helpful to look at the graph of the function, especially when we know how the graph should look. The graph of $g(x) = x^2 + 3$ is a parabola that opens upward. The parabola is symmetric about the y axis and shifted up 3 units. This means the vertex is at $(0, 3)$.

We are looking for the set of all outputs. The output value at the vertex is the smallest output value. Therefore, the range of the function g is all real numbers greater than or equal to 3. We can write this symbolically as $y \geq 3$.

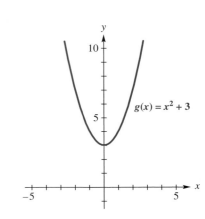

b. Looking at the function $f(x) = \sqrt{x + 3}$, we know that we cannot take the square root of a negative number in the real number system. Therefore, we must make sure that the value of the expression under the radical, $x + 3$, is not negative. That is,

$$x + 3 \geq 0$$
$$x \geq {}^-3$$

Therefore, the domain is all real numbers greater than or equal to $^-3$; that is, $x \geq {}^-3$.

Next, we determine the range of $f(x) = \sqrt{x + 3}$. The range is the set of all output values of $f(x)$, or $\sqrt{x + 3}$. Let's start at the border of our domain, $x = {}^-3$. At $x = {}^-3$, $f(^-3) = 0$. As x increases, $x + 3$ increases. Therefore, the output $\sqrt{x + 3}$ continues to increase. This means that the range of f is all real numbers greater than or equal to 0; that is, $y \geq 0$.

We can visualize these results graphically.

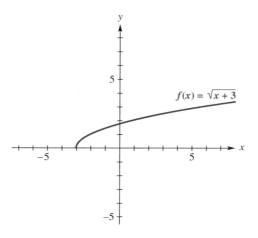

We can see from our graph in the preceding figure that the graph of the function begins at the point $(^-3, 0)$ and extends to the right. This supports our claim that the domain is $x \geq {}^-3$. We can also see that the graph begins at the point $(^-3, 0)$ and extends upward. This supports our claim that the range is all nonnegative real numbers.

As we have seen, there are examples in which the mathematical model has a restricted domain. The two situations we must avoid are inputs that cause us to have zero in a denominator or a negative number inside an even radical.

When writing algebraic functions to model real-life situations, we may need to restrict our models further to fit the problem situation. We call the set of all reasonable input values for a given problem situation the **domain of the problem situation.**

Example 5

Pollution Index.   For a certain large city, the pollution increases in an approximately linear fashion between 8:00 A.M. and 4:00 P.M. each day. The function $P(t) = 12.5t + 18$ models the pollution P in parts per million (ppm), in terms of t, the number of hours after 8:00 A.M.

a. What is the domain of the function? What is the domain of this problem situation, and what is the range for this problem situation? Draw a graph of the problem situation.

b. Is it reasonable to expect the pollution index to continue to increase in a linear fashion after 4:00 P.M.? Why or why not? If not, sketch one possible graph of the function between 8:00 A.M. one day and 8:00 A.M. the following day.

Solution   a. All values of t are allowable in the linear function $P(t) = 12.5t + 18$, so the domain is all real numbers. However, the problem situation is defined for hours between 8:00 A.M. and 4:00 P.M., so allowable values of t in this problem situation are between 0 and 8. The *domain of the problem situation* is $0 \leq t \leq 8$.

The function $P(t) = 12.5t + 18$ is linear with a positive slope, which means that the function increases as the input t increases. Thus, the minimum value of the function occurs at $P(0)$, and the maximum value of the function occurs at $P(8)$. From this we can see that the range of the problem situation is

$$P(0) \leq P(t) \leq P(8)$$
$$12.5 * 0 + 18 \leq P(t) \leq 12.5 * 8 + 18$$
$$18 \leq P(t) \leq 118$$

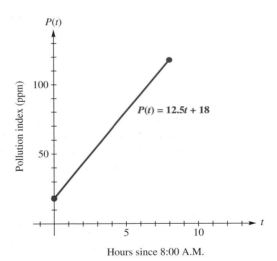

b. It is not reasonable to conclude that the graph of the pollution index continues to increase. At some time between 4:00 P.M. and 8:00 A.M. the following day the pollution index decreases from 118 ppm back to approximately 18 ppm.

One *possible* graph of the problem situation is shown in the following figure.

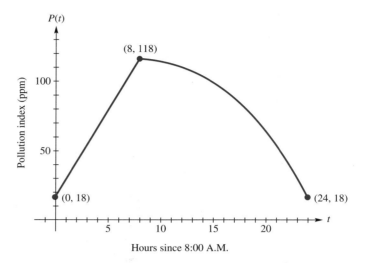

Hours since 8:00 A.M.

We do not have enough information from the statement of the problem to know the exact shape of the model between 4:00 P.M. and 8:00 A.M. the following day. This graph is one possible guess but there are many other possibilities.

In this section we introduced the notation $f(x)$ to represent the value of the function f at x. The **domain** of a function is the set of all inputs for the function that result in real number outputs. The **range** is the set of all output values of the function. In many problem situations, the domain is more restricted than the algebraic model. This restricted domain is called the **domain of the problem situation.**

Problem
Set
9.1

1. Use the following tables to answer these questions.
   a. Determine $f(2)$, $f(4)$, and $f(^-1)$.
   b. Determine $g(2)$, $g(4)$, and $g(^-1)$.

x	f(x)
$^-1$	3
0	4
1	-2
2	6
3	2
4	-1

x	g(x)
$^-1$	3
0	1
1	0
2	-3
3	-5
4	6

2. a. Given $f(x) = x^2 - 5$, determine $f(3)$.

   b. Given $g(x) = 2x + 7$, determine $g(^-5)$.

   c. Given $h(x) = \sqrt{2x - 5}$, determine $h(15)$.

   d. Given $F(x) = \dfrac{x}{x + 1}$, determine $F(3)$.

   e. Given $G(x) = 3x^3 - 7x^2 + 5$, determine $G(^-2)$.

   f. Given $H(x) = \sqrt{x^2 - 13}$, determine $H(7)$.

3. The function $S(r) = 2\pi r^2 + 12\pi r$ gives us the surface area for a 6-foot-tall cylinder given the radius r in feet.
   a. Determine the surface area if the radius is 3.2 feet.
   b. Determine $S(5.1 \text{ ft})$.

4. $P(t)$ is a function,
   where $P =$ the population of Little Rock, AR, in ten thousands, and
   $t =$ the number of years after 1980.
   a. Interpret the meaning of the equation $P(10) = 17.6$.
   b. What does $P(17)$ represent?

5. $S(n)$ is a function,
   where $S =$ an American shoe size, and
   $n =$ a person.
   a. Interpret the meaning of $S(\text{Ian}) = 10\frac{1}{2}$.
   b. What does $S(\text{Penny})$ represent?

6. $C(m)$ is a function,
   where $C =$ cost of a trip in dollars, and
   $m =$ number of miles driven
   a. Interpret the meaning of $C(1000) = 650$.
   b. What does $C(450)$ represent?

7. Given $F(x) = \dfrac{x^2}{x+1}$, determine each of the following.

    a. $F(2)$ **d.** $F(t)$

    b. $F(4)$ **e.** $F(m+1)$

    c. $F(0)$ **f.** $F(^-1)$

8. Solve each equation.

    a. $f(x) = 12$, where $f(x) = 2x + 4.5$ **d.** $F(x) = 24$, where $F(x) = 3 * 2^x$

    b. $g(x) = 7$, where $g(x) = 3x^2 + x$ **e.** $G(x) = 36$, where $G(x) = (x+5)^2$

    c. $h(x) = 5$, where $h(x) = \dfrac{3}{x-2}$ **f.** $H(x) = 8$, where $H(x) = |2x + 5|$

9. Use the graphs to answer the following questions. Assume that a dot at an endpoint indicates the function stops, and a line without a dot indicates the function continues in that direction.

    a. Determine $f(3), f(^-5), f(0),$ and $f(^-6)$.

    b. Determine $g(^-3), g(4), g(0), g(^-5),$ and $g(8)$.

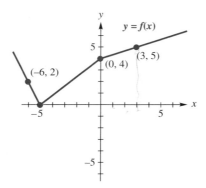 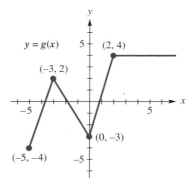

10. Identify the domain and range of the functions $f(x)$ and $g(x)$ from Problem 9.

11. Use the graphs to answer the following questions.

    a. Determine $h(^-4), h(^-1),$ and $h(2)$.

    b. Determine $k(2), k(4),$ and $k(^-1)$.

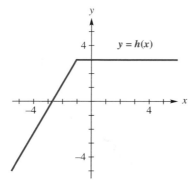 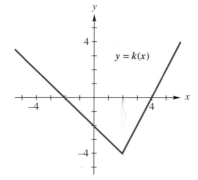

12. Identify the domain and range of the functions $h(x)$ and $k(x)$ from Problem 11.

13. Find the domain and range for the following functions.

    a. $f(x) = 2x + 14$ **d.** $k(x) = {}^-x^2 - 10$

    b. $g(x) = x^2 - 135$ **e.** $F(t) = \sqrt{t - 5}$

    c. $h(t) = |t|$ **f.** $G(x) = x^2 + 18x$

14. Find the domain and range for the following functions.

    a. $F(x) = |x - 3|$ **d.** $H(t) = t^{1/6}$

    b. $K(x) = \sqrt{{}^-x}$ **e.** $g(t) = t^{2/3}$

    c. $f(x) = \dfrac{1}{x - 2}$

15. Suppose we define a relation so that the input is *someone in your family* and the output is *date of his or her birthday*.

    a. Is this relation a function? Explain.

    b. Suppose you and your brother both share a birthday of August 14. Is this still a function? Explain.

16. Determine which of the following relations are functions. Explain your decision.

    a. The input is a *name entry in the phone directory,* and the output is the *phone number for that name.*

    b. The input is a *student in this class,* and the output is *shoe size.*

    c. The input is a *month of the year,* and the output is *number of days in the month.*

17. Cory's Cardboard Company.   A cardboard company makes cardboard boxes from sheets of board 2 feet by 3 feet. The boxes are made by cutting squares of equal size from each corner and folding up the sides. Write an equation for the volume of the box in terms of the length of the side of the square removed. What is the domain of the problem situation? Explain how you decided.

9.2 Inverse Functions

Activity Set 9.2

Informally, two functions are inverses if they undo each other. We have already seen several functions that are inverses. For example, to undo a cube we must apply a cube root. Therefore, $y = x^3$ and $y = \sqrt[3]{x}$ are inverse functions. (*Recall:* Another notation for $\sqrt[3]{x}$ is $x^{1/3}$.) Similarly, $y = 2x$ and $y = \frac{1}{2}x$ are inverse functions.

1. a. Complete the following tables.

TABLE A	
x	$y = x^3$
-2	
-1	
0	
1	
2	

TABLE B	
x	$y = \sqrt[3]{x} = x^{1/3}$
-8	
-1	
0	
1	
8	

TABLE C	
x	$y = 2x$
-1	
0	
1	
2	
3	
4	

TABLE D	
x	$y = \frac{1}{2}x$
-2	
0	
2	
4	
6	
8	

TABLE E	
x	$y = \sin(x)$
$0°$	0
$30°$	
$45°$	
$60°$	

TABLE F	
x	$y = \arcsin(x)$
0	
0.50	
0.71	
0.87	

b. Describe the relationship between Tables A and B.

c. Describe the relationship between Tables C and D.

d. Describe the relationship between Tables E and F.

2. Table G is a table of values for some function g. Table H is a table of values for the inverse of g. Use your observations from Activity 1 to complete Table H.

TABLE G	
x	$y = g(x)$
$^-2$	$^-3$
$^-1$	$^-1$
0	1
1	3
2	5
3	7
4	9

TABLE H	
x	Inverse of g
$^-3$	
$^-1$	
1	
3	
5	
7	
9	

3. On the same set of axes, sketch the graphs of $y = x^3$ and $y = \sqrt[3]{x}$ by plotting the points from Table A and the points from Table B in Activity 1. Graph the line $y = x$ on this same set of axes. Describe any relationship that you see.

4. On the same set of axes, sketch the graphs of $y = 2x$ and $y = \frac{1}{2}x$ by plotting the points from Table C and the points from Table D in Activity 1. Graph the line $y = x$ on this same set of axes. Describe any relationship that you see.

5. a. Complete the following tables.

TABLE I	
x	$h(x) = 2^x$
$^-2$	$\frac{1}{4}$
$^-1$	
0	
1	
2	
3	
4	

TABLE J	
x	Inverse of h
$\frac{1}{4}$	
$\frac{1}{2}$	
1	
2	
4	
8	
16	

b. On the same set of axes, graph h and the inverse of h by plotting the points. Graph the line $y = x$ on this same set of axes. Describe any relationship that you see.

c. What are the domain and range of h? What are the domain and range of the inverse of h? Describe the relationship between the domains and ranges of inverse functions, and explain why this relationship must always hold.

6. Following is the graph of some function $y = f(x)$. Sketch the graph of the inverse of the function f, and describe your process.

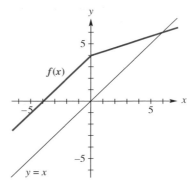

Discussion 9.2

Informally, the inverse function of a function f is a function that undoes f. For example $y = x^{1/3}$ is the inverse of $y = x^3$; the cube root undoes the cube. In the activities, we saw that the input and output values of inverse functions are interchanged. We observed that the graph of the inverse is the reflection of the graph of the original function about the line $y = x$. We also observed that if (a, b) is on the graph of a function f, then (b, a) is on the graph of the inverse function. This leads us to a formal definition.

Definition

The **inverse function** of a function f is the function whose graph consists of all points (b, a), where (a, b) is on the graph of f, provided that the inverse is itself a function. The inverse function of a function f is denoted by f^{-1}.

NOTE: The notation f^{-1} represents the inverse function of f, *not the reciprocal* of f. That is, if $f(x) = x^3$, then $f^{-1}(x) = x^{1/3}$, which is *not the same as* $\frac{1}{x^3}$.

We can look at inverse functions numerically, graphically, and algebraically.

Example 1

The functions g, h, and k are defined by the following tables.

a. Make a table of values for g^{-1}, for h^{-1}, and for k^{-1}, if the inverse function exists.

b. Determine $g(2)$, $g^{-1}(18)$, and $h^{-1}(3)$.

x	g(x)
−2	6
0	12
2	18
4	24
6	30

x	h(x)
−1	1.5
0	3
1	6
2	12
3	24

x	k(x)
−2	4
−1	1
0	0
1	1
2	4

Solution   a. We know that to determine g^{-1}, we can interchange the inputs and outputs of g. Similarly for h^{-1}. However, notice that $k(^-2)$ and $k(2)$ are both equal to 4. Therefore, if we interchange the inputs and outputs of k we produce a relation in which an input of 4 produces two different outputs: $^-2$ and 2. A relationship with more than one output for a given input is not a function. Therefore, k does not have an inverse function. Following are the tables of values for g^{-1} and h^{-1}.

x	$g^{-1}(x)$
6	$^-2$
12	0
18	2
24	4
30	6

x	$h^{-1}(x)$
1.5	$^-1$
3	0
6	1
12	2
24	3

b. The expression $g(2)$ represents the output of the function g for an input of 2. From the table for function g, we can see that $g(2) = 18$.

The expression $g^{-1}(18)$ represents the output of the function g^{-1} for an input of 18. From the table for function g^{-1}, we can see that $g^{-1}(18) = 2$. Alternatively, we can think of $g^{-1}(18)$ as representing the input of g that produces an output of 18 and use the table for function g to reach the same conclusion. Notice, using the second method, we are reading the table backwards.

The expression $h^{-1}(3)$ represents the output of the function h^{-1} for an input of 3. From the table for the function h^{-1}, we can see that $h^{-1}(3) = 0$. Alternatively, we can think of $h^{-1}(3)$ as representing the input of h, which produces an output of 3, and use the table for the function h to reach the same conclusion.

In Example 1, we saw that not all functions have inverse functions. Given a function defined by a table of values, if two inputs have the same output, then the function does not have an inverse function.

> *Existence of Inverse Functions*
>
> A function has an inverse function if and only if each input has a unique output.

In the activities, you may have found that the graph of the inverse of a function is the mirror image, or reflection, of that function about the line $y = x$. We can use this idea to graph inverse functions.

Example 2

Following are the graphs of functions $g(x) = 1.5^x$ and $h(x) = x^2$.

a. Determine the domain and range of g and h.

b. Sketch g^{-1} and h^{-1}, if they exist.

c. Determine the domain and range of g^{-1}.

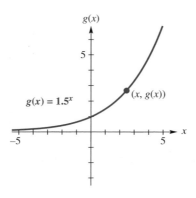

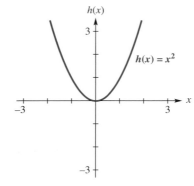

Solution   a. The domain of g is all real numbers, and the range of g is all positive real numbers, that is, $y > 0$.

The domain of h is all real numbers, and the range of h is all nonnegative real numbers, that is, $y \geq 0$.

b. To graph the inverse of g we reflect the graph of g across the line $y = x$. In doing so, we obtain the graph seen in the following figure.

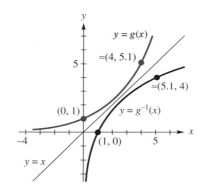

When we reflect h about the line $y = x$, we obtain a graph that is not a function. A function must pass the vertical line test. That is, any vertical line drawn only intersects the graph of a function at most once. Therefore, h does not have an inverse *function*.

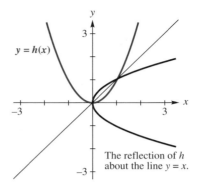

$y = h(x)$

The reflection of h about the line $y = x$.

c. The domain of g^{-1} is all positive real numbers, $x > 0$, and the range of g^{-1} is all real numbers.

A couple of interesting observations can be made from Example 2. First, because the input and output values of inverse functions are interchanged, the domain and range of inverse functions are interchanged. That is, the domain of f^{-1} is the range of f, and the range of f^{-1} is the domain of f. Second, we found from the previous two examples that not all functions have inverse *functions*. If a graph passes the vertical line test then it is a function. Because the inverse of a function f is obtained by reflecting f across the line $y = x$, the resulting graph passes the vertical line test if and only if the original function f passes a horizontal line test. A graph passes a **horizontal line test** if every horizontal line intersects the graph at most once.

Definition

A function is **one-to-one** if each input produces a unique output.

Therefore, we say that a function has an inverse function if and only if it is one-to-one. We use the horizontal line test to determine if a function is one-to-one.

One-to-One Functions Have Inverses

A function has an inverse function if and only if it is one-to-one.

Example 3

Write an equation for the inverse of each of the following functions.

a. $y = (x + 2)^3$

b. $y = 2x - 5$

Solution

a. There are two approaches we can take. We can use the idea that inverse functions undo each other, or we can use algebra. We first think of how to undo a function.

The function $y = (x + 2)^3$

- Adds 2 to its input.
- Then cubes the result.

To undo this process, we must undo the operations that were done and perform these in reverse order.

Therefore, the inverse must

- First apply a cube root.
- Then subtract 2 from the result.

We can write this inverse function as $y = x^{1/3} - 2$.

To verify our results we can make a table of values for $y = (x + 2)^3$, then use the outputs of this function as inputs into $y = x^{1/3} - 2$.

x	$y = (x + 2)^3$
$^-1$	1
0	8
1	27
2	64
3	125

x	$y = (x^{1/3} - 2)$
1	$1^{1/3} - 2 = ^-1$
8	$8^{1/3} - 2 = 0$
27	$27^{1/3} - 2 = 1$
64	$64^{1/3} - 2 = 2$
125	$125^{1/3} - 2 = 3$

Because the output values of $y = x^{1/3} - 2$ are the same as the input values of $y = (x + 2)^3$, we conclude that $y = x^{1/3} - 2$ is the inverse of $y = (x + 2)^3$.

Alternatively, because the inputs and outputs are interchanged in inverse functions, the role of the dependent and independent variable are reversed. To accomplish this algebraically, we interchange x and y in the original function. Because we want the new function expressed in terms of the dependent variable, we must then solve for y.

$y = (x + 2)^3$ This is the original function.

$x = (y + 2)^3$ Interchange x and y.

$x^{1/3} = ((y + 2)^3)^{1/3}$ To solve for y, first apply the one-third power to both sides.

$x^{1/3} = y + 2$

$x^{1/3} - 2 = y$ Next, subtract 2 from both sides.

This produces the same inverse function as before, $y = x^{1/3} - 2$.

b. The function $y = 2x - 5$

- Doubles its input.
- Then subtracts 5.

To undo this function we must undo the operations in reverse order. Therefore, the inverse function must

- First add 5 to the input.
- Then take half of the result.

Therefore, the inverse function of $y = 2x - 5$ is the function

$$y = \frac{x + 5}{2}$$

We could also find this by using the algebraic process of interchanging x and y and solving for y.

$x = 2y - 5$ Interchange x and y.

$x + 5 = 2y$

$\dfrac{x + 5}{2} = y$ Solve for y.

Because we found the same inverse by two methods, we can skip the numerical verification that was done in part a.

Therefore, we conclude that the inverse function of $y = 2x - 5$ is the function

$$y = \frac{x + 5}{2}$$

We may not always be able to find simple algebraic formulas for inverse functions. For example, in Activity 4 at the beginning of this section we found both a graphical and numerical representation of the inverse function of $y = 2^x$. However, algebraically, the inverse of $y = 2^x$ is the function defined by $x = 2^y$. At this time, we do not have any means to solve for y because y appears in the exponent. This topic will be discussed further in Section 10.1.

In this section, we looked at **inverse functions.** Informally, the inverse function of a function f is a function that undoes f. The inverse function of f is denoted f^{-1}. Numerically, the inverse function of f is the function that results by interchanging the input and output values of f, provided the result is a function. Graphically, the inverse function of f is obtained by reflecting f across the line $y = x$, provided the result passes the **vertical line test.** A function f has an inverse function if f passes the **horizontal line test.** Algebraically, the inverse function of a function f is the function obtained by interchanging the input and output variables and then solving for the dependent variable, if possible.

Problem
Set
9.2

1. Which of the following functions have inverses that are functions? Explain why or why not, and include any assumptions you are making. For those that have inverse functions, identify the input and output of the inverse function.

   a. *R*(number of music CDs sold) = amount earned in royalties

   b. *H*(name of a player on the college softball team) = percentage of hits

   c. *M*(number of calories you want to burn) = number of minutes you need to spend on an exercise bike

2. Which of the following functions have inverses that are functions? Explain why or why not, and include any assumptions you are making. For those that have inverse functions, identify the input and output of the inverse function.

   a. *S*(number of minutes after leaving home) = speed of your car

   b. *A*(*w*) = area of a square with width *w*

   c. *P*(number of kilowatt hours) = cost of your power bill

   d. *S*(a student in this class) = the student's shoe size

   e. *W*(*t*) = weight of Rick *t* years after birth

   f. *N*(*t*) = number of bacteria in a colony that is doubling every 15 minutes, at time *t*.

3. The functions *g, h,* and *k* are defined by the following tables.

   a. For each function, decide if it has an inverse that is a function. Explain.

   b. For each function that has an inverse function, make a table of values for the inverse.

   c. Determine $g(5)$, $g^{-1}(5)$, $h(-1)$, $k(3)$, and $k^{-1}(3)$.

x	y = g(x)
−5	−2.5
0	0
5	2.5
10	5
15	7.5

x	y = h(x)
−8	15
−1	6
0	−1
1	6
8	15

x	y = k(x)
$\frac{1}{3}$	−1
1	0
3	1
9	2
27	3

4. *P*(*t*) is a one-to-one function,

   where *P* = the population of Little Rock, AR, in ten thousands and

   t = the number of years after 1980

   Describe what each expression represents.

   a. $P(0)$ **b.** $P(20)$ **c.** $P^{-1}(25)$

5. *C*(*s*) is a one-to-one function,

   where *C* = the cost to build a house in thousands of dollars and

   s = the square footage of the house

   Describe what each expression represents.

   a. $C(800)$ **b.** $C^{-1}(100)$ **c.** $C^{-1}(200)$

6. $D(m)$ is a one-to-one function,

where $D =$ cost of a trip in dollars and

$m =$ number of miles driven

Describe what each expression represents.

a. $D(50)$ **b.** $D^{-1}(100)$ **c.** $D(100)$

7. $C(y)$ is a one-to-one function,

where $C =$ the predicted number of centenarians (persons 100 years or older) living in Florida and

$y =$ the year

Describe what each expression represents.

a. $C(2050)$ **b.** $C^{-1}(2100)$ **c.** $C^{-1}(4000)$

8. $V(s)$ is a one-to-one function,

where $V =$ velocity of a rock dropped from a bridge in feet per second and

$s =$ number of seconds since the rock was dropped

Describe what each expression represents.

a. $V(10)$ **b.** $V^{-1}(21)$ **c.** $V(60)$

9. Write an equation for the inverse function of each function. Verify your responses.

a. $y = x + 2$ **c.** $y = \dfrac{x}{2} + 5$

b. $y = 3x - 4$ **d.** $y = x + \dfrac{5}{2}$

10. Write an equation for the inverse function of each function. Verify your responses.

a. $y = \sqrt[3]{x}$ **c.** $y = 3(x + 6)$

b. $y = x^3 - 4$ **d.** $y = (1 + x)^{1/5}$

11. Suppose f is a one-to-one function whose domain is all real numbers and whose range is all positive real numbers. What are the domain and range of f^{-1}? Explain your answer.

12. The functions *f, g, h,* and *k* are defined by the following graphs.

   a. For each function that has an inverse function, sketch a graph of the inverse. For each function that does not have an inverse function, explain why not.

   b. Approximate $f(-1), f(4), g(2), h(4), h(9)$, and $k(4)$.

   c. Approximate $f^{-1}(-1), f^{-1}(1.5), h^{-1}(2), h^{-1}(3)$, and $k^{-1}(-4)$.

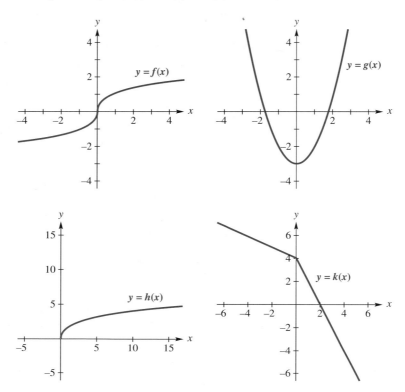

13. **a.** Does $y = x^2$ have an inverse function? Explain why or why not.

   b. Does $y = \sqrt{x}$ have an inverse function? Explain why or why not.

   c. What are the domain and range of x^2?

   d. What are the domain and range of $\sqrt{x}$?

   e. Sketch a graph of the inverse function of $y = \sqrt{x}$.

   f. How does your graph in part e compare with a graph of $y = x^2$?

Chapter Nine Summary

In this chapter we revisited the concept of functions. A **function** is a rule that assigns a *single output value* to every input value. The input is the independent variable and the output is the dependent variable. We looked at what values could be used as the input for specific functions and what the output would be. The set of all input values that produce a real number output is the **domain** of the function. The **range** is the set of all output values of the function.

In this course, we determine the domain of a function by considering two conditions:

- The denominator cannot be equal to zero.
- We cannot have an even root of a negative number.

To determine the range of a function, we often *begin* by graphing the function. Then we need to analyze the graph mathematically. It is not sufficient to use tracing features on your calculator to decide what the range of a function is because this method produces results that are only approximate.

In addition to looking at the mathematical expressions to determine the domain and range, we saw that application problems may place more restrictions on these values. For example, suppose we have a function for which the input variable is the height of a box and mathematically the domain of the function is all real numbers. Clearly the height of a box cannot be negative, and in most situations there is also an upper limit for what is reasonable for the height. Then the **domain of this problem situation** is not all real numbers, but is restricted to values that make sense for this problem.

In this chapter we introduced **function notation.** The notation $f(x)$ represents the value of the function f at x. In this notation, f is the **name** of the function, and x is called the **argument** of the function.

We found that a function f has an inverse denoted f^{-1} if and only if the original function is one-to-one. A function is one-to-one if each input has a unique output.

NOTE: The notation f^{-1} represents the inverse function of f, *not the reciprocal* of f. That is, if $f(x) = x^3$, then $f^{-1}(x) = x^{1/3}$, which is *not the same* as $\frac{1}{x^3}$.

Determining the Existence of an Inverse Function

- To decide whether a **function has an inverse function from a table of values,** we look to see if each input value produces a unique output. If the same output value is paired with two or more different input values, then the function does not have an inverse function.
- To decide whether a **function has an inverse function from a graph,** we use the **horizontal line test.**
- To determine whether a **function has an inverse function from an equation,** we interchange x and y and solve the resulting equation for y. If solving for y creates more than one output value, the function does not have an inverse function.

If a function is one-to-one, then we can find the inverse by the following methods.

> **How to Create the Inverse Function**
>
> - If a function is given as a **table of values,** then the inverse can be determined by interchanging the input and output values.
>
> - If a function is given as a **graph,** the inverse can be drawn by reflecting the graph across the line $y = x$.
>
> - If a function is given as an **equation,** then the inverse can be found by one of two methods.
>
>   1. The inverse of a function can be written by writing a relation that undoes each of the steps of the original function but in reverse order.
>   2. Alternatively, we can write the inverse of a function by interchanging x and y and then solving the resulting equation for y, if possible.

Chapter Ten

Logarithmic Functions and Equations

In Chapter 6, we studied exponential functions. Later we studied the inverse functions of many different functions. We found that we did not have an algebraic function that worked as the inverse of the exponential function. In this chapter, we will be introduced to that important inverse of the exponential function, the logarithmic function. We will learn how to graph logarithmic functions. Algebraic techniques for solving exponential and logarithmic equations will be applied.

10.1 Logarithmic Functions

Activity Set

10.1

1. a. For each of the following functions, sketch the graph of the function and its inverse.

$$y = 2^x \qquad y = 5^x \qquad y = 10^x$$

b. Describe how the graph of the inverse of $y = b^x$ for $b > 1$ looks.

c. What is the domain and range of $y = b^x$ for $b > 1$? Explain.

d. What is the domain and range of the inverse of $y = b^x$ for $b > 1$? Explain.

2. a. What power of three produces 27? _____

What power of 3 produces 9? _____

What power of 3 produces 3? _____

What power of 3 produces 1? _____

What power of 3 produces $\frac{1}{3}$? _____

b. Complete Table A and Table B.

TABLE A	
Input w	**The output is the power of 3 that produces the input w.**
27	
9	2
3	
1	
$\frac{1}{3}$	

TABLE B	
Input x	**Output** 3^x
3	
2	
1	
1	
$^-1$	

c. When the input and output values of a function are interchanged, what is the relationship between the original function and the resulting function?

d. What is the relationship between Table A and Table B? Explain.

3. a. Complete Tables C and D.

TABLE C	
Input w	**The output is the power of 10 that produces the input w.**
$\frac{1}{100}$	$^-2$
$\frac{1}{10}$	
1	
10	
1000	
$1.0 * 10^6$	

TABLE D	
Input w	**The output is the power of 5 that produces the input w.**
$\frac{1}{25}$	
1	
5	
25	
125	
625	

b. Using Table C, answer the following questions.

Given an input of 1000, what power of 10 produces this input?

Given an input of 10, what power of 10 produces this input?

Given an input of $\frac{1}{100}$, what power of 10 produces this input?

Given an input of w, describe in words what the output is of the inverse of $y = 10^x$.

c. Using Table D, answer the following questions.

Given an input of 25, what power of 5 produces this input?

Given an input of 5, what power of 5 produces this input?

Given an input of 125, what power of 5 produces this input?

Given an input of w, describe in words what the output is of the inverse of $y = 5^x$.

d. Given an input of w, describe in words what the output is of the inverse of $y = b^x$ for $b > 1$.

Discussion 10.1

In Chapter 6, we explored exponential functions of the form $y = b^x$, where $b > 0$ and $b \neq 1$. In this section, we will be looking at a function, called the **logarithmic function,** which is the inverse of the exponential function. Recall, an inverse relation is formed by interchanging input and output values.

When we write $y = b^x$, it is assumed that x is the input variable and y is the output variable. To see this more visually, we can write an exponential function as

$$\text{output} = b^{\text{input}} \text{ where } b > 0 \text{ and } b \neq 1$$

Then to create the inverse of the exponential function, we can interchange the input and output. This means that the inverse of the exponential function is defined by the equation

$$\text{input} = b^{\text{output}} \text{ where } b > 0 \text{ and } b \neq 1$$

Therefore, the output of the inverse of the exponential function $y = b^x$ is the *power* of b that produces the input. Typically, we like to solve our equations for the output variable (dependent). Because we currently do not have a function that produces this power, we create one, and it is called the **logarithmic function with base b,** denoted $y = \log_b(x)$. Therefore, the output of the logarithmic function with base b is the *power* of b that produces the input x. The expression $\log_b(x)$ is read "log base b of x." For example, the inverse of the exponential function $y = 3^x$, is the logarithmic function with base 3, denoted $y = \log_3(x)$. In words, the output of $y = \log_3(x)$ is the *power* of 3 that produces the input x.

Example 1

Determine the value of each expression.

a. $\log_3(9)$ b. $\log_3\left(\frac{1}{9}\right)$ c. $\log_2(16)$

Solution

a. The output of $\log_3(9)$ is the *power* of 3 that produces 9. Therefore, we need to determine the exponent in the equation $3^? = 9$. Because $3^2 = 9$, $\log_3(9) = 2$.

b. The output of $\log_3\left(\frac{1}{9}\right)$ is the *power* of 3 that produces 1/9. Therefore, we need to determine the exponent in the equation $3^? = \frac{1}{9}$. Because $3^{-2} = \frac{1}{9}$, $\log_3\left(\frac{1}{9}\right) = {}^{-}2$.

c. The output of $\log_2(16)$ is the *power* of 2 that produces 16. Therefore, we need to determine the exponent in the equation $2^? = 16$. Because $2^4 = 16$, $\log_2(16) = 4$.

To summarize:

> For $b > 0$ and $b \ne 1$, the **logarithmic function with base b** is the inverse of the exponential function with base b.
>
> The statement
>
> $$\text{output} = \log_b(\text{input}) \qquad \text{is equivalent to} \qquad \text{input} = b^{\text{output}}.$$
>
> This can also be written as
>
> $$y = \log_b(x) \quad \text{is equivalent to} \quad x = b^y$$

The equation $\text{output} = \log_b(\text{input})$ is called a logarithmic equation or the logarithmic form of the equation. The equivalent equation $\text{input} = b^{\text{output}}$ is called an exponential equation or the exponential form of the equation. Notice that the exponent in the exponential equation is the output of the logarithmic equation. The logarithmic function outputs an *exponent*.

Example 2

Rewrite each logarithmic equation in its exponential form, then solve the resulting equation.

    a. $\log_3(27) = p$ b. $\log_5(m) = {}^-2$ c. $\log_4(k) = \frac{4}{3}$

Solution

a. The logarithmic equation $\log_3(27) = p$ is equivalent to the exponential equation $3^p = 27$. Because $3^3 = 27$, $p = 3$.

b. The logarithmic equation $\log_5(m) = {}^-2$ is equivalent to the exponential equation $5^{-2} = m$. Because $5^{-2} = \frac{1}{25}$, $m = \frac{1}{25}$.

c. The logarithmic equation $\log_4(k) = \frac{4}{3}$ can be written in the exponential form, $4^{4/3} = k$. Because $4^{4/3} \approx 6.35$, $k \approx 6.35$.

Graphs of Logarithmic Functions

Using what we now know about logarithmic functions, we graph two particular logarithmic functions.

Example 3

Complete the following table, and sketch a graph.

x	$y = \log_2(x)$
16	4
	3
	2
	1
	0
	$^-1$
	$^-2$

Solution   In the table we are given the output and must find the input. In the second row $y = 3$, which means that $3 = \log_2(x)$. This logarithmic equation is equivalent to the exponential equation $2^3 = x$. Therefore, when $y = 3$, $x = 8$. Similarly, $2 = \log_2(x)$ is equivalent to $2^2 = x$, or $x = 4$, and so on. The completed table is shown here.

x	$y = \log_2(x)$
16	4
8	3
4	2
2	1
1	0
$\frac{1}{2}$	-1
$\frac{1}{4}$	-2

We can then plot these points to obtain the following graph.

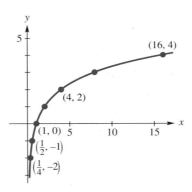

Example 4

Complete the following table, and sketch a graph.

x	$y = \log_3(x)$
$\frac{1}{3}$	
1	
3	
9	
27	
6	
60	

Solution   In the table we are given the input and must find the output. In the first row $x = \frac{1}{3}$, which means that $y = \log_3\left(\frac{1}{3}\right)$. This logarithmic equation is equivalent to the exponential equation $3^y = \frac{1}{3}$. Therefore, we must determine the power of 3 that produces $\frac{1}{3}$. Because $3^{-1} = \frac{1}{3}$, $y = {}^-1$. Similarly, from the second row we have $y = \log_3(1)$, which is equivalent to $3^y = 1$. Hence, $y = 0$. We can complete the first five rows in this fashion to obtain the table.

x	$y = \log_3(x)$
$\frac{1}{3}$	$^-1$
1	0
3	1
9	2
27	3
6	
60	

In the sixth row, we have $y = \log_3(6)$, which is equivalent to $3^y = 6$. We know that $3^1 = 3$ and $3^2 = 9$; therefore, y must be between 1 and 2. Using trial and error on a calculator, we can approximate the value of y to the nearest tenth to be approximately 1.6 because $3^{1.6} \approx 5.8$. Similarly, $y = \log_3(60)$ is equivalent to $3^y = 60$. Because $3^3 = 27$ and $3^4 = 81$, y must be between 3 and 4. To the nearest tenth y is approximately 3.7 because $3^{3.7} \approx 58.3$. With this we can complete the table and sketch a graph.

x	$y = \log_3(x)$
$\frac{1}{3}$	$^-1$
1	0
3	1
9	2
27	3
6	≈ 1.6
60	≈ 3.7

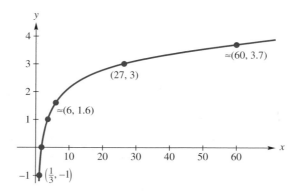

The graphs of the logarithmic functions in Examples 3 and 4 have very similar shapes. Let's now compare that shape with the shape of the graphs of exponential functions.

Recall that the graphs of a function and its inverse are reflections across the line $y = x$. Therefore, we can sketch a graph of a logarithmic function by reflecting the graph of an exponential function about the line $y = x$. The following graph is the graph of the exponential function $y = b^x$ for $b > 1$.

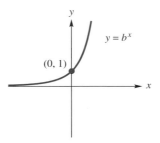

In the next figure, we have the graph of $y = b^x$ for $b > 1$ and its reflection about the line $y = x$. This reflection must be the graph of the inverse function $y = \log_b(x)$. This graph looks like we expected from the graphs we saw in the previous two examples.

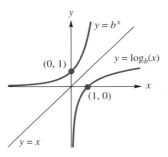

The following graph shows just the logarithmic function with the base b.

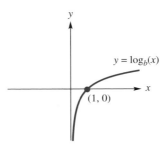

Every exponential function of the form $y = b^x$ passes through the point $(0, 1)$. Therefore, every logarithmic function of the form $y = \log_b(x)$ passes through the point $(1, 0)$.

Recall that the exponential function $y = b^x$ is asymptotic to the line $y = 0$, which is the x-axis. For very small values of x, the function can get arbitrarily close to an output of zero. Because the reflection about the line $y = x$ of the line $y = 0$ is the line $x = 0$, the logarithmic function $y = \log_b(x)$ is asymptotic to the line $x = 0$, the y-axis. We can also come to this conclusion by thinking about interchanging the input and output between the function and its inverse.

The domain of the exponential function $y = b^x$ is all real numbers, and the range is all positive real numbers. Because the input and output values of inverse functions are interchanged, the domain of the logarithmic function $y = \log_b(x)$ is all positive real numbers, and the range is all real numbers. Look at the graph of the function $y = \log_b(x)$ to see that this is reasonable.

This information is summarized in the following box.

> *Domain and Range of Exponential and Logarithmic Functions*
>
> The exponential function $y = b^x$ has the following domain and range.
>
> > domain = all real numbers
> >
> > range = all positive real numbers
>
> The logarithmic function $y = \log_b(x)$ has the following domain and range.
>
> > domain = all positive real numbers
> >
> > range = all real numbers

We looked only at logarithmic functions with a base greater than one. This is because in practice, the base of a logarithmic function is usually greater than one. Recall that the exponential function $y = \left(\frac{1}{2}\right)^x$ can also be written as $y = 2^{-x}$. We rewrote the exponential function with a base less than one as an exponential function with base greater than one. The same applies to logarithmic functions.

Two bases are used more often than others: the natural base e and the base 10. The logarithmic function with the base e is called the natural logarithmic function, denoted $y = \ln(x)$, and the logarithmic function with the base 10 is called the common logarithmic function, denoted $y = \log(x)$. Most calculators have the natural and common logarithmic functions built in.

Definition

The function $y = \log_{10}(x)$ is the **common logarithmic function** and is denoted $y = \log(x)$.

Definition

The function $y = \log_e(x)$ is the **natural logarithmic function** and is denoted $y = \ln(x)$.

Being able to change equations from exponential form to logarithmic form or vice versa can be useful in solving exponential and logarithmic equations.

Example 5

Algebraically solve each of the following equations. Rewriting the equation may be helpful.

a. $\log(t) = 2$ 　　　　　　　 b. $\ln(x) = 3.2$ 　　　　　　　 c. $e^{t+1} = 15$

Solution

a. Because the base is not explicitly written in this logarithmic equation, the base is understood to be 10.

$$\log(t) = 2$$
$$10^2 = t \qquad \text{Rewrite in exponential notation using base 10.}$$
$$100 = t$$

b. The base of the function $\ln(x)$ is e.

$$\ln(x) = 3.2$$
$$e^{3.2} = x \qquad \text{Rewrite in exponential notation using base } e.$$
$$24.5 \approx x$$

c. Because the variable is in the exponent, rewriting the equation may be helpful.

$$e^{t+1} = 15$$
$$\log_e(15) = t + 1 \qquad \text{Rewrite in logarithmic notation using base } e.$$
$$\ln(15) = t + 1 \qquad \text{Recall that ln is the notation used for base } e.$$
$$2.71 \approx t + 1$$
$$1.71 \approx t$$

In this section, we began to explore **logarithmic functions.** For $b > 0$ and $b \neq 1$, the logarithmic function with the base b is the inverse of the exponential function with the base b. Because the input of an exponential function is the exponent, and inputs and outputs of inverse functions are interchanged, the output of the logarithmic function is an exponent. More specifically, the output of the logarithmic function $y = \log_b(x)$ is the power of b that produces the input x.

We looked at the graphs of logarithmic functions for which the base was restricted to be greater than one. These graphs were the reflections about the line $y = x$ of the graphs of exponential functions for bases greater than one. The domain of logarithmic functions is all positive real numbers, and the range is all real numbers. The graph of $y = \log_b(x)$ contains the point $(1, 0)$ and is asymptotic to the line $x = 0$, the vertical axis.

Problem
Set
10.1

1. Without using a calculator, evaluate each logarithmic expression.

   a. $\log_6(36)$ **c.** $\log_4\left(\frac{1}{16}\right)$ **e.** $\log_5(1)$

   b. $\log_5(25)$ **d.** $\log_{1/2}(8)$ **f.** $\log(1000)$

2. Without using a calculator, evaluate each logarithmic expression.

   a. $\log_2(8)$ **d.** $\log_3\left(\frac{1}{9}\right)$ **g.** $\log_3(81)$

   b. $\log(10{,}000)$ **e.** $\ln(e^3)$ **h.** $\log_4(16)$

   c. $\log_{13}(1)$ **f.** $\log_{1/2}(8)$

3. **a.** Explain how you know that $\log_3(100)$ is between 4 and 5.

   b. Explain how you know that $\log_3\left(\frac{1}{100}\right)$ is between $^-5$ and $^-4$.

4. Using a calculator, evaluate each logarithmic expression. Round approximate values to three significant digits. Explain why each result is reasonable.

   a. $\ln(3.45)$ **d.** $\ln(10)$

   b. $\log(13{,}250)$ **e.** $\log(256)$

   c. $\log(4)$ **f.** $\log(0.01)$

5. Rewrite each logarithmic equation in exponential form and each exponential equation in logarithmic form.

   a. $\log_2(x) = 5$ **d.** $\log_4(x^2) = -\frac{1}{2}$

   b. $4^m = 16$ **e.** $\ln(^-x) = 4$

   c. $\log_x(5) = \frac{1}{2}$ **f.** $10^{-m} = 12{,}000$

6. Solve each equation analytically. You may find rewriting some equations in alternative form helpful. Round approximate solutions to four significant digits.

   a. $\log_3(x) = 4$ **e.** $\ln(^-2x) = 7$

   b. $2^t = 16$ **f.** $10^k = 10{,}000$

   c. $\log_b(3) = \frac{1}{2}$ **g.** $10^k = 700$

   d. $e^{-x} = 5.78$

7. Solve each equation analytically. Round approximate solutions to four significant digits.

   a. $\log_7(x) = 2$ **d.** $\ln(x) = 1.4$

   b. $e^t = 17.6$ **e.** $10^{2x} = 785$

   c. $2^n = 0.25$ **f.** $\log(x + 1) = -2.4$

8. Complete the following table, and sketch a graph.

x	$y = \log_5(x)$
$\frac{1}{5}$	
1	
5	
25	
10	
60	
200	

9. For each of the following parts, assume that x represents the independent variable and y represents the dependent variable.

   a. For each function on the left, identify an *equivalent* exponential function on the right.

   b. For each function on the left, identify its *inverse* exponential function on the right.

$$y = \log_2(x) \qquad\qquad y = 10^x$$
$$x = 2^y$$
$$y = \log(x) \qquad\qquad x = e^y$$
$$y = 2^x$$
$$y = \ln(x) \qquad\qquad x = 10^y$$
$$y = e^x$$

10. Identify each statement as true or false.

   a. $\log_2\left(\frac{1}{8}\right) = \frac{1}{3}$ **c.** $\log_2(^-8) = {}^-3$

   b. $\log_8(2) = \frac{1}{3}$ **d.** $\log_5(2) = 25$

11. **a.** Graph $y = \log(10^x)$ using your graphing calculator. What other function have you seen that has this same graph?

   b. Graph $y = \ln(e^x)$ using your graphing calculator. What other function have you seen that has this same graph?

   c. Rewrite $\log_b(b^x)$ for $b > 1$ as an equivalent expression using your observations from parts a and b. What is the domain of the function $y = \log_b(b^x)$ for $b > 1$? Explain.

   d. Graph $y = 10^{\log(x)}$ using your graphing calculator. What other function have you seen that has this same graph?

   e. Graph $y = e^{\ln(x)}$ using your graphing calculator. What other function have you seen that has this same graph?

   f. Rewrite $b^{\log_b(x)}$ for $b > 1$ as an equivalent expression using your observations from parts d and e. What is the domain of the function $y = b^{\log_b(x)}$ for $b > 1$? Explain.

12. Graphically solve each of the following inequalities.

   a. $\log(m) < 3$

   b. $\ln(p) \leq {}^-4$

13. a. Is $\ln(4)$ less than, greater than, or equal to $\log(4)$? Explain.

　b. Is $\ln(2)$ less than, greater than, or equal to $\log(2)$? Explain.

　c. Is $\ln(1)$ less than, greater than, or equal to $\log(1)$? Explain.

　d. Is $\ln\left(\frac{1}{2}\right)$ less than, greater than, or equal to $\log\left(\frac{1}{2}\right)$? Explain.

　e. Is $\ln\left(\frac{1}{10}\right)$ less than, greater than, or equal to $\log\left(\frac{1}{10}\right)$? Explain.

　f. Using your results from parts a–e, solve the inequality $\ln(x) < \log(x)$.

14. a. Is $\log_2(10)$ less than, greater than, or equal to $\log_5(10)$? Explain.

　b. Is $\log_2(6)$ less than, greater than, or equal to $\log_5(6)$? Explain.

　c. Is $\log_2(1)$ less than, greater than, or equal to $\log_5(1)$? Explain.

　d. Is $\log_2\left(\frac{1}{2}\right)$ less than, greater than, or equal to $\log_5\left(\frac{1}{2}\right)$? Explain.

　e. Is $\log_2\left(\frac{1}{5}\right)$ less than, greater than, or equal to $\log_5\left(\frac{1}{5}\right)$? Explain.

　f. Using your results from parts a–e, solve the inequality $\log_2(x) < \log_5(x)$.

15. Henderson–Hasselbach.　The Henderson–Hasselbach equation can be used to relate a patient's blood pH (or acidity) to the concentrations of bicarbonate and carbonic acid in the blood plasma by the following equation.

$$pH = 6.1 + \log\left(\frac{\text{bicarbonate}}{\text{carbonic acid}}\right)$$

　a. In a healthy person the concentration of bicarbonate ranges from 24 to 28 mEq/L, the concentration of carbonic acid ranges from 1.2 to 1.4 mEq/L, and the ratio between bicarbonate and carbonic acid should be 20 to 1. What is the concentration of bicarbonate in a healthy person if the concentration of carbonic acid is 1.3 mEq/L? What is the pH of this person's blood plasma?

　b. In a patient with lung disease the carbonic acid concentration increases. If this concentration reaches 2 mEq/L the patient is considered to have pulmonary failure. Is a patient with a bicarbonate concentration of 26 mEq/L and a blood pH of 7.1 at risk of pulmonary failure? Explain your response.

10.2 Exponential and Logarithmic Equations

1. Without using a calculator, determine the value of each of the following.
   a. $\log_2 32$
   b. $\log_3 27$
   c. $\log 100{,}000$

2. Without using a calculator, determine the value of each expression in the following table. For the first column, refer to Activity 1.

$\log_b(x*y)$	$\log_b x$	$\log_b y$
$\log_2(8*4) =$	$\log_2 8 =$	$\log_2 4 =$
$\log_3(3*9) =$	$\log_3 3 =$	$\log_3 9 =$
$\log(100*1000) =$	$\log 100 =$	$\log 1000 =$

   a. Look for a pattern that relates $\log_b(x*y)$ to $\log_b x$ and $\log_b y$.
   b. Use the pattern you have observed to make the following equation true by changing the ? to an arithmetic operation.

$$\log_b(x*y) = \log_b x \ ? \ \log_b y.$$

3. Without using a calculator, determine the value of each expression in the following table.

$\log_b\left(\frac{x}{y}\right)$	$\log_b x$	$\log_b y$
$\log_2\left(\frac{8}{4}\right) =$	$\log_2 8 =$	$\log_2 4 =$
$\log_3\left(\frac{81}{3}\right) =$	$\log_3 81 =$	$\log_3 3 =$
$\log\left(\frac{10{,}000}{100}\right) =$	$\log 10{,}000 =$	$\log 100 =$

   a. Look for a pattern that relates $\log_b\left(\frac{x}{y}\right)$ to $\log_b x$ and $\log_b y$.
   b. Use the pattern you have observed to make the following equation true by changing the ? to an arithmetic operation.

$$\log_b\left(\frac{x}{y}\right) = \log_b x \ ? \ \log_b y.$$

4. In Activity 2, you found that $\log_b(x*y) = \log_b x + \log_b y$. We can use this property to simplify $\log(a^4)$.

$$\log(a^4) = \log(a*a*a*a)$$
$$= \log(a) + \log(a) + \log(a) + \log(a)$$
$$= 4*\log(a)$$

   a. Based on this rewriting of $\log(a^4)$, how would you rewrite $\log(a^p)$?
   b. Verify your conjecture from part a.

Discussion 10.2

Solving Logarithmic Equations

An equation in which the variable appears in the argument of a logarithm is called a **logarithmic equation.** Translating the logarithmic equation into its equivalent exponential form can solve many logarithmic equations. Recall that the logarithmic equation $y = \log_b(x)$ is equivalent to the exponential equation $x = b^y$.

Example 1

Solve the equation $\log(x - 3) + 5.6 = 7.3$. Round the solution to the nearest tenth.

Solution   The equation $\log(x - 3) + 5.6 = 7.3$ is a logarithmic equation because the variable x is part of the argument of the log function.

$$\log(x - 3) + 5.6 = 7.3$$

$$\log(x - 3) = 1.7 \qquad \text{Isolate the term involving the logarithm.}$$

$$10^{1.7} = x - 3 \qquad \text{Translate the common log to its equivalent exponential form with base 10.}$$

$$10^{1.7} + 3 = x \qquad \text{Use a calculator to evaluate.}$$

$$53.1 \approx x$$

CHECK

$$\log(53.1 - 3) + 5.6 \overset{?}{\approx} 7.3$$

$$7.2998 \approx 7.3 \qquad ✔$$

The solution is $x \approx 53.1$.

Some logarithmic equations cannot be solved by simply translating to an equivalent exponential equation. We will now look at other techniques that can be used to solve logarithmic equations.

In the activities you discovered equivalent logarithmic expressions. These equivalent expressions are actually properties of logarithms and can be used to rewrite expressions containing several terms involving logarithms. We will formalize these properties now and look at why they work.

> *Product Property for Logarithms*
>
> For positive real numbers x, y, and b with $b \neq 1$
>
> $$\log_b(xy) = \log_b(x) + \log_b(y)$$
>
> The log of the product of x and y is equivalent to the sum of the log of x and the log of y.

Let's see why this logarithmic property is true. Assign variables to $\log_b(x)$ and $\log_b(y)$.

$\log_b(x) = u$ and $\log_b(y) = w.$

$b^u = x$ and $b^w = y$	Translate each equation to exponential form.
$xy = b^u * b^w$	Write the product of x and y.
$xy = b^{u+w}$	Apply the property for multiplying powers.
$\log_b(xy) = u + w$	Translate the equation to logarithmic form.
$\log_b(xy) = \log_b(x) + \log_b(y)$	Replace u and w by their equivalent expressions.

We have shown that the product property for logarithms is true.

> *Quotient Property for Logarithms*
>
> For positive real numbers x, y, and b with $b \neq 1$
>
> $$\log_b\left(\frac{x}{y}\right) = \log_b(x) - \log_b(y)$$
>
> The log of the quotient of x and y is equivalent to the difference of the log of x and the log of y.

The quotient property for logarithms can be justified using an argument similar to the one used to justify the product property for logarithms. This is left for you to do as a problem at the end of the section.

> *Power Property for Logarithms*
>
> For positive real numbers x, y, and b with $b \neq 1$
>
> $$\log_b(x^n) = n * \log_b(x)$$
>
> The log of the nth power of x is equivalent to the product of n and the log of x.

We show why this logarithmic property is true in the same manner as the first property.

$\log_b(x) = u$	Assign a variable to represent $\log_b(x)$.
$b^u = x$	Translate the equation from logarithmic to exponential form.
$x^n = (b^u)^n$	Apply the nth power to each side.
$x^n = b^{n*u}$	Apply the property for taking a power of a power.
$\log_b(x^n) = n * u$	Translate from exponential to logarithmic form.
$\log_b(x^n) = n * \log_b(x)$	Substitute the equivalent expression for u.

We have justified the power property for logarithms.

We will find these properties useful in solving logarithmic equations. This is demonstrated in the following example.

Example 2

Solve each logarithmic equation. Round approximate solutions to the nearest hundredth.

    a. $\log(2^x) = 12$

    b. $\ln(t) + \ln(t + 1) = 2.31$

Solution   a. Translating the logarithmic equation $\log(2^x) = 12$ to exponential form gives us $10^{12} = 2^x$. This does not get us any closer to a solution. Instead, we try applying one of the properties of logarithms.

$$\log(2^x) = 12$$

$$x * \log(2) = 12 \qquad \text{Apply the power property of logarithms.}$$

$$x = \frac{12}{\log(2)} \qquad \text{Divide by } \log(2).$$

$$x \approx 39.9 \qquad \text{Use a calculator to evaluate.}$$

CHECK

$$\log(2^x) = 12$$

$$\log(2^{39.9}) \overset{?}{=} 12$$

$$12.0111 \approx 12 \qquad ✔$$

The solution to $\log(2^x) = 12$ is approximately 39.9.

b. The equation $\ln(t) + \ln(t + 1) = 2.31$ cannot be translated to an exponential equation because there are two terms that are logarithmic expressions.

$$\ln(t) + \ln(t + 1) = 2.31$$

$$\ln[t(t + 1)] = 2.31 \qquad \text{Apply the product property of logarithms.}$$

$$t(t + 1) = e^{2.31} \qquad \text{Rewrite the equation in logarithm form,}$$
$$\qquad\qquad\qquad\quad \text{recalling that } \ln(t) = \log_e(t).$$

$$t^2 + t \approx 10.07$$

$$t^2 + t - 10.07 \approx 0$$

$$t = \frac{-1 \pm \sqrt{1^2 - 4 * 1 * {}^-10.07}}{2 * 1} \qquad \text{Use the quadratic formula.}$$

$$t \approx 2.71 \qquad \text{or} \qquad t \approx {}^-3.71 \qquad \text{We obtain two possible solutions.}$$

CHECK

Check in the original equation.

$t \approx 2.71$

$$\ln(2.71) + \ln(2.71 + 1) \overset{?}{\approx} 2.31$$

$$2.30798 \approx 2.31 \qquad ✔$$

$t \approx {}^-3.71$

$$\ln({}^-3.71) + \ln({}^-3.71 + 1) \overset{?}{\approx} 2.31$$

$$\text{undefined} \neq 2.31 \qquad ✗$$

Because logarithmic functions are not defined for input values that are negative, $t \approx {}^-3.71$ is an extraneous solution. Our only solution is $t \approx 2.71$.

We now summarize a strategy for solving logarithmic equations.

Strategy

Solving Logarithmic Equations

1. If there are two or more terms in which the variable appears in the argument of a logarithm, use the properties of logarithms to rewrite with only one term involving a logarithm. If the variable appears as a power in the argument of the logarithm, use the power property to rewrite it.

2. If the variable is still in the argument of the logarithm, isolate the factor that includes the logarithm.

3. Rewrite the equation in exponential form.

4. Solve for the variable.

5. Check all possible solutions in the original equation. Remember that the argument of a logarithm must be nonnegative.

Solving Exponential Equations

An equation in which the variable appears in the exponent is called an **exponential equation.** Using the inverse of the exponential function, the logarithmic function, can solve most exponential equations. Because log base 10 and log base e are built-in functions in most technology, we restrict our solution processes to using one of these two logarithms and then use the properties of logarithms to help us solve exponential equations involving other bases.

Example 3

Solve the following two exponential equations. Round approximate solutions to the nearest thousandth.

    a. $10^{x-2} = 860$

    b. $3 * 5^n = 123$

Solution

a. To solve the exponential equation $10^{x-2} = 860$, we can translate it into a logarithmic equation.

$$10^{x-2} = 860$$

$$\log(860) = x - 2 \qquad \text{Rewrite in logarithmic form.}$$

$$x = \log(860) + 2$$

$$x \approx 4.934$$

CHECK

$$10^{x-2} = 860$$

$$10^{(4.934-2)} \stackrel{?}{=} 860$$

$$859 \approx 860 \qquad ✔$$

The solution to $10^{x-2} = 860$ is $x \approx 4.934$.

b. To solve the exponential equation $3 * 5^n = 123$, we must first isolate the factor containing the exponent.

$$3 * 5^n = 123$$

$$5^n = 41$$

$$\log_5(41) = n \qquad \text{Translate to logarithmic form.}$$

This equation is solved for n, but we have no way to evaluate $\log_5(41)$. Instead of translating the equation $5^n = 41$ to a logarithmic equation, we take the logarithm of both sides of the equation. We can use either the common logarithm (log) or the natural logarithm (ln) because our calculator can evaluate these logarithmic functions.

$$5^n = 41$$

$$\log(5^n) = \log(41) \qquad \text{Take the common log of both sides.}$$

$$n * \log(5) = \log(41) \qquad \text{Apply the power property of logarithms.}$$

$$n = \frac{\log(41)}{\log(5)}$$

$$n \approx 2.307$$

CHECK   Check the solution in the original equation.

$$3 * 5^n = 123$$

$$3 * 5^{2.307} \stackrel{?}{=} 123$$

$$122.93 \approx 123 \qquad ✔$$

The solution to $3 * 5^n = 123$ is $n \approx 2.307$.

In Example 3 we saw that if we have an exponential equation with a base of e or 10, we can translate to a logarithmic equation that we can solve. If the base is neither e nor 10, we need to apply either the natural logarithm (ln) or the common logarithm (log) to both sides of the equation. We can summarize this strategy.

Strategy

Solving Exponential Equations

1. Isolate the factor involving the exponent.
2. If the base of the exponent is e or 10, translate to a logarithmic equation and solve.
3. If the base of the exponent is neither e nor 10, apply either the common logarithm or the natural logarithm to both sides of the equation. Then apply the power property of logarithms and solve.
4. Check all possible solutions in the original equation.

Example 4

Doubling the Population. The formula for the growth of the population of a South American country is

$$P = P_o e^{0.032t}$$

where P_o is the population today, and P is the predicted population in t years. Determine when the population doubles.

Solution The population doubles when the population is twice what it is today. That is, when $P = 2P_0$. We can then substitute $2P_0$ for P in our equation.

$$2P_0 = P_0 e^{0.032t} \qquad \text{Substitute } 2P_0 \text{ for } P.$$

$$\frac{2P_0}{P_0} = e^{0.032t} \qquad \text{Isolate } e^{0.032t}.$$

$$2 = e^{0.032t}$$

Notice that P_0 is canceled from both sides of the equation. The time it takes the population to double does not depend on the original amount. Next, rewrite the equation in logarithmic form.

$$\ln(2) = 0.032t \qquad \text{Rewrite in logarithmic form.}$$

$$\frac{\ln(2)}{0.032} = t \qquad \text{Solve for } t.$$

$$21.66 \approx t$$

CHECK

$$2P_0 \overset{?}{\approx} P_0 e^{0.032 * 21.66}$$

$$2P_0 \approx 1.99995 P_0 \qquad \checkmark$$

We conclude that the population doubles in the 22nd year.

In this section we looked at how to analytically solve **logarithmic equations** and **exponential equations.** Logarithmic equations are those in which the variable appears in the argument of a logarithm. Exponential equations are those in which the variable appears in the exponent. Strategies were presented for solving each of these types of equations. Because logarithmic functions have restricted domains, some solutions may be extraneous. Therefore, checking possible solutions in the original equation is always necessary.

Problem
Set
10.2

1. Identify each of the following equations as representing a linear function, a quadratic function, a power function, a logarithmic function, an exponential function, or other.

   a. $f(x) = 3.5^{x/2}$

   b. $y = 2x^2 - x + 5$

   c. $g(x) = \dfrac{x+6}{5}$

   d. $h(x) = \dfrac{x+6}{x}$

   e. $y = \ln(x+4)$

   f. $y = 3x^5$

2. Solve each logarithmic equation algebraically. Round approximate solutions to the nearest hundredth. Check your solutions.

   a. $\ln(x) = 1.2$

   b. $\log(3x) = 0.35$

   c. $\log(2x - 1) = 2.5$

   d. $\ln\left(\dfrac{x}{2}\right) = -1$

3. Solve each logarithmic equation algebraically. Round approximate solutions to the nearest hundredth.

   a. $\ln(t) + 5 = 6.25$

   b. $\log_5(2x - 1) = -1$

   c. $\log(x^3) = 4$

   d. $\log_3(2m - 7) = 3$

   e. $\ln(x - 2) - \ln(x) = 1.4$

   f. $\log_2(x) - \log_2(x - 1) = -2$

   g. $\ln(n + 5) = 3.2 - \ln(3.2)$

   h. $\log(3x) + \log(x + 5) = 1$

4. Solve each exponential equation algebraically. Round approximate solutions to the nearest hundredth. Check your solutions.

   a. $e^{x-1} = 15$

   b. $3(10^t) + 15 = 21$

   c. $10e^{2x} = 400$

   d. $10^{x/2} = \dfrac{1}{2}$

   e. $200(1 + 0.05)^x = 400$

   f. $10{,}000 = 600 * 5^{(m/4)}$

5. Solve each exponential equation. Round approximate solutions to the nearest hundredth. Check your solutions.

   a. $7^x = 561$

   b. $4 * 5^n = 673$

   c. $3^{(6+x)} = 15.7$

   d. $2^k + 348.9 = 698.5$

   e. $200(1 + 0.05)^x = 400$

   f. $10{,}000 = 600 * 5^{(m/4)}$

6. Solve for the indicated variable.

   a. $R = \log(I)$ for I

   b. $P = P_o e^{rt}$ for t

   c. $M = \ln(x) - \ln(x - 1)$ for x

   d. $L = 10 * \log\left(\dfrac{I}{I_0}\right)$ for I

   e. $A = A_0 e^{-kt}$ for k

   f. $2 = \left(1 + \dfrac{r}{n}\right)^{nt}$ for t

   g. $P_0(2)^{t/4} = P_0 e^{kt}$ for k

7. Show, using the properties of exponents, that the quotient property of logarithms is true. Looking at the process used in the text to show that the product property of logarithms is true may be helpful.

8. *The Henderson–Hasselbach Equation* can be used to relate a patient's blood pH (acidity) to the concentrations of bicarbonate and carbonic acid in the blood plasma by the following equation.

$$pH = 6.1 + \log\left(\frac{\text{bicarbonate}}{\text{carbonic acid}}\right)$$

Solve the formula for the carbonic acid concentration in terms of the pH and the bicarbonate concentration.

9. Annual Compounding of Interest.   How long will it take for a $1000 investment to double if it is in an account that earns 6.5% annual interest compounded annually? Algebraically solve this problem.

10. Compounding Continuously.   How long will it take for a $1000 investment to double if it is in an account that earns 6.5% annual interest compounded continuously? Algebraically solve this problem to the nearest tenth of a year.

11. **a.** Complete the table, and sketch the graph of $y = \log_5(x)$.

x	$\frac{1}{5}$	1	5	25	125
y					

   b. From the graph estimate $\log_5(50)$.
   c. Rewrite the equation $y = \log_5(50)$ in exponential form.
   d. Solve the equation from part c for y using either $\log_{10}$ or ln. Compare this result with your estimate in part b.
   e. Rewrite the equation $y = \log_b x$ in exponential form. Solve the resulting equation for y using either the common logarithm or the natural logarithm.

12. A Savings Plan.   For many people, a savings plan is based on regular deposits, not one large deposit. If you plan to deposit $100 monthly into a savings account that pays 0.5% monthly interest, the formula

$$B = A\frac{(1 + i)^n - 1}{i}$$

   where B = balance after n months
   A = amount deposited each month
   i = interest rate per month
   n = number of months
   can be used to compute the balance in the account after n months.

   a. Using this model, how much is in the account after one year? After five years?
   b. How long will it take to reach $2400?

13. Half-Life.   A highly unstable nuclear element has a half-life of 38 hours. This means that after 38 hours half of the original element remains in its original form. Write an algebraic model for the amount of substance left after t hours if 100 grams is present originally. Use your model to determine when 40% of the original element remains.

14. Noise Levels.   The formula for decibel readings is $D = 10 \log(I)$, where I is the number of times more intense a sound is than the sound on the threshold of hearing.

   a. A local rock club had a decibel reading of 95 dB. What was the intensity level at the rock club?

   b. Use resources such as the library, health professionals, and so on to find the decibel level at which damage to a person's ears can occur.

   c. If equipment at a local mill registers 60.2 dB, should the mill be required to provide protection for workers' ears? Explain.

15. Solve each of the equations. Round approximate answers to four significant digits.

   a. $3^{5x} = \dfrac{1}{27}$

   b. $10e^{0.25x} = 20$

   c. $(5x)^3 = \dfrac{1}{27}$

   d. $150 = 400 * \left(\dfrac{x}{9}\right)^5$

   e. $\left(\dfrac{5}{3}\right)^{-x} = 20$

   f. $\sqrt[3]{5x} = 10$

   g. $5 + 4^{3x} = 1.2$

   h. $5 * 4^{-x^2} = 1.2$

16. Inflation Problem.   One gallon of milk cost $2.09 in 2000. Assume a 5% annual rate of inflation. The formula for the amount A an item costs after t years is

$$A = P(1 + r)^t$$

Assume that the rate of inflation remains constant for the next five years and has been close to 5% for the last five years.

   a. What will a gallon of milk cost in 2002? 2005?

   b. What did a gallon of milk cost in 1997 according to our model?

   c. Make a table of values for A as t ranges from -5 to 5.

   d. Draw a graph by plotting the points from your table and connecting the points with a smooth curve.

17. Inflation Problem, Generalized.   If the rate of inflation is such that the cost of living is doubling every ten years, what is the average rate of inflation per year as a percent?

Chapter Ten Summary

The inverse of an exponential function is a **logarithmic function.**

Each exponential function $y = b^x$ has a corresponding logarithmic function $y = \log_b(x)$ as its inverse. The output of the logarithmic function with base b is the *power* of b that produces the input x. For example, $\log_5(25)$ is the power that must be applied to 5 to get 25. In other words, $\log_5(25) = 2$. This relationship is summarized in the following box.

> For $b > 0$ and $b \neq 1$, the **logarithmic function with base b** is the inverse of the exponential function with base b.
>
> The statement
>
> $$output = \log_b(input) \qquad \text{is equivalent to} \qquad input = b^{output}$$
>
> This could also be written as
>
> $$y = \log_b(x) \qquad \text{is equivalent to} \qquad x = b^y$$

It is very important to distinguish clearly between the exponential equation that is the inverse of a logarithmic equation and the exponential equation that is equivalent to a logarithmic equation.

$$y = \log_b(x) \qquad \text{and} \qquad y = b^x \qquad \text{are inverses.}$$
$$y = \log_b(x) \qquad \text{and} \qquad x = b^y \qquad \text{are equivalent.}$$

The equivalence is used most often. For example, $4 = \log_5(x)$ can be written in its equivalent exponential form $x = 5^4$.

Thinking of the shape of the exponential function $y = b^x$, $b > 1$, can be helpful in figuring out the shape of its inverse, the logarithmic function $y = \log_b(x)$. Because they are inverse functions, the graph of the logarithmic function is the reflection across the line $y = x$ of the exponential function.

> *The General Shape of a Graph of the Logarithmic Function for $b > 1$*
>
> - The domain of the logarithmic function is all positive real numbers.
> - The range of the logarithmic function is all real numbers.
> - The horizontal intercept of the graph is $(1, 0)$.
> - The graph is asymptotic to the line $x = 0$, the vertical axis.
>
>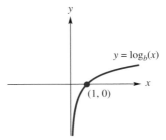

Two bases are used more often than others: the natural base e and the base 10. The logarithmic function with the base e is called the **natural logarithmic function,** denoted $y = \ln(x)$. The logarithmic function with the base 10 is called the **common logarithmic function,** denoted $y = \log(x)$. These logarithmic functions are found on scientific calculators.

An equation in which the variable appears in the argument of a logarithm is called a **logarithmic equation.** Translating the logarithmic equation into its equivalent exponential form can solve many logarithmic equations.

Sometimes translating to an equivalent form is not enough to allow us to solve an equation. In these cases, we can use the following properties to rewrite the equations.

Properties for Logarithms
For positive real numbers x, y, and b with $b \neq 1$

$$\log_b(xy) = \log_b(x) + \log_b(y)$$ The log of a product is the sum of the logs.

$$\log_b\left(\frac{x}{y}\right) = \log_b(x) - \log_b(y)$$ The log of a quotient is the difference of the logs.

$$\log_b(x^n) = n * \log_b(x)$$ The log of a power is the power times the log.

The following strategy was presented for solving logarithmic equations.

$\mathsf{Strategy}$

Solving Logarithmic Equations

1. If there are two or more terms in which the variable appears in the argument of a logarithm, use the properties of logarithms to rewrite with only one term involving a logarithm. If the variable appears as a power in the argument of the logarithm, use the power property to rewrite it.

2. If the variable is still in the argument of the logarithm, isolate the factor that includes the logarithm.

3. Rewrite the equation in exponential form.

4. Solve for the variable.

5. Check all possible solutions in the original equation. Remember that the argument of a logarithm must be nonnegative.

An equation in which the variable appears in the exponent is called an **exponential equation.** Previously, we had been able to solve these equations numerically or graphically. With the introduction of the logarithmic function, we are now able to solve them analytically. Using the inverse of the exponential function, the logarithmic function, we can solve most exponential equations. Because log base 10 and log base e are built in functions in most calculators, we restrict our solution processes to using one of these two logarithms. We use the properties of logarithms to help us solve exponential equations involving other bases.

The strategy for solving exponential equations is repeated here.

$\mathsf{Strategy}$

Solving Exponential Equations

1. Isolate the factor involving the exponent.

2. If the base of the exponent is e or 10, translate to a logarithmic equation and solve.

3. If the base of the exponent is neither e nor 10, apply either the common logarithm or the natural logarithm to both sides of the equation. Then apply the power property of logarithms and solve.

4. Check all possible solutions in the original equation.

Chapter
Review

7–10

1. Look over the following list of functions. Suppose you are asked to graph each function without using your calculator. Without actually graphing these functions, rank them in order from easiest to hardest. (Label the easiest #1, second easiest #2, and so on.)

   a. $y = 4x(x - 15)$ _____

   b. $y = {}^-2.5x^2 + 40$ _____

   c. $y = x^2$ _____

   d. $y = 10$ _____

   e. $y = 2x - 14$ _____

   f. $y = (x + 2)(x - 15)$ _____

   g. $y = x$ _____

   h. $y = x^2 - 14x + 25$ _____

2. Suppose you are asked to solve each of the following equations algebraically. Without actually solving them, rank them in order from easiest to hardest. (Label the easiest #1, second easiest #2, and so on.)

   a. $(x - 2)^2 = 49$ _____

   b. $(x - 6)(x + 5) = 0$ _____

   c. $x^2 - x - 9 = {}^-7$ _____

   d. $(x - 2)^2 = {}^-49$ _____

   e. $(x - 6)(x + 5) = 12$ _____

   f. $x^2 - 3x - 9 = 0$ _____

   g. $(3x - 5) = 25$ _____

3. Suppose you are asked to solve each of the following equations for the indicated variable. Without actually solving them, rank them in order from easiest to hardest. (Label the easiest #1, second easiest #2, and so on.)

   a. $52 = \pi r^2 h$ for h _____

   b. $0 = AP - P^2$ for A _____

   c. $8 = AP - P^2$ for P _____

   d. $0 = AP - P^2$ for P _____

4. Suppose you are asked to find the horizontal intercepts for each function algebraically. Without actually finding them, rank them in order from easiest to hardest. (Label the easiest #1, second easiest #2, and so on.)

   a. $y = 5x + 15$ _____

   b. $y = 2x^2 - 4x + 5$ _____

   c. $4x - y = 28$ _____

   d. $y = {}^-0.5x^2 - 4$ _____

5. Suppose you are asked to find the vertex of each function. Without actually finding them, rank them in order from easiest to hardest. (Label the easiest #1, second easiest #2, and so on.)

   a. $y = 3x^2 - 8$ _____

   b. $y = |15 - 2x|$ _____

   c. $y = 3(x - 2)(x - 10)$ _____

   d. $y = {}^-0.5x^2 - 4x$ _____

6. Describe how to use a table to solve the following equation.

$$-3x^2 - 5x + 10 = 0.5x + 4$$

7. Describe how to use a graph to solve the following equation.

$$-3x^2 - 5x + 10 = 0.5x + 4$$

8. Describe how to solve the following equation algebraically.

$$-3x^2 - 5x + 10 = 0.5x + 4$$

9. Consider the function $y = ax^2 + bx + c$. Draw one possible graph of this function if we know that $a > 0$, $c > 0$, and $b^2 - 4ac = 0$.

10. Write the equation of the line passing through $(8, 15)$ with slope $^-3$.

11. Write the equation of the line shown.

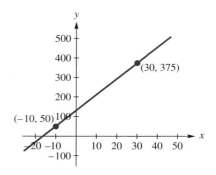

12. Write the equation of the following line. Check your equation.

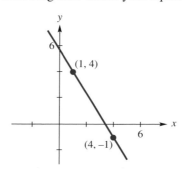

13. For each equation, indicate whether the equation graphs as a line, a parabola, a V-shaped graph, or other. You should be able to make your decision without graphing the equation.

a. $y + \dfrac{x}{2} = \dfrac{2}{3}$

b. $\sqrt{7}x = 2y$

c. $y = \dfrac{x^2}{2} + 3$

d. $y = |x - 10|$

e. $y = \sqrt{3x}$

f. $y = \sqrt[3]{9}$

g. $y = 2x(x - 15)$

14. Graph each linear function. Your graph should clearly show all intercepts.

a. $y = x$

b. $y = {}^-5x + 100$

c. $x = {}^-4$

d. $y = 5.4$

e. $4.6x + 0.25y = 24$

15. Evaluate the following expression in one step on your calculator.

$$\frac{12.4 - \sqrt{({}^-12.4)^2 - 4 * 6.2 * {}^-3.7}}{2 * 6.2}$$

16. For each equation, indicate whether the equation graphs as a line, a parabola, a V-shaped graph, a circle, an exponential function, or other.

a. $y = 2x - 5$

b. $x^2 + y^2 = 25$

c. $y = x^2 + 15$

d. $y = 2^x$

e. $y = x^4$

f. $y = 13x^2 - 20x + 48$

g. $y = 0.428 * 0.5^{-x}$

h. $y = x^{1/3}$

i. $y = 4x(28 - 5x)$

17. Predict how each graph will look. Then find a window that shows a complete graph on your calculator. Your graph should clearly show all intercepts.

a. $y = 2x^2 + 15$

b. $y = {}^-0.25x^2$

c. $y = 20 - \dfrac{x^2}{5}$

d. $y = x^2 - 15x$

e. $y = {}^-9.8x^2 + 49.6x$

f. $y = {}^-2x^2 + 140x - 2300$

18. Solve the 2×2 linear system graphically and algebraically.

$$\begin{cases} 2x + 5y = 12 \\ 2y = x - 30 \end{cases}$$

19. Solve each equation. Approximate numerical results to three significant digits. Check your solutions.

a. ${}^-46 = 0.025x^{5/3}$

b. $250(1 + r)^5 = 300$

c. $40x^4 - 100 = 75$

d. $\sqrt[3]{2x + 5} + 50 = 45$

20. Write the equations of three different lines that are parallel to the line $y = 4 - 7x$.

21. Write the equation of the line that passes through the point $({}^-2, 5)$ and is perpendicular to the line $y = 4x - 1$.

22. Write the equations of three different lines that are perpendicular to

$$y = \frac{{}^-2x - 6}{3}$$

23. Solve the following system graphically and algebraically.

$$\begin{cases} 2x + y = 15 \\ y = 2x^2 + 30x - 50 \end{cases}$$

24. Solve each system. Round approximate solutions to the thousandths place.

a. $\begin{cases} y = 10 - x^2 \\ y = -2x + 2 \end{cases}$

c. $\begin{cases} x^2 + y^2 = 10 \\ y = \dfrac{x + 3}{2} \end{cases}$

b. $\begin{cases} y = x^2 + 12x \\ 5x + y = 25 \end{cases}$

d. $\begin{cases} y = 0.5x^2 - 10x + 10 \\ y = -0.1x^2 + 6.4 \end{cases}$

25. Solve each inequality graphically.

a. $-0.15x^2 - 2x + 8 < 0$

b. $-10 - 2x \le x^2 - 20x + 104$

26. Solve each equation.

a. $3k^2 = k$

b. $(2t - 3)^2 = 4$

c. $\dfrac{2M + 5}{3} = \dfrac{4T - 2}{T}$ for T

d. $3k^2 + m = k$ for k

Write an inequality or compound inequality to describe each of the following graphs.

27. a.

b.

c.

d.

28. *Without using a calculator,* evaluate the following numerical expressions.

a. $27^{2/3}$

b. $(-27)^{2/3}$

c. $-4^{3/2}$

d. $25^{-3/2}$

e. $\left(\dfrac{1}{81}\right)^{-1/2}$

29. Using your calculator in one step, evaluate the expression $7x^2 - y^3 + y^{-2}$ when $x = -3$ and $y = -5$. Show your substitution.

30. Using your calculator in one step, evaluate $-k^2 + k^{2/3}$ when $k = 10$. Show your substitution.

31. Simplify $(25w)^{1/2}$. Verify your result.

32. Simplify $5x^{1/3} * 10x^{-1/2}$. Verify your result.

33. Graph $x^2 + y^2 = 64$. Is $x^2 + y^2 = 64$ a function?

34. Solve the following right triangles.

a.

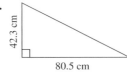

42.3 cm

80.5 cm

b. 142 mm

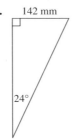

24°

c.

3.05 m

2.12 m

d.

65°

12 in.

35. Find the height of the tree.

120 ft

22°

Height of tree

36. Find the height h of the tree in the following figure.

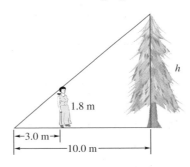

1.8 m

3.0 m

10.0 m

h

37. Find the area of the following trapezoid.

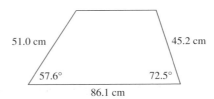

51.0 cm

45.2 cm

57.6°

72.5°

86.1 cm

38. Heading Back to Camp.   A group of campers left the campsite to explore the surrounding forest. They left the campsite heading N28°E and traveled about 1.5 miles. At this point they turned and headed N34°W and traveled an additional 4 miles. Now they want to return directly to their campsite. Without making a scale drawing, determine the bearing and distance to their campsite.

39. Pentagon.   The following regular pentagon is inscribed in a circle with radius 6.8 cm. Determine the perimeter and area of the pentagon.

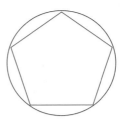

40. Pipe in the Corner.   A corner in a basement forms an angle of 84°. A tank with a diameter of 2.5 feet is placed in the corner. The walls of the basement touch the tank as shown and are perpendicular to the radius at these points. Determine the largest size pipe that will fit between the tank and the corner.

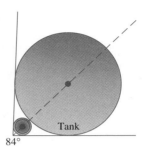

Tank

84°

41. Find the domain and range for the following functions.

    a. $f(x) = 5x - 16$

    b. $g(x) = x^2 + 12$

    c. $h(t = |t - 2|$

    d. $k(t) = \dfrac{5}{t - 3}$

    e. $F(x) = \sqrt{x + 1}$

42. Solve each equation for x.

    a. $f(x) = 2$, where $f(x) = 5x - 4.5$

    b. $g(x) = 5$, where $g(x) = 2x^2 + 3x$

    c. $h(x) = 5$, where $h(x) = \dfrac{5}{x + 2}$

43. Given that $G(x) = x^2 + 4x$,

    a. Find the value(s) of x for which $G(x) = 5$.

    b. Find the value of $G(5)$.

44. Write an equation for the inverse function of each of the following functions.

    a. $y = 2x + 9$

    b. $y = x^3 + 6$

    c. $y = \dfrac{x}{2} - 13$

45. $B(n)$ is a one-to-one function,

where B = Benson's monthly power bill in dollars

n = number of kilowatt hours of electricity used during the month

Describe what each expression represents.

a. $B^{-1}(200)$

b. $B(200)$

c. $B^{-1}(165)$

46. $T(m)$ is a one-to-one function,

where T = length of a trip in minutes

m = distance traveled in miles

Describe what each expression represents.

a. $T(50)$

b. $T^{-1}(100)$

c. $T^{-1}(117)$

47. $y(x)$ is a one-to-one function,

where x = temperatures in degrees Celsius

y = temperature in degrees Fahrenheit

Describe what each expression represents.

a. $y(32)$

b. $y^{-1}(32)$

c. $y(-40)$

48. Use the following table and graph to determine the values $F(-1)$, $F^{-1}(4)$, $G(-2)$, $G^{-1}(8)$, and $G^{-1}(-2)$.

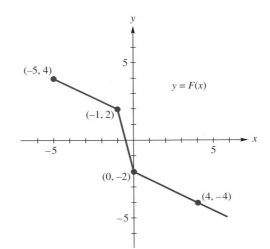

x	$G(x)$
-2	5
0	-2
2	4
4	0
6	8
8	-10

49. Determine the domain and range of function F from the previous problem.

50. Determine whether each of the following tables represents a linear function, an exponential function, or neither. Explain.

a.

Input	Output
0	0.25
1	1
2	4
3	16
4	64

c.

Input	Output
1	$^-3$
2	$^-12$
3	$^-27$
4	$^-48$
5	$^-75$

b.

Input	Output
5	45
10	37.5
15	30
20	22.5
25	15

d.

Input	Output
1	96
2	72
3	54
4	40.5
5	30.375

51. For each table, write an equation that models the data. Check your equation.

a.

t	C
0	5.75
1	7.00
2	8.25
3	9.50

c.

m	P
0	1458
1	486
2	162
3	54

b.

n	B
2	10.08
3	12.096
4	14.5152
5	17.41824

d.

n	B
10	$^-56$
12	$^-50$
14	$^-44$
16	$^-38$

52. Predict how the graph of each of the following exponential functions will look. Verify your predictions by graphing.

a. $y = 5^x$

b. $y = 0.25^x$

c. $y = 6\left(\frac{1}{2}\right)^x$

d. $y = 3^{-x}$

e. $y = 3.7 * 2^{3x}$

f. $y = e^x$

53. Ants.   Your wooden deck has been infested with a colony of 50 carpenter ants. The colony can triple in size every five days. How long before there are 500,000 ants in the colony?

54. Write an equation for the following tables.

a.

t	N
-2	$\frac{5}{4}$
-1	$\frac{5}{2}$
0	5
1	10
2	20
3	40

b.

P	D
-6	$\frac{-33}{4}$
-3	$\frac{-9}{2}$
0	$\frac{-3}{4}$
3	3
6	$\frac{27}{4}$
9	$\frac{21}{2}$

55. Without using a calculator, match each expression in the left column with an equivalent expression in the right column.

$\log_2(3x)$ $\qquad\qquad\qquad\qquad\qquad\qquad$ $\log_5(1) - \log_5(5)$

$\log_2(8x)$ $\qquad\qquad\qquad\qquad\qquad\qquad$ $\log_2(x) + \log_2(3)$

$\log_5\left(\frac{1}{5}\right)$ $\qquad\qquad\qquad\qquad\qquad\qquad$ $\dfrac{x\log(2)}{5}$

$\log(2^{x/5})$ $\qquad\qquad\qquad\qquad\qquad\qquad$ $\log_2(x) + 3$

56. Rewrite each logarithmic equation in exponential form.

a. $\log_2(x) = 5$ $\qquad\qquad\qquad$ **c.** $\log(2x + 50) = 2$

b. $\log_x(64) = 2$ $\qquad\qquad\qquad$ **d.** $\log_{x^2}(64) = 3$

57. Without using a calculator, evaluate each logarithmic expression.

a. $\log_2(8)$ $\qquad\qquad\qquad\qquad$ **e.** $\log_6(1)$

b. $\log(100)$ $\qquad\qquad\qquad\qquad$ **f.** $\log_3(81)$

c. $\log_5\left(\frac{1}{25}\right)$ $\qquad\qquad\qquad$ **g.** $\ln(e^4)$

d. $\log_8(8^5)$

58. Solve each logarithmic equation analytically. Round approximate solutions to the nearest hundredth.

a. $\ln(t) + 3 = 6.5$ $\qquad\qquad\qquad$ **d.** $\log(2m - 465) = 3$

b. $\log_2(5x - 9) = 3$ $\qquad\qquad\qquad$ **g.** $\ln(k) = 4.6 - \ln(2)$

c. $\log_8(x^3) = 4$ $\qquad\qquad\qquad$ **h.** $\log(2x) + \log(x + 5) = 2$

59. Solve each logarithmic equation anlytically. Round approximate solutions to the nearest hundredth. Check your solutions.

a. $\ln(x) = 1.2$ $\qquad\qquad\qquad\qquad$ **c.** $\log(2x - 1) = 2.5$

b. $\log(3x) = 0.35$ $\qquad\qquad\qquad$ **d.** $\ln\left(\frac{x}{2}\right) = -1$

60. Solve each of the following equations. Round approximate solutions to the nearest thousandth.

   a. $4x^5 + 5.1 = 52$

   b. $6x^2 - 5 = 3x$

   c. $360 = 3.7 * 2^x$

   d. $5x(x - 3) = 3(8 - 5x)$

   e. $78(1 + 0.05)^x = 215$

   f. $\sqrt{x + 18} = 3 + x$

61. Noise.   Decibels are used to measure the intensity of sound. The relationship between the intensity of sound and the power of sound is given as $D = 10 * \log\left(\dfrac{P}{10^{-16}}\right)$, where D is the intensity of sound in decibels, and P is power of the sound in watts per centimeter squared (watt/cm^2).

   a. Normal conversation has a power of about $3.16 * 10^{-11}$ watt/cm^2. Determine the intensity of sound for normal conversation.

   b. A piece of farm machinery produces noise with sound intensity of 100 decibels. Determine the power of sound for this machinery.

   c. The threshold of pain occurs when the intensity of sound level reaches 140 decibels. Determine the power of sound at the threshold of pain. How many times higher is the power of this noise than the farm machinery described in part b?

62. Hydrogen.   The pH scale ranges from 0 to 14. It measures the concentration of hydrogen ions in a solution. This indicates the level of acidity or basicity of a solution. A pH of 7 is a neutral solution. Pure water has pH $= 7$. A pH of less than 7 is acidic and a pH of more than 7 is basic. The formula for computing pH is given as pH $= {}^{-}\log[\text{H}^+]$, where $[\text{H}^+]$ is the concentration of hydrogen ions in a solution.

   a. When growing camellias, the desired concentration of hydrogen ions is 10^{-8}; what is the pH of this soil?

   b. The pH of rain in northern Norway has been measured to be about 4.4. Is this acidic or basic?

   c. Unpolluted rain has a pH of about 5.6. How many times higher is the concentration of hydrogen ions in the rain from northern Norway compared with unpolluted rain?

   d. Lemon juice has a pH of 2.3. How many times higher is the concentration of hydrogen ions in lemon juice than the rain in northern Norway?

63. Predict how the graph of each of the following functions will look. Verify your predictions by graphing.

   a. $y = 2 * 5.1^x$

   b. $y = \ln(x)$

   c. $y = 150 * 0.5^x$

   d. $y = \log(x)$

   e. $y + 7x = 14$

   f. $y = 4 * 3^{-x}$

64. Solve each exponential equation algebraically. Round approximate solutions to the nearest hundredth. Check your solutions.

   a. $10^{x-1} = 560$

   b. $3(10^t) + 15 = 21$

   c. $9e^{3x} = 4500$

   d. $e^{x/2} = \dfrac{7}{2}$

   e. $350(1 + 0.08)^x = 600$

   f. $4.2^{(3+t)} = 150.7$

Appendices

The topics covered in the appendices should be familiar to you. We include them here for a refresher and as a reference. Depending on your background, any part of this material may be new to you. It is up to you to read it and to get help as needed. We assume you have a working knowledge of these topics.

Appendix A The Real Number System

Discussion A

It is often important to know what part or subset of our number system we are working in. We will review the real number system and its subsets in this section. Historically, numbers were developed to help people count things. We will start with this first set of numbers and then expand the set to include the numbers that will be used in this course.

Definition _____

The set of **natural numbers** is also known as the counting numbers. It consists of the numbers we use to count objects: 1, 2, 3, 4, . . .

One of the first additions made to this set was the number zero. This may seem like an unimportant distinction, but historically, the addition of zero was an important step in mathematical sophistication.

Definition _____

The set of **whole numbers** includes all of the natural numbers and zero. It consists of the numbers 0, 1, 2, 3, . . .

If we are subtracting whole numbers, the result is not always a whole number. For example, we need negative numbers to evaluate $6-10$, so we expand our numbers to include these negatives.

Definition _____

The set of **integers** includes the natural numbers, zero, and the opposites of all of the natural numbers. The integers are the numbers . . . , $^-3$, $^-2$, $^-1$, 0, 1, 2, 3, . . .

We can now add, subtract, or multiply any two integers and the result will be an integer. However, if we divide any two integers, the result is not always an integer. We need to add fractions to our set of numbers to be able to evaluate $5 \div 3$.

Definition _____

The set of **rational numbers** consists of all numbers that can be written as ratios of two integers, provided the divisor is not zero. That is, all numbers that can be written in the form $\frac{a}{b}$, where a and b are integers and $b \neq 0$, are rational numbers.

The rational numbers include all of the integers, because any integer can be written as a ratio, for example, $4 = \frac{4}{1}$. The rational numbers include numbers represented by common fractions, improper fractions, and mixed numbers because these can be written as ratios also. It can be shown that any ratio of integers can be written as either a terminating or a repeating decimal. So the rational numbers include all terminating and repeating decimals as well. Because integers, common fractions, improper fractions, and mixed numbers can all be expressed as terminating or repeating decimals, the rational numbers can be defined as those numbers that can be represented as terminating or repeating decimals.

Within the set of rational numbers, we can now do any of the four basic operations—addition, subtraction, multiplication, or division (except by zero)—and get an answer still in the set. Some additional numbers are not rational numbers. One type of number is the result we get when taking the square root of a number, if the number is not a perfect square. The decimal for $\sqrt{6}$ does not terminate or repeat; therefore, it is not a rational number. You probably know that π, the ratio of the circumference to the diameter of a circle, is also a nonterminating, nonrepeating decimal. We need a new set to include these numbers.

Notice that the sets mentioned so far are all nested; each one contains all of the numbers from the previous set. The following set does not include any of the previous sets.

Definition _____

An **irrational number** is any number that can be written as a nonterminating, nonrepeating decimal. In other words, any real number that *cannot* be written exactly in the form of a fraction is an irrational number.

This set includes such numbers as π and $\sqrt{2}$. When you enter $\sqrt{2}$ in your calculator, you get a decimal that may appear to be a rational number. What you can see on your calculator display is, in fact, a rational number but it is only an *approximation* of the irrational number you entered. The approximation is sufficient for most applications.

The number 2.020020002 . . . is a nonrepeating nonterminating decimal; therefore, it is irrational. Although we see a pattern to the decimal 2.020020002 . . . , it is not a repeating decimal. In a repeating decimal the exact same digits must repeat.

Notice that the irrational numbers are defined as those numbers that cannot be written in the form of a fraction; that is, they cannot be written as a rational number. We can think of irrational as meaning not rational.

Definition _____

The **real numbers** consist of all the rational numbers together with all the irrational numbers.

The following figure gives us a visual representation of the real number system. The picture can help us see which sets are contained in other sets.

A Visual Model of the Real Number System and Its Subsets

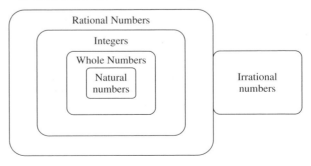

The following table lists the real number system and its subsets along with some examples of numbers belonging to the subsets. Because the entire table represents the real number system, any number in the table is also a real number.

THE REAL NUMBER SYSTEM	
Natural Numbers	1, 9, 27, 1080
Whole Numbers	0, 4, 56, 200
Integers	9, $^-$10, $^-$100, 0, 54
Rational Numbers	0.124, 2.3333 . . . , $^-$0.5, 17, $\frac{1}{2}$, $^-3\frac{5}{7}$, $^-$250
Irrational Numbers	π, $\sqrt{10}$, 0.010010001 . . .

Example 1

Which of the following are whole numbers?

$$\sqrt{8} \qquad 0 \qquad 0.5 \qquad \frac{8}{4} \qquad 2\frac{1}{3} \qquad {}^-10 \qquad 5 \qquad \sqrt{36}$$

Solution   The whole numbers consist of the numbers 0, 1, 2, 3, . . . We can easily see that 0 and 5 are whole numbers. Fractions are not part of the set, so 0.5 and $2\frac{1}{3}$ are not whole numbers. But what about the fraction $\frac{8}{4}$? Because this fraction is equal to 2, it is a whole number. Negative numbers are not included in the set, so $^-$10 is not a whole number. We need to be careful when deciding whether $\sqrt{8}$ and $\sqrt{36}$ are whole numbers. Because 8 is not a perfect square, $\sqrt{8}$ is an irrational number. Because $\sqrt{36}$ equals 6, $\sqrt{36}$ is a whole number. We conclude that 0, 5, $\frac{8}{4}$, and $\sqrt{36}$ are all whole numbers.

Notice from the example that when we are deciding what set a number belongs to, we need to think about the value of the number, not the form it is written in.

Problem Set A

1. For each part, a–f, identify all of the numbers from the list on the right that belong to the set.

   a. Natural numbers

   b. Whole numbers

   c. Integers

   d. Rational numbers

   e. Irrational numbers

   f. Real numbers

$2\frac{3}{5}$

0

$\sqrt{7}$

2.5

$-\frac{1}{4}$

15

0.667

$^-$10

0.151515 . . .

$^-\sqrt{9}$

0.010120123 . . .

2. For each of the following numbers, list *all* of the number systems to which it belongs.

   a. 0.5

   b. $-6\frac{1}{2}$

   c. $\sqrt{43}$

   d. 100

   e. $\frac{2}{7}$

   f. $\sqrt{64}$

   g. $^-17$

   h. π

   i. $\frac{23}{5}$

   j. $^-2.25$

3. Decide if each of the following statements is true or false.

   a. Irrational numbers are not real numbers.

   b. All integers are also rational numbers.

   c. Every rational number is a real number.

   d. Every real number is a rational number.

   e. Zero is both a rational number and an irrational number.

   f. Every repeating decimal number is a rational number.

   g. $\sqrt{89}$ lies between the integers 8 and 9.

   h. $\sqrt{16}$ is a rational number.

   i. $\frac{6}{2}$ is a natural number.

4. Without using a calculator, locate the following numbers on a number line.

   a. 2.25   **b.** $-\frac{5}{4}$ **c.** $\frac{1}{3}$ **d.** $\sqrt{7}$ **e.** $^-2\frac{7}{8}$ **f.** $^-3$

   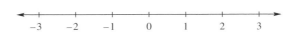

5. How many natural numbers are there between $-\frac{3}{2}$ and $\frac{5}{2}$?

   a. None   **b.** 2   **c.** 4   **d.** Infinitely many

6. The numbers $\frac{22}{7}$, 3.14, and π are all

   a. integers   **b.** rationals   **c.** irrationals   **d.** reals

7. Which of the following is not a rational number?

   a. $\frac{22}{7}$ **b.** $\sqrt{11}$ **c.** 3.14   **d.** 0.7

Appendix B The Language of Mathematics

Discussion B

A comfortable understanding of the language of mathematics is very important in solving problems and communicating ideas. Our assumption is that you are already familiar with the basics of this language. This section reviews basic definitions, notation, terminology, and usage.

In mathematics, we use numbers and letters that represent numbers. Sometimes a letter is a variable. A **variable** is a letter that is able to take on different numerical values. Other times a letter may be used to represent a particular number—a **constant.** An example of this is our use of the Greek letter π to represent the constant ratio of the circumference to the diameter of a circle.

An **operation** acts on one or more numbers and gives a result. The six basic operations of arithmetic are addition, subtraction, multiplication, division, exponentiation, and root extraction. Notice that each of these operations has an inverse operation. For example, multiplication and division are inverse operations. You are probably familiar with using inverse operations to solve basic equations. You can find a review of solving equations in Appendix E.

The following table summarizes the basic operations, symbols used to express them, and terminology associated with each.

Operation	Symbols Used	Parts	Result
addition	$2 + 3$	term $+$ term	sum
subtraction	$5 - 2$	term $-$ term	difference
multiplication	$2 * 3, 2 \cdot 3, 2(3), 2 \times 3$	factor $*$ factor	product
division	$\frac{6}{2}, 6 \div 2$	$\frac{\text{dividend}}{\text{divisor}}$	quotient
exponentiation	2^3	$\text{base}^{\text{exponent}}$	power
root extraction	$\sqrt[3]{8}, \sqrt{9}$	$\sqrt[\text{index}]{\text{radicand}}$	root

The following definitions help us understand the meaning of the collections of mathematical symbols that we encounter.

Definition _____

The symbols [ ], ( ), — (fraction bar), and $\sqrt{}$ are **grouping symbols** and are used to denote the result of the operation inside the grouping symbols.

Definition _____

An **algebraic expression** is any single number, single variable, or numbers and variables combined with mathematical operations.

The following all represent algebraic expressions.

5 $\qquad\qquad\qquad$ $2\pi r(h + 4)$

$3n$ $\qquad\qquad\qquad$ $5x^{-4} - 0.5y$

$3n + 8ay - 5$ $\qquad\qquad$ $\dfrac{4x}{2 + n} + \dfrac{5}{n}$

Definition _____

Terms of an expression are the parts of sums or differences.

Strategy

Identifying the Terms of an Expression

Because terms are the parts of a sum or difference, *addition and subtraction* symbols *outside of grouping symbols* separate terms. Therefore, to identify the terms of an expression, find all of the addition and subtraction symbols that are not in grouping symbols; the collections of symbols on either side of these are the terms.

Example 1

Underline the terms in the following expressions.

a. $3n + 8ay - 5$ b. $2\pi r(h + 4) + 3r$

Solution

a. The expression $3n + 8ay - 5$ contains one addition symbol and one subtraction symbol.

$$\underline{3n} + \underline{8ay} - \underline{5}$$

These symbols separate the terms of the expression. Therefore, $3n$ is a term, $8ay$ is a term, and 5 is a term.

b. The expression $2\pi r(h + 4) + 3r$ contains one addition symbol that is not in grouping symbols.

$$\underline{2\pi r(h + 4)} + \underline{3r}$$

This addition symbol separates the terms of the expression. Therefore, $2\pi r(h + 4)$ is a term and $3r$ is a term.

In Example 1, $2\pi r(h + 4)$ is a single term because the only addition symbol is inside of grouping symbols. Terms are separated by addition or subtraction symbols that are not within grouping symbols. In the single term

$$\frac{2x + a}{3y - 6}$$

the + and − are within the groupings implied by the fraction bar. In the single term $\sqrt{b^2 - 4ac}$, the − is within the grouping symbol $\sqrt{}$.

In the expression $3n + 8ay + 5$ the third term, 5, is called the **constant term.** It is a term that has a constant value. In the first term, the **coefficient** of n is 3. In the term $3n$, the 3 is also the **numerical coefficient.** In the second term, the numerical coefficient is 8, the coefficient of y is $8a$ and the coefficient of a is $8y$. In the expression $2\pi r(h + 4)$, the numerical coefficient is 2π, and the coefficient of r is $2\pi(h + 4)$.

An **equation** is the equality of two expressions. Solving an equation means finding the value or values of a particular variable to make the two expressions equal. Expressions cannot be solved, only simplified.

Equality is an example of a relation. **Relations** compare two quantities. Besides equality (=), you should be familiar with inequalities ($\neq, <, >, \leq, \geq$) and approximate equality ($\approx$). We use = and $\approx$ to distinguish between exactly equal and approximately equal and expect you to be careful to do the same.

Two extremely important distinctions in mathematics are between expressions and equations and between factors and terms. Be sure you are clear on these differences.

In mathematics we use the symbol "$-$" in two ways. It can mean the opposite or negative of a number, as in ^-8x or $^-(^-6)$. It can also mean subtraction, as in $2t - 3$. The meaning should be clear in context. This distinction is important in reading mathematical expressions and also in entering expressions into a calculator. There are two different keys on the calculator for these two meanings.

You should be able to read mathematical expressions and equations using the vocabulary reviewed here. This ensures that you understand clearly the meaning of an expression and also that you can communicate your ideas clearly. The mathematical symbolism is precise in its meaning but the English language makes it possible for us to read mathematics in ways that are ambiguous unless we are careful and use the correct terminology. For example, if I am trying to read the expression $6 * (2 + 3)$ and say *six times two plus three,* someone might think that I mean $6 * 2 + 3$. If I read the expression $6 * (2 + 3)$ as *six times the sum of two and three,* then it is clear. The word *sum* groups the two and the three. There is no confusion in what is meant.

Knowing how many terms are in a numerical expression and how to read the expression can be helpful when using a calculator to evaluate an expression in one step. For example, suppose you are trying to find the average of 22, 32, 35, and 31. The numerical expression is

$$\frac{22 + 32 + 35 + 31}{4}$$

One common incorrect calculator entry for this expression is $22 + 32 + 35 + 31/4$, and the result is 96.75. Our list of numbers ranges from 22 to 35. Clearly, 96.75 is not the average of these numbers.

Let's see what happened. The expression $22 + 32 + 35 + 31/4$ has four terms, each separated by plus signs.

$$\underline{22} + \underline{32} + \underline{35} + \underline{31/4} \qquad \text{Underline the terms.}$$

When the calculator evaluates this expression, it divides 31 by 4, rather than dividing the sum by 4. Then the other three terms are added to this result.

The expression

$$\frac{22 + 32 + 35 + 31}{4}$$

has one term because the fraction bar acts as a grouping symbol. When entering this expression in the calculator, we need to use parentheses as the grouping symbol to indicate that the addition should be performed first. Let's try entering the expression in the calculator again with grouping symbols, $(22 + 32 + 35 + 31)/4$. The result is 30. This answer is reasonable.

This expression is not too difficult, and some students might be tempted to evaluate it in two steps instead of entering it in one step. However, throughout the course, you will be evaluating many complex numerical expressions. Being able to enter these in one step is critical to your success.

Problem Set B

1. Use the following word list to fill in the blanks in the following sentences with the best choice. (Use plurals as needed.)

addition	equation	irrational	reciprocal
approximated	exponent	less than	root
approximately	exponentiation	multiplication	root extraction
base	expression	opposite	square (squared)
cube (cubed)	factor	power	subtraction
difference	greater than	product	sum
division	grouping symbols	quotient	term
equal to	inverse	rational	whole numbers

a. Any repeating or terminating decimal is a _____ number.

b. $2xy - \frac{3xy}{2z} + (5x)^7$ is an example of an _____.

c. The _____ $2xy - \frac{3xy}{2z} + (5x)^7 = x(x + 2y)(3x - y)$ has three _____ on the left side of the equation and three _____ on the right side.

d. The expression 4^3 can be read as the third _____ of four, or as four _____.

e. The expression $(xy)^2$ can be read as the _____ of the _____ of x and y.

f. The number 0.1020030004 . . . is an _____ number.

g. The _____ number $\sqrt{3}$ can be _____ by the _____ number 1.732.

h. The inverse of a power is a _____.

i. $x^2 + y^2$ is read as the _____ of the _____ of x and the _____ of y.

j. $(r + s)^2$ is read as the _____ of the _____ of r and s.

k. _____ are used to indicate an operation that is to be done first.

l. Multiplication is a shortcut for doing repeated _____.

2. Translate each of the following phrases or sentences into correct symbolic mathematics.

a. the product of five and x

b. the sum of six and w

c. the sixth power of five

d. four times the sum of n and fifteen

e. the opposite of the sum of p and t

f. the square of the sum of p and t

g. the sum of the square of p and the square of t

h. the product of six cubed and k squared

i. the reciprocal of five

j. the reciprocal of the sum of five and x

k. the square root of m

l. the square root of the sum of m and 25

m. The sum of three and twice p is the same as the square of p.

n. The product of t and q is equal to the sum of t and q.

o. The sum of five and the cube of x is more than 5.6.

p. The quotient of 7.2 and the sum of y and 3.56 is at least as big as the difference between y and 91.3.

3. Write the following expressions or equations in English to communicate them clearly. (There may be more than one correct answer.)

   a. $2 - x^2$

   b. $(2 - x)^2$

   c. $\dfrac{b}{2a} = b + 2$

   d. $7(x + 5)$

   e. $\sqrt{t - 6.5}$

   f. $\dfrac{1}{x + 6.5}$

   g. $\dfrac{1}{x} + 6.5$

   h. $Y = 5(x - 3)^6$

4. **a.** How many terms are in the expression $^{-}wRP + 7RP^2 - 4w(R + P) + \dfrac{P}{5}$?

   b. What are the simple factors of the term $7RP^2$?

   c. What are the simple factors of the term $4w(R + P)$?

   d. What is the coefficient of R in the term $7RP^2$?

   e. What is the numerical coefficient in the term $7RP^2$?

   f. What is the numerical coefficient in the term ^{-}wRP?

   g. What is the numerical coefficient in the term $\dfrac{P}{5}$?

5. First evaluate each of the following expressions by hand. Then enter the expression *in one step* on your calculator. Compare the two results. If they do not agree, try to find your error.

   a. $\sqrt{16 + 9}$

   b. $\sqrt{16} + 9$

   c. $\dfrac{1}{2 + 3}$

   d. $\dfrac{14 - 4}{4 + 1}$

   e. $\dfrac{24}{2 * 4}$

6. For each numerical expression, underline the terms and then evaluate the expression showing at least one intermediate step.

   a. $7 - 5(4 - 7)$

   b. $24 - \dfrac{8}{4 * 2}$

   c. $5 * 2^2$

   d. $(5 * 2)^2$

   e. $\sqrt{5 * 9 - 3^2}$

   f. $(^{-}4)^2 - 7 + 12$

   g. $^{-}4^2 - 7 + 12$

   h. $\dfrac{^{-}12}{3}(3 + {}^{-}5)$

   i. $\dfrac{^{-}12}{3(3 + {}^{-}5)}$

   j. $\left(\dfrac{4 - 12}{4}\right)^3$

7. Evaluate each expression from Problem 6 *in one step* using your calculator. Compare your results to what you got before. If your results are different, try to find where you made an error.

Appendix C The Cartesian Coordinate System

You should already have been introduced to the Cartesian coordinate system and some basic graphing of data. We include here the basic terminology and definitions. If you need a more detailed review, see your instructor for additional references.

The Cartesian coordinate system is designed to locate and name any point in a plane, a two-dimensional flat surface. This system is formed by intersecting two real number lines at right angles as shown in the following figure. The **horizontal axis** is often called the **x-axis** and is oriented with zero in the center, positive numbers to the right and negative numbers to the left. The **vertical axis** is often called the **y-axis** with zero in the middle, positive numbers above the horizontal axis and negative numbers below the horizontal axis. The point of intersection of the axes is called the **origin.** The four sections of the plane are referred to as the **quadrants** (I through IV) and are numbered counterclockwise beginning in the upper right.

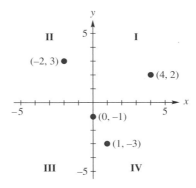

Each point is identified by an **ordered pair** (a, b). The a and b are called the **coordinates** of the point. The first coordinate is the directed horizontal distance between the point and the origin. The second coordinate is the directed vertical distance between the point and the origin. The order of the coordinates is critical!

On the previous graph, each of the equal markings on the axes represented one unit. With some data this is not appropriate. We need to select our scale to work with the data we are graphing.

In 1997, Keiko, the killer whale that starred in several movies, was moved from an aquarium in Mexico City to a rehabilitation facility in Newport, Oregon. During his first 12 months in Newport, Keiko gained approximately 1000 pounds. Keiko's health has raised concern about all whales held in captivity. The following table shows data collected on seven killer whales in captivity over the last 12 months.

Length of whale (ft)	11	17	24	21	8	10	20
One year's weight gain (or loss) (lbs)	1000	405	$^-$27	35	800	$^-$98	$^-$140

We graph the length of the whales on the horizontal axis and the weight gain or loss on the vertical axis. Our horizontal axis needs only positive values and the largest value is 24. We could scale our horizontal axis with each division representing 5 units. (Another good choice might be each unit representing 2 units.)

The values for the vertical axis range from ⁻140 to 1000. One reasonable choice is to scale this axis from ⁻200 to 1000, with each increment representing 100 units.

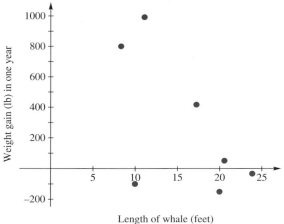

Length of whale (feet)

Notice that we chose different scales for the two axes. This is allowed. However, along a given axis the scale must remain the same. It cannot, for example, be 100 units per division for positive values and 20 units per division for negative values. This would distort the presentation of the data.

Because the data in the previous graph is in context, the graph must include a clearly marked scale and labels. The **scale** shows the numerical increments. The scale should be clearly labeled even if each tick mark on the axis represents 1 unit. The **labels** on the axes tell us what is being represented. In the previous example the horizontal axis was labeled as the length of the whale in feet, and the vertical axis was labeled as the weight gain in pounds in the last year. Whenever we graph information in context, each axis must be labeled, including units, to indicate what the axes represent. Look at the graphs in the examples of Chapter 1 for more examples of good scaling and graph labeling.

Problem Set C

1. Identify the coordinates of each point graphed here. Assume points have integer coordinates.

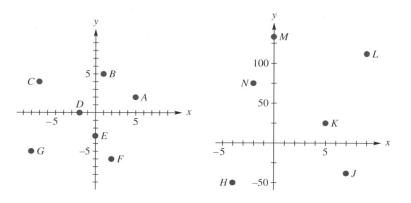

2. Draw a coordinate system, and plot the following points.

   a. (4, ⁻5)   **b.** (⁻2, 6)   **c.** (⁻3, ⁻4)   **d.** (4, 0)   **e.** (0, 8)

3. The following graph shows Sally's distance away from home after leaving on a bike trip.

   a. Discuss Sally's bike ride from the time she left home until its end. You might consider when she was going fastest, when she stopped, when she turned around, and so forth. Was this a fast trip or a leisurely ride?

   b. When was Sally approximately two blocks from home?

   c. How far from home was Sally 10 minutes after leaving home?

   d. How fast was Sally going after 7 minutes?

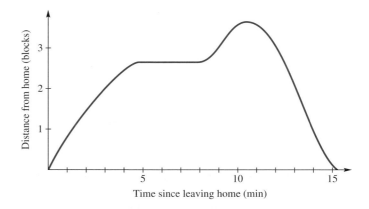

4. *WILDFIRE DAMAGE*

   a. During what years was more acreage damaged by wildfire in Oregon than in Washington state?

   b. Approximately how many acres of wildfire damage occurred in 1992 in Oregon? In Washington?

   c. During what year were approximately 8700 acres damaged by wildfire on Oregon state lands?

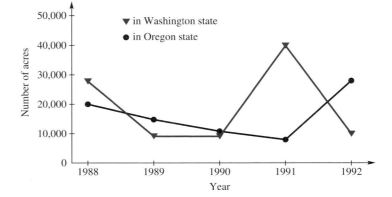

Appendix D Graphing Equations

Throughout this course we will be looking at information given in a table of values, presented on a graph, and in equations. All three of these models are related. From a table of values we can draw a graph and sometimes write an equation, from a graph we can make a table and sometimes write an equation, and from an equation we can make a table and draw a graph. In this section we will focus on drawing the graph of an equation.

Suppose we start with the equation $x + 2 = 5$. When an equation contains a variable, we often want to know what value(s) of this variable make the equation true. The value(s) of the variable that make the equation true are called **solution(s)** to the equation. In our example $x + 2 = 5$, we can see that if $x = 3$, then the equation is true. Therefore, $x = 3$ is the solution to the equation. To **solve** an equation means to find all of the solutions.

Example 1

Identify the solutions to each equation by inspection.

    a. $m + 5 = 15$ b. $k - 3 = 20$

Solution

a. We want to know what number we add to 5 to obtain 15. We can see that 10 works because $10 + 5 = 15$. Therefore, the solution to the equation $m + 5 = 15$ is $m = 10$.

b. Similarly, the solution to $k - 3 = 20$ is $k = 23$ because $23 - 3 = 20$.

As you know, not all equations can be solved by inspection. In Appendix E, we will review an algebraic process for solving some equations.

Next let's consider an equation that contains two different variables, for example, $y = x + 2$. To find the solution to an equation in two variables, we need to find a value for x and a value for y that together make the equation true. For example, if $x = 3$ and $y = 5$, the equation $y = x + 2$ is true. Does this equation have other solutions? What if $x = 1$? Then $y = 1 + 2$, or $y = 3$. Therefore, $x = 1$ and $y = 3$ is another solution to the equation.

One way to identify solutions to an equation that contains two variables is to pick a value for one of the variables and determine the value of the other. In our equation $y = x + 2$, we pick values for x and determine the corresponding values for y. We can record our results in a table.

x	$y = x + 2$
$^-2$	$^-2 + 2 = 0$
$^-1$	$^-1 + 2 = 1$
0	$0 + 2 = 2$
1	$1 + 2 = 3$
2	$2 + 2 = 4$
3	$3 + 2 = 5$

Therefore, $x = ^-2$ and $y = 0$ is a solution, $x = ^-1$ and $y = 1$ is a solution, and so on. Because each solution is a value of x and y, we often list the solutions as ordered pairs. For example, we say $(^-2, 0)$ is a solution, $(^-1, 1)$ is a solution, and so on. As you can imagine from this table, we can continue this process indefinitely. The equation $y = x + 2$ has an infinite number of solutions, each of which is an ordered pair. Therefore, it is not possible to list all of the solutions to the equation $y = x + 2$. Instead, we graph the solutions by plotting the ordered pairs from our table and connecting the points with a smooth curve. This is shown on the following graph.

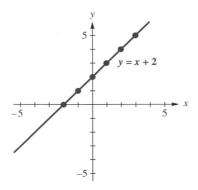

As you can see, the points we plotted for this equation appear to lie in a straight line, and in fact they do.

The line drawn is a representation of *all* of the solutions to the equation $y = x + 2$. For example, if we pick any point that lies on this line, it should be a solution to the equation. Let's try it.

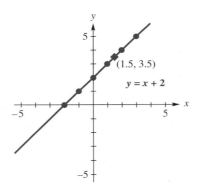

The point (1.5, 3.5) appears to lie on the line. Let's see if it is a solution.

$$3.5 = 1.5 + 2$$
$$3.5 = 3.5$$

We can see that (1.5, 3.5) is a solution.

To recap:

- Equations in two variables have an infinite number of solutions.
- Each solution to an equation in two variables is an ordered pair.
- We visually represent the solutions to an equation in two variables by graphing the equation. At this point, we create the graph by plotting some of the ordered-pair solutions and connecting the points with a smooth curve.

Therefore, if we need to graph an equation in two variables, we first must find some solutions to the equation. One way to do this is to make a table of values. When making a table of values, it is best to choose both positive and negative values for one of the variables because we do not know how the graph is going to look.

Example 2

Draw a graph of the solutions for each equation by plotting points.

a. $y = 4 - 3x$ b. $y = x(x + 2)$

Solution a. To plot points, we first need to make a table. In making our table, we pick values for x because this allows us easily to determine the values for y.

x	$y = 4 - 3x$
$^-3$	$4 - 3(^-3) = 13$
$^-2$	$4 - 3(^-2) = 10$
$^-1$	7
0	4
1	1
2	$^-2$
3	$^-5$

Now we can plot these points. The points appear to lie in a straight line, so we draw a line through these points. The line we draw represents all of the ordered pairs that are solutions to the equation $y = 4 - 3x$. We assume when drawing the line that it extends in both directions so that it represents all of the points that solve $y = 4 - 3x$.

We can graph this equation in a calculator to gain confidence that our hand-drawn graph is correct. To do this, we should set our calculator window to match our hand-drawn graph.

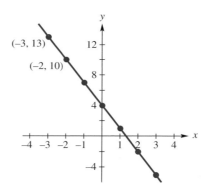

b. Similarly, we make a table of values for $y = x(x + 2)$ and plot these points. Remember that it is a good idea to pick both positive and negative numbers for x.

x	$y = x(x + 2)$
$^-3$	$^-3(^-3 + 2) = 3$
$^-2$	$^-2(^-2 + 2) = 0$
$^-1$	$^-1$
0	0
1	3
2	8
3	15

Next, we plot these points and draw the graph. These points clearly do not lie in a straight line. It is a good idea to check the shape of this graph using your graphing calculator before trying to draw the graph with these few points. Set the window the same as the graph you have drawn by hand.

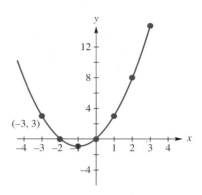

In parts a and b of this example, we used a table to find several points on the graph of the given equations. We then connected the points with a straight line or a smooth curve. We do not know enough about the graphs of these equations to be certain that we graphed them completely. A complete graph of an equation includes all places where the graph changes direction. As you study more mathematics you will learn how the graphs of these and other equations should look. For now, when asked to graph an equation be sure to pick several values for the input variable, some that are positive and some that are negative. This gives you the best chance of seeing how the complete graph looks.

Problem
Set
D

1. Identify the solution to each equation by inspection. Check your solution.

   a. $x - 2 = 18$ **b.** $p + 5 = 9$ **c.** $3 + x = 9$

2. List three solutions to each equation.

   a. $y = x - 3$ **b.** $y = 20 - x^2$ **c.** $y = 2\sqrt{x}$

3. **a.** Complete the following table for the equation $y = 12 - x$.

   b. Graph the equation $y = 12 - x$ by plotting the points from your table.

x	$y = 12 - x$
$^-3$	
$^-2$	
$^-1$	
0	
1	
2	
3	

4. a. Complete the following table for the equation

$$y = \frac{x + 1}{2}$$

b. Graph the following equation by plotting the points from your table.

$$y = \frac{x + 1}{2}$$

x	$y = \dfrac{x + 1}{2}$
-5	
-3	
-1	
0	
1	
3	
5	

5. a. Make a table of values for the equation $y = (x - 4)(x + 4)$.

b. Draw a graph of the equation $y = (x - 4)(x + 4)$.

6. Draw a graph that represents the solutions to the equation $y = x^3$.

Appendix E Solving Linear and Rational Equations

Discussion E

Solving Linear Equations in One Variable

Can you solve the following number puzzle?

> If you multiply an unknown number by $^-3$ and add 64 you get 22. What is the unknown number?

One strategy for solving this puzzle is to work backwards, "undoing" each of the operations until you know the number.

> Start with 22 and *subtract* 64, then *divide* the result by $^-3$. The number is 14.

It is easy to check if 14 solves the puzzle.

> Start with 14, multiply by $^-3$, and add 64. The result is 22, so our answer is correct.

Notice that when we worked backwards, we used subtraction to undo addition, and division to undo multiplication. These operations are called inverses and are useful when solving equations.

Solving equations is similar to solving a puzzle. As we solve an equation symbolically, we need to think about how to undo the operations so we can isolate the variable. It is also important to think of the equation as a balance. If we change an equation by adding or subtracting a number or by multiplying or dividing by a number on one side, we must do the same thing to the other side so that the equation remains balanced.

Let's look at the preceding puzzle symbolically. Our first step is to translate the statement "If you multiply an unknown number by $^-3$ and add 64, you get 22" into symbols. We let N represent the unknown number.

$$^-3N + 64 = 22$$

$$^-3N + 64 - \mathbf{64} = 22 - \mathbf{64} \qquad \text{Subtract 64 from both sides to undo the addition. This isolates the term containing } N.$$

$$\frac{^-3N}{^-\mathbf{3}} = \frac{^-42}{^-\mathbf{3}} \qquad \text{Divide both sides by } ^-3 \text{ to undo the multiplication. This isolates } N.$$

$$N = 14$$

Example 1

Solve the following equations. Check your solutions.

     a. $\dfrac{x}{4} + 24 = 76$ b. $30.25 = 5C - 12.25$

Solution   a.

$$\frac{x}{4} + 24 = 76$$

$$\frac{x}{4} + 24 - \mathbf{24} = 76 - \mathbf{24} \qquad \text{Isolate the term containing the variable } \tfrac{x}{4} \text{ by subtracting 24 from both sides.}$$

$$\frac{x}{4} = 52 \qquad \text{Simplify both sides.}$$

$$\mathbf{4} * \frac{x}{4} = 52 * \mathbf{4} \qquad \text{Multiply both sides by 4 to undo the division. This isolates the variable.}$$

$$x = 208$$

CHECK

$$\frac{208}{4} + 24 \overset{?}{=} 76 \qquad \text{Substitute } x = 208 \text{ into the original equation.}$$

$$76 = 76 \qquad ✔ \qquad \text{Because both sides match, our solution works.}$$

Therefore, $x = 208$ is the solution to the equation $\frac{x}{4} + 24 + 76$.

b.
$$30.25 = 5C - 12.25$$

$$30.25 + \mathbf{12.25} = 5C - 12.25 + \mathbf{12.25} \qquad \text{Isolate the term containing the variable } 5C \text{ by adding 12.25 to both sides.}$$

$$\frac{42.5}{\mathbf{5}} = \frac{5C}{\mathbf{5}} \qquad \text{Divide both sides by 5 to undo the multiplication.}$$

$$8.5 = C$$

CHECK

$$30.25 \overset{?}{=} 5 * 8.5 - 12.25 \qquad \text{Substitute } C = 8.5 \text{ into the original equation.}$$

$$30.25 = 30.25 \qquad ✔$$

Therefore, $C = 8.5$ is the solution to the equation $30.25 = 5C - 12.25$.

When we check solutions, our first step is to substitute the solution into the original equation. Then, we evaluate each side in one step using a calculator. If both sides of the equation agree, the solution is correct. It is *not* a good idea to work the arithmetic out step by step when checking a solution. If you make a mistake solving the equation and then check the answer by going through a step-by-step process, you are very likely to make the same mistake. Your incorrect answer seems to work, even though the answer is wrong!

Sometimes we need to simplify one or both sides of an equation first before beginning to solve an equation by undoing.

Example 2

Chelsey is planning for some elective eye surgery. She must pay the first $50 of the treatment and then her insurance company pays 90% of the costs after that initial $50. If Chelsey wants to pay no more than $200 for the eye surgery, how expensive could the total bill be?

Solution

Chelsey's costs are $50 plus 10% of the remaining bill. We want to find a total cost such that Chelsey's cost is $200. Our unknown in this problem is the total cost of the eye surgery. We begin by defining this as our variable; let T = total cost in dollars. Next, we need to translate the problem situation into an equation; that is, translate "$50 plus 10% of the *remaining cost* is equal to $200" into symbols. The *remaining cost* is the total bill less $50. We need to find 10% of the remaining cost, so the expression $T - 50$ needs to be written in grouping symbols.

$$50 + 0.10(T - 50) = 200$$

Now we can solve for T.

$$50 + 0.10T - 5 = 200 \qquad \text{Use the distributive property to simplify the left-hand side of the equation. We are not changing the equation, so we do not need to do anything to the right-hand side.}$$

$$45 + 0.10T = 200 \qquad \text{Simplify the left side of the equation.}$$

$$45 + 0.10T - \mathbf{45} = 200 - \mathbf{45} \qquad \text{Subtract 45 from both sides to undo the addition.}$$

$$0.10T = 155$$

$$\frac{0.10T}{\mathbf{0.10}} = \frac{155}{\mathbf{0.10}} \qquad \text{Divide both sides to undo the multiplication.}$$

$$T = 1550$$

We check our solution in the problem situation because we may have made an error in writing the equation. If the bill is $1550, Chelsey pays the first $50. This leaves a balance of $1500 and she must pay 10% of that, which is $150. This means that Chelsey pays $50 + $150, or $200. This is the amount she was able to pay, so our solution is correct.

We conclude that the total cost of the eye surgery could go as high as $1550.

The previous two examples showed how to solve some types of linear equations in one variable. We now clarify what makes an equation linear in one variable.

Characteristics of a Linear Equation in One Variable

An equation is called a **linear equation in one variable** if the following conditions are met.

- It contains only one variable.
- When the equation is written in simplified form, the variable occurs only to the first power.
- The variable does not appear in the denominator or under a radical.

Some situations are modeled by equations that need to be simplified before we can solve them. Other situations are modeled by equations in which the variable occurs on both sides of the equation. The strategy for solving these equations is to first simplify both sides, then add or subtract to move the terms containing the variable to just one side. When all of the terms containing the variable are on one side of the equation, we can proceed to isolate the variable.

The following box shows a strategy that summarizes how to solve these equations.

Strategy

Solving Linear Equations in One Variable

1. Simplify both sides of the equation by applying the distributive property or combining like terms (or both). If the equation contains fractions, you may want to multiply both sides of the equation by the least common denominator to eliminate the fractions.

2. Add or subtract terms from both sides of the equation to collect all terms containing the variable on one side and all other terms on the opposite side. Again combine like terms if possible.

3. Divide both sides by the coefficient of the variable.

4. Check your solution in the original equation.

Example 3

Solve the following equations. Check your solutions.

a. $3w - 5(15 - 2w) = 12w + 32$

b. $\dfrac{p}{5} + \dfrac{3p}{4} = 1$

Solution a.

$3w - 5(15 - 2w) = 12w + 32$	Begin by simplifying both sides of the equation.
$3w + {}^-5(15 + {}^-2w) = 12w + 32$	Rewrite subtraction as "adding the opposite."
$3w - 75 + 10w = 12w + 32$	Use the distributive property to simplify the second term on the left-hand side.
$13w - 75 = 12w + 32$	Combine like terms.
$13w - 75 - \mathbf{12w} = 12w + 32 - \mathbf{12w}$	Subtract $12w$ from both sides to collect all terms containing w on the left side of the equation.
$w - 75 = 32$	
$w - 75 + \mathbf{75} = 32 + \mathbf{75}$	Add 75 to both sides to isolate w.
$w = 107$	

CHECK

$$3 * 107 - 5(15 - 2 * 107) \overset{?}{=} 12 * 107 + 32$$

$$1316 = 1316 \quad ✔$$

Therefore, $w = 107$ is the solution to $3w - 5(15 - 2w) = 12w + 32$.

b. If an equation contains fractions, it is usually easier to multiply both sides of the equation by the least common denominator to eliminate the fractions in the beginning.

$\dfrac{p}{5} + \dfrac{3p}{4} = 1$	
$\mathbf{20}\left(\dfrac{p}{5} + \dfrac{3p}{4}\right) = 1 * \mathbf{20}$	The least common denominator for 5 and 4 is 20. Multiply both sides by 20 to eliminate the fractions.
$20 * \dfrac{p}{5} + 20 * \dfrac{3p}{4} = 20$	Distribute 20 over the sum on the left-hand side.
$4p + 15p = 20$	Simplify the terms on the left-hand side.
$19p = 20$	Combine like terms.
$\dfrac{19p}{\mathbf{19}} = \dfrac{20}{\mathbf{19}}$	Divide both sides by 19.
$p = \dfrac{20}{19} \approx 1.053$	

To check, substitute $\frac{20}{19}$ for p into the original equation.

CHECK

$$\frac{p}{5} + \frac{3p}{4} = 1$$

$$\frac{20/19}{5} + \frac{3(20/19)}{4} \overset{?}{=} 1$$

$$1 = 1 \quad ✔$$

Let's see what happens if we check our approximate solution of 1.053 instead of the exact solution.

CHECK

$$\frac{p}{5} + \frac{3p}{4} = 1$$

$$\frac{1.053}{5} + \frac{3 * 1.053}{4} \stackrel{?}{=} 1$$

$$1.00035 \approx 1 \qquad ✔$$

Notice that when we check an approximate solution the values do not match exactly, but they are very close.

Solving Rational Equations in One Variable

We now look at how to solve rational equations that, when simplified, result in linear equations.

Recall that rational numbers are numbers that can be written as a ratio of two whole numbers with the denominator not equal to zero. We are now going to look at a type of equation called a rational equation. The rational equations we are working with contain at least one term that is the ratio of two linear expressions with a variable in the denominator.

The first step in solving rational equations is to simplify the equation by eliminating the fractions. To do this, we multiply both sides of the equation by the least common denominator. Eliminating the fractions allows us to solve the resulting equation using the strategy presented earlier.

Example 4

Solve the following equations and check your solutions.

a. $\dfrac{1}{4} + \dfrac{5}{2m} = \dfrac{1}{6}$ b. $\dfrac{3}{x} - \dfrac{5}{x+4} = \dfrac{2}{x+4}$

Solution   a. The least common denominator for 4, $2m$, and 6 is $12m$.

$$\frac{1}{4} + \frac{5}{2m} = \frac{1}{6}$$

$$\mathbf{12m}\left(\frac{1}{4} + \frac{5}{2m}\right) = \left(\frac{1}{6}\right)\mathbf{12m} \qquad \text{Multiply both sides by the least common denominator.}$$

$$12m\left(\frac{1}{4}\right) + 12m\left(\frac{5}{2m}\right) = \left(\frac{1}{6}\right)12m \qquad \text{Distribute } 12m \text{ over the sum and cancel the common factors in each term.}$$

$$3m + 30 = 2m \qquad \text{Simplify both sides of the equation.}$$

$$3m + 30 - \mathbf{3m} = 2m - \mathbf{3m} \qquad \text{Subtract } 3m \text{ from both sides.}$$

$$30 = {}^-m$$

$${}^-1 * 30 = {}^-1 * {}^-m \qquad \text{Multiply both sides by } {}^-1.$$

$${}^-30 = m$$

To solve for m when you have ${}^-m$ in an equation, you can *either* divide both sides by ${}^-1$ or you can multiply both sides by ${}^-1$. This only works if the numerical coefficient is ${}^-1$.

CHECK

$$\frac{1}{4} + \frac{5}{2m} = \frac{1}{6}$$

$$\frac{1}{4} + \frac{5}{2*-30} \overset{?}{=} \frac{1}{6}$$

$$0.16667 = 0.16667 \quad \checkmark$$

Therefore, $m = {}^-30$ is the solution to the equation

$$\frac{1}{4} + \frac{5}{2m} = \frac{1}{6}$$

b. The least common denominator for x, $x + 4$, and $x + 4$ is $x(x + 4)$.

$$\frac{3}{x} - \frac{5}{x + 4} = \frac{2}{x + 4}$$

$$x(x + 4)\left(\frac{3}{x} - \frac{5}{x + 4}\right) = \left(\frac{2}{x + 4}\right)x(x + 4) \qquad \text{Multiply both sides by the least common denominator.}$$

$$x(x + 4)\left(\frac{3}{x}\right) - x(x + 4)\left(\frac{5}{x + 4}\right) = \left(\frac{2}{x + 4}\right)x(x + 4) \qquad \text{Distribute the } x(x + 4) \text{ over the difference on the left-hand side.}$$

$$(x + 4)3 - x*5 = 2*x \qquad \text{Cancel common factors in each term.}$$

$$3x + 12 - 5x = 2x \qquad \text{Simplify both sides of the equation.}$$

$$^-2x + 12 = 2x \qquad \text{Continue to simplify by combining like terms on the left-hand side.}$$

$$^-2x + 12 + \mathbf{2x} = 2x + \mathbf{2x} \qquad \text{Add } 2x \text{ to both sides of the equation.}$$

$$12 = 4x$$

$$\frac{\mathbf{12}}{\mathbf{4}} = \frac{\mathbf{4x}}{\mathbf{4}} \qquad \text{Divide both sides by 4.}$$

$$3 = x$$

CHECK

$$\frac{3}{3} - \frac{5}{3 + 4} \overset{?}{=} \frac{2}{3 + 4}$$

$$0.2857 = 0.2857 \quad \checkmark$$

Therefore, $x = 3$ is the solution to the equation

$$\frac{3}{x} - \frac{5}{x + 4} = \frac{2}{x + 4}$$

Solving Proportions

A special type of rational equation is a proportion.

Definition _____

A proportion is an equation stating that two ratios are equal. That is, $\frac{a}{b} = \frac{c}{d}$, where $b \neq 0$ and $d \neq 0$.

Because proportional equations are a type of rational equation, all proportional equations can be solved using the techniques just discussed. However, there is a shortcut that works with proportional equations but not with other rational equations. You are very likely to encounter more proportional equations than other rational equations because proportions have many applications. Learning the shortcut makes solving these faster and easier.

Let's look at a proportion and see if we can discover the shortcut.

$$\frac{a}{b} = \frac{c}{d}$$

$$bd * \frac{a}{b} = \frac{c}{d} * bd \qquad \text{Multiply both sides by the least common denominator to clear the fractions.}$$

$$ad = bc \qquad \text{Cancel common factors in each term.}$$

Notice that after clearing the fractions and canceling the common factors we get a product on each side of the equation. To be more specific, the product of the numerator of the left side of the proportion and the denominator of the right side is equal to the product of the denominator of the left side and the numerator of the right side. The result is always the same when we are simplifying proportions, so we can skip the step when we multipy both sides by the least common denominator and just write the product. That is, if $\frac{a}{b} = \frac{c}{d}$, then $ad = bc$. This shortcut for solving proportional equations is often called **cross multiplication** because we are multiplying the numerator of each side by the denominator of the other. If we connect these paths the lines "cross."

$$\frac{a}{b} \diagdown\!\!\!\!\diagup \frac{c}{d}$$

$$ad = bc$$

Example 5

Solve the equation

$$\frac{x}{3x - 4} = \frac{5}{7}$$

Solution

$$\frac{x}{3x - 4} = \frac{5}{7} \qquad \text{Because each side of this equation is a single fraction, this is a proportion, and we can use cross multiplication.}$$

$$7 * x = 5 * (3x - 4) \qquad \text{Cross multiply to simplify the proportion.}$$

$$7x = 15x - 20$$

$$7x - \mathbf{15x} = 15x - 20 - \mathbf{15x} \qquad \text{Subtract } 15x \text{ from both sides.}$$

$$^{-}8x = {}^{-}20$$

$$\frac{^{-}8x}{\mathbf{-8}} = \frac{^{-}20}{\mathbf{-8}} \qquad \text{Divide both sides by } {}^{-}8.$$

$$x = 2.5$$

CHECK

$$\frac{2.5}{3 * 2.5 - 4} \overset{?}{=} \frac{5}{7}$$

$$0.71429 = 0.71429 \qquad ✔$$

Therefore, $x = 2.5$ is the solution to the equation

$$\frac{x}{3x - 4} = \frac{5}{7}$$

Try solving Example 5 by multiplying both sides of the original equation by $7(3x - 4)$, the least common denominator. Compare this process to the solution process in Example 5. Trying this example both ways may help you decide which of the two methods works better for you.

In this section, we reviewed techniques for solving **linear equations in one variable.** The techniques for solving a linear equation were summarized in a strategy box. We also found that if we multiply both sides of a **rational equation** by the least common denominator, the fractions are eliminated. Then we can apply the strategy for solving a linear equation to the result. It is important to point out that multiplying a rational equation by the least common denominator does *not* always result in a linear equation. In Chapter 7, we learn how to solve rational equations that result in quadratic equations.

A special type of rational equation is a **proportional equation.** Cross multiplication is a special technique for solving proportional equations. It is a shortcut for multiplying by a common denominator that works in the special case of proportional equations.

Problem
Set
E

1. Solve the following equations. Check your solutions.

   a. $4z + 9 = 31$

   b. $^-7A - 12 = 30$

   c. $^-24 = 16G - 336$

   d. $11y - 5y = 15$

   e. $\dfrac{m}{3} + 12 = 70$

   f. $21 = 4(w - 3) + 5w$

   g. $\dfrac{2x - 9}{3} = 11$

   h. $2D - 7 + 5D = 9$

   i. $5x - 3(x - 2) = {}^-10$

   j. $\dfrac{2}{3}t + 12 = {}^-2t$

   k. $4(a + 12) + 7a = 26$

   l. $2p = 14p - 18$

   m. $4(k + 5) + 5k = 20.3$

2. Solve the following equations. Check your solutions.

   a. $(n - 2)180 = 450$

   b. $14 = \dfrac{s}{2} - 8$

   c. $\dfrac{a}{3} - 10 = \dfrac{3}{2}$

   d. $3y - 7 = 10 - 4y$

   e. $6 + 0.6m + 1.2m = 18.6$

   f. $\dfrac{5m + 2}{8} = 60$

   g. $^-14 = 5y + 8(y + 1)$

   h. $\dfrac{w}{4} + \dfrac{w}{7} = 16.5$

   i. $\dfrac{x + 10}{5} = \dfrac{x}{4}$

3. Shown here is a student's solution process for solving the equation

$$\frac{1}{5} - \frac{x-3}{4} = 12$$

Can you find where this student made a mistake in solving the equation? Explain why the solution appears to check even though the student made a mistake.

SOLVE

$$\frac{1}{5} - \frac{x-3}{4} = 12$$

$$20\left(\frac{1}{5} - \frac{x-3}{4}\right) = 12 * 20$$

$$20 * \frac{1}{5} - 20 * \frac{x-3}{4} = 240$$

$$4 - 5 * x - 3 = 240$$

$$1 - 5x = 240$$

$$\frac{^-5x}{^-5} = \frac{239}{^-5}$$

$$x = {^-}47.8$$

CHECK

$$\frac{1}{5} - \frac{^-47.8 - 3}{4} \stackrel{?}{=} 12$$

$$20\left(\frac{1}{5} - \frac{^-47.8 - 3}{4}\right) \stackrel{?}{=} 12 * 20$$

$$20 * \frac{1}{5} - 20 * \frac{^-47.8 - 3}{4} \stackrel{?}{=} 240$$

$$4 - 5 * {^-}47.8 - 3 \stackrel{?}{=} 240$$

$$4 + 239 - 3 \stackrel{?}{=} 240$$

$$240 = 240 \qquad ✔$$

4. Which of the following equations are proportions? Do not solve these yet.

a. $\dfrac{3}{x} + 2 = 7$

b. $\dfrac{R}{R+1} = 9$

c. $5 + \dfrac{1}{x+3} = \dfrac{8}{x+3}$

d. $\dfrac{2}{5} = \dfrac{8}{2x+3}$

e. $\dfrac{^-8}{D+3} - \dfrac{5}{D} = \dfrac{22}{D}$

f. $\dfrac{10}{b} = \dfrac{15}{36}$

g. $\dfrac{x+7}{2x+5} = \dfrac{3}{5}$

h. $\dfrac{10}{x-2} - \dfrac{5}{x} = \dfrac{^-3}{x}$

5. Solve the following equations. Check your solutions.

a. $\dfrac{3}{x} + 2 = 7$

b. $\dfrac{R}{R + 1} = 9$

c. $5 + \dfrac{1}{x + 3} = \dfrac{8}{x + 3}$

d. $\dfrac{2}{5} = \dfrac{8}{2x + 3}$

e. $\dfrac{^-8}{D + 3} - \dfrac{5}{D} = \dfrac{22}{D}$

f. $\dfrac{10}{b} = \dfrac{15}{36}$

g. $\dfrac{x + 7}{2x + 5} = \dfrac{3}{5}$

h. $\dfrac{10}{x - 2} - \dfrac{5}{x} = \dfrac{^-3}{x}$

6. Solve the following equations. Check your solutions.

a. $^-4m - 12 - 5m = 13$

b. $^-37 = \dfrac{5t}{6} - 19$

c. $7.49 + 0.75H = 1.39H$

d. $\dfrac{375}{2} = \dfrac{6}{x}$

e. $\dfrac{T + 8}{2} = \dfrac{5T}{6}$

f. $^-k = 7(k + 45)$

g. $\dfrac{3x}{4x + 5} = \dfrac{3}{2}$

h. $\dfrac{3}{m} - \dfrac{5}{4m} = 1.5$

i. $^-44 = 53 - \dfrac{97}{5y}$

7. College Fees. Cory's school charges a tuition of \$47 per credit, \$2.50 per credit technology fee, and a student activities fee of \$75. The activity fee is a flat fee, not per credit hour. Cory has budgeted \$900 for tuition and fees. How many credits can Cory take and stay within budget?

8. Which Long-Distance Plan?   Cory's parents are trying to decide on a long-distance phone option. One option is an 800-number that charges 25¢ a minute with a monthly charge of \$2.50. The other option is a plan that would charge $37\frac{1}{2}$¢ a minute.

a. If Cory talks to her parents for 10 minutes in a month, which plan is cheaper? What if she talks for 45 minutes?

b. Determine the number of minutes for which the two plans are equal.

9. The Computer.   Your company just bought a computer for \$2499. Each year its value decreases by \$350.00 due to depreciation. If the company plans to replace the computer when its value falls below \$1000, how long will the company keep the computer?

Appendix F Literal Equations and Formulas

Discussion F

Many mathematics applications are based on formulas.

Distance:	distance = rate $*$ time
Area of a circle:	$A = \pi r^2$
Perimeter of a rectangle:	$P = 2L + 2W$
Simple interest:	$I = prt$
Electronics:	$E = IR + Ir$
Temperature conversion:	$F = \frac{9}{5}C + 32$

Formulas are written with numbers, variables, and constants. Variables and some constants are represented by letters or entire words that we call **literal numbers.** The conversion formula $F = \frac{9}{5}C + 32$ contains the numbers $\frac{9}{5}$ and 32, and the literal numbers F and C. A formula or mathematical model is valuable if you need to solve a certain problem many times.

Example 1

A formula used in the health industry to determine an individual's cost for major medical insurance is

$$A = (T - D)(1.00 - P) + D$$

where T = total bill

D = amount of deductible

P = percent paid by insurance, as a decimal

A = amount you must pay

Suppose your health insurance has a $200 deductible and then pays 90% of the remaining bill for all major medical. Rewrite the formula with this given information. Use your new formula to complete the following table.

Total Bill	$2000	$2500	$3000	$3500
Amount You Must Pay				

Solution The new formula is

$$A = (T - 200)(1.00 - 0.90) + 200$$

Now evaluate the formula for each value of T to complete the table.

Total Bill	$2000	$2500	$3000	$3500
Amount You Must Pay	$380	$430	$480	$530

In Example 1, we saw how a formula can be used over and over to solve for a specific value. In some applications, the formula is not written in the best form for your needs. For example, the formula for converting between degrees Fahrenheit and degrees Celsius is $F = \frac{9}{5}C + 32$, where F is the temperature in degrees Fahrenheit and C is temperature in degrees Celsius. Suppose that you need to convert 72° Fahrenheit to degrees Celsius. Substituting this information into the formula we get

$$72 = \frac{9}{5}C + 32$$

Now, instead of just evaluating an expression, we need to solve the equation for C. Actually, this is not too hard, but suppose that you had to convert *several* temperatures from Fahrenheit to Celsius. Then it is simpler to solve the original formula for C once and use the resulting formula to determine the Celsius temperatures. Solving a literal equation is similar to solving an equation in one variable. Let's solve both $72 = \frac{9}{5}C + 32$ and $F = \frac{9}{5}C + 32$ to see how the steps match.

One Variable Equation	Formula
$72 = \frac{9}{5}C + 32$	$F = \frac{9}{5}C + 32$
$\mathbf{5} * 72 = \left(\frac{9}{5}C + 32\right) * \mathbf{5}$	$\mathbf{5} * F = \left(\frac{9}{5}C + 32\right) * \mathbf{5}$
$360 = \frac{9}{5}C * 5 + 32 * 5$	$5F = \frac{9}{5}C * 5 + 32 * 5$
$360 = 9C + 160$	$5F = 9C + 160$
$360 - \mathbf{160} = 9C + 160 - \mathbf{160}$	$5F - \mathbf{160} = 9C + 160 - \mathbf{160}$
$200 = 9C$	$5F - 160 = 9C$
$\dfrac{200}{9} = \dfrac{9C}{9}$	$\dfrac{5F - 160}{9} = \dfrac{9C}{9}$
$22.2 \approx C$	$\dfrac{5F - 160}{9} = C$

So now we know that if Fahrenheit reads 72°, then the temperature is approximately 22.2°C.

We also have a new formula for converting from Fahrenheit temperatures to Celsius. This formula can be used to convert many different values. However, we should check our formula before we use it. Normally, when we solve an equation for C, we substitute the value for C into the original equation to check our result. We can see that in solving the literal equation, C is equal to an expression, not a numerical value. Therefore, to verify our result we pick values for all of the variables except C to obtain a value to use in our check.

Strategy

Verifying the Results to a Literal Equation

1. Obtain a value for the variable that you solved for.
   a. Choose values for all of the variables *except* the one that you solved for. Do not choose 0, 1, or 2.
   b. Substitute your chosen values into your *result,* and calculate the value of the variable that you solved for.
2. Substitute the values of all variables (this includes the calculated value) into the *original* equation and check to see if the statement is true.

To verify our result, we need to find a value for C. We start with our results and choose a value for F. So, we arbitrarily let $F = 25$.

$$\frac{5F - 160}{9} = C \qquad \text{Start with the result.}$$

$$\frac{5 * 25 - 160}{9} = C \qquad \text{Substitute the chosen value into the result.}$$

$$^-3.89 \approx C \qquad \text{Evaluate the expression to obtain a value for the variable we solved for.}$$

Next, we substitute *both values* into the *original equation* and check to see that the statement is true for these values.

CHECK

$$F = \frac{9}{5}C + 32 \qquad \text{Return to the original equation.}$$

$$25 \stackrel{?}{=} \frac{9}{5} * {}^-3.89 + 32 \qquad \text{Substitute all values into this equation.}$$

$$25 \approx 24.998 \qquad ✔ \qquad \text{Evaluate each side to determine if the statement is true.}$$

We know that the formula works when $F = 25$ and $C \approx {}^-3.89$. We have not proven that the formula works for all numbers, but because we did not choose one of the numbers with special properties, we can feel confident that the solution to the formula is correct. Because this process does not check the solution for all values, we say that we are **verifying** the solution, not checking it. However, if the process of numerically verifying a literal solution leads to an incorrect statement, we know the solution does not work and we need to find the mistake.

We now list the steps to solve a linear literal equation, but remember that these steps are very similar to what you do if you are solving an equation with just one variable.

Strategy

Solving a Linear Literal Equation

1. Simplify both sides of the equation by applying the distributive property or combining like terms. If the equation contains fractions, you may want to multiply both sides of the equation by a common denominator to eliminate the fractions.

2. Identify the terms containing the variable you are solving for.

3. Add or subtract terms from both sides of the equation to collect all terms containing the variable you are solving for on one side and all other terms on the opposite side. Again combine like terms if possible.

4. If the variable for which you are solving occurs in more than one term, factor out the common variable.

5. Divide both sides by the coefficient of the variable for which you are solving.

6. Verify your solution.

Example 2

Solve for the indicated variable. Verify your solutions.

a. $E = IR + Ir$,   for R b. $E = IR + Ir$,   for I c. $ax + b = cx + d$,   for x

Solution   a.

$$E = \underline{IR} + Ir$$ We begin by identifying the terms that contain the variable we are solving for, R.

$$E - \mathbf{Ir} = IR + Ir - \mathbf{Ir}$$ Isolate the term containing R.

$$E - Ir = IR$$

$$\frac{E - Ir}{I} = \frac{IR}{I}$$ Divide by the coefficient of R.

$$\frac{E - Ir}{I} = R$$

VERIFY   We first choose values for all of the variables except the one we solved for, in this case E, I, and r. We let $E = 30$, $I = 4$, and $r = 5$.

$$\frac{E - Ir}{I} = R$$ Start with the result.

$$\frac{30 - 4 * 5}{4} = R$$ Substitute the chosen values into the result.

$$2.5 = R$$ Evaluate the expression to obtain a value for R.

Next, substitute the values of E, I, r, and R into the original equation.

$$E = IR + Ir$$ Return to the original equation.

$$30 \overset{?}{=} 4 * 2.5 + 4 * 5$$ Substitute all values into this equation.

$$30 = 30 \qquad ✔$$

Because the values we determined using the solution to the literal equation satisfy the original equation, we conclude that the literal solution is

$$R = \frac{E - Ir}{I}$$

b. To solve $E = IR + Ir$ for I, we begin by identifying the terms that contain the variable I.

$$E = \underline{IR} + \underline{Ir}$$ There are two terms containing the variable I.

$$E = I(R + r)$$ Because I occurs in more than one term, we factor I out on the right side of the equation.

$$\frac{E}{(R + r)} = \frac{I(R + r)}{(R + r)}$$ Divide by the coefficient of I.

$$\frac{E}{(R + r)} = I$$

VERIFY   To verify, let $E = 36$, $R = 3$, and $r = 5$.

$$\frac{E}{(R + r)} = I$$

$$\frac{36}{3 + 5} = I$$

$$4.5 = I$$

Substitute the values for E, I, r, and R into the original equation.

$$E = IR + Ir$$
$$36 \overset{?}{=} 4.5 * 3 + 4.5 * 5$$
$$36 = 36 \quad \checkmark$$

Because the values we determined using the solution to the literal equation satisfy the original equation, we conclude that the literal solution is

$$\frac{E}{(R + r)} = I$$

c. To solve $ax + b = cx + d$ for x, we begin by identifying the terms that contain the variable x.

$$\underline{ax} + b = \underline{\boldsymbol{cx}} + d \qquad \text{The terms } ax \text{ and } cx \text{ contain the variable } x.$$

Two terms contain x, and they are on opposite sides of the equation. Our next step is to subtract terms to isolate all of the terms containing x to one side of the equation.

$$ax + b - cx = cx + d - \boldsymbol{cx} \qquad \text{Subtract } cx \text{ from both sides to get all of the terms}$$
$$\text{containing } x \text{ on one side.}$$

$$ax + b - cx = d$$
$$ax + b - cx - \boldsymbol{b} = d - \boldsymbol{b} \qquad \text{Subtract } b \text{ to isolate the terms containing } x.$$
$$ax - cx = d - b$$
$$x(a - c) = d - b \qquad \text{Factor } x \text{ out on the left-hand side.}$$
$$\frac{x(a - c)}{(\boldsymbol{a - c})} = \frac{d - b}{(\boldsymbol{a - c})} \qquad \text{Divide by the coefficient of } x.$$
$$x = \frac{d - b}{a - c}$$

VERIFY   Let $a = 7$, $b = 10$, $d = 5$, and $c = 3$, then

$$x = \frac{5 - 10}{7 - 3}$$
$$x = {}^-1.25$$

Substitute all of the values into the original equation.

$$ax + b = cx + d$$
$$7 * {}^-1.25 + 10 \overset{?}{=} 3 * {}^-1.25 + 5$$
$$1.25 = 1.25 \quad \checkmark$$

We conclude that

$$x = \frac{d - b}{a - c}$$

As we saw from the examples, solving literal equations is very similar to solving equations in one variable. The one exception is when the variable we are solving for occurs in more than one term. Then we need to factor the variable out to determine its coefficient. Verifying results is somewhat different from checking solutions. When we solve literal equations, our result is not a number. Therefore, to verify, the first step is to obtain a value for the variable we solved for by picking numbers for all of the other variables and substituting these into the literal solution. Then substitute all of the values, including the one we just found, into the original equation. If this results in a true statement, we gain confidence that our solution is correct. If it does not work, we must check for where we made our error.

Problem
Set
F

1. Solve for the indicated variable and verify.

   a. $y = mx + b,$ for x

   b. $P = \pi r + 2W + L,$ for W

   c. $ab + k = 5ab - 10k,$ for a

   d. $T(1 + r_1) = A + Tr_2,$ for T

   e. $\dfrac{M}{2} + N = Y$ for M

2. The following is a student's solution and verification. Find the error the student made. Explain why the solution appears to check even though the student made a mistake in the solution process.

   Solve

   $$\frac{M}{2} + N = Y \quad \text{for } M$$

   $$2 * \frac{M}{2} + N = Y * 2$$

   $$M + N = 2Y$$

   $$M = 2Y - N$$

   VERIFY   Let $N = 5$ and $Y = 4$.

   $$\frac{M}{2} + N = Y$$

   $$\frac{M}{2} + 5 = 4$$

   $$2 * \frac{M}{2} + 5 = 4 * 2$$

   $$M + 5 = 8$$

   $$M = 3$$

   Then substitute into $M = 2Y - N$.

   $$3 \stackrel{?}{=} 2 * 4 - 5$$

   $$3 = 3 \quad ✔$$

3. Solve for the indicated variable. Verify your solutions.

   a. $3(1 - x) + 3a = ax,$ for x

   b. $P = \dfrac{94 + 89 + 78 + x}{4},$ for x

   c. $T_d = 3(T_2 - T_3),$ for T_2

   d. $V = \frac{1}{3}\pi r^2 h,$ for h

   e. $M = \dfrac{ab}{a + b},$ for a

   f. $R = \dfrac{C - S}{t},$ for t

   g. $\dfrac{(x - 2)^2}{9} + \dfrac{y}{4} = 1,$ for y

   h. $LCM^2 + RCM + 1 = 0,$ for C

4. The Hospital Bill Revisited.   Recall the formula for determining the amount you must pay for major medical from example 1.

$$A = [(T - D)(1.00 - P)] + D$$

where T = total bill

D = amount of deductible

P = proportion paid by insurance, as a decimal

A = amount you must pay

a. Solve the hospital bill formula for T, the total amount of the bill.

b. Use your formula to determine the total hospital bill you can afford to pay if you budget $500 toward paying for a hospital bill after insurance has been paid. Assume your health insurance has a $200 deductible and then pays 90% of the remaining bill for all major medical.

5. a. A geometric figure is made up of a semicircle, rectangle, and triangle. Write a formula for the area and a formula for the perimeter of the figure in terms of a, b, c, and d.

b. Solve your perimeter formula for the variable a. Verify your results numerically.

c. Solve your area formula for the variable a. Verify your results numerically.

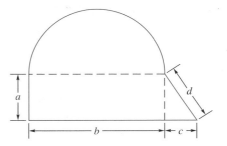

6. Two students are trying to come up with a formula to convert hours:minutes:seconds to decimal hours.

One student decides to test the following formula: where S is seconds, M is minutes, H is hours, and DH is decimal hours. Note DH is a single variable, it does not represent a product.

$$\frac{\dfrac{S}{60} + M}{60} + H = DH$$

The other student's formula looks like this:

$$\frac{S}{3600} + \frac{M}{60} + H = DH$$

Use each of the formulas to convert 2:30:00 and 6:30:45 to decimal hours. What can you say about the two formulas?

Appendix G Unit Conversion

We can measure dimensions such as length, area, weight, time, speed, and so on. In measuring these dimensions we use units. For example, we can measure length in inches or meters, and we can measure weight in pounds or kilograms. It is important to record both the numerical value and its units of measurement. In addition, the units in a problem situation can be useful in solving the problem. In this section, we will be using unit analysis and unit fractions to convert between different units of measurement.

Before we continue, we need to discuss the different uses of the word "unit." A unit is any fixed quantity or amount used as a standard. For example, when we measure someone and record their height as 6 ft, the unit for this measurement is feet because the standard measure for this was 1 foot. Similarly, gallons, miles, kilograms, and miles per hour are all examples of units of measurement.

The word *unit* is also used to represent one whole. We will see in the next few definitions both uses of the word unit.

Definition _____

Unit analysis or **dimensional analysis** involves analyzing the units (feet, liters, grams, and so on) in a problem situation so that the dimensions match that of the desired answer.

Definition _____

Unit fractions are fractions in which the numerator and denominator are equivalent measurements and have units associated with them. Therefore, unit fractions are always equal to 1.

For example, because 12 inches is equal to 1 foot, both of the fractions $\frac{12 \text{ in.}}{1 \text{ ft}}$ and $\frac{1 \text{ ft}}{12 \text{ in.}}$ are unit fractions. When multiplying factors, the units cancel just as numerical factors cancel.

When we ask you to convert between different units of measurement, we want you to use unit analysis together with unit fractions to show your work. We will discuss this process in the remainder of this section.

Suppose a car is traveling at an average speed of 45 miles per hour. How far does the car travel in 3 hours? We know that to determine the distance we calculate the product of 45 and 3, but how do the units work? What do units of miles per hour look like? To travel at a rate of 45 miles per hour means that in 1 hour the car travels a distance of 45 miles. This rate is the ratio of 45 miles to 1 hour, which we write as the fraction $\frac{45 \text{ miles}}{1 \text{ hour}}$. So the units on 45 are $\frac{\text{miles}}{\text{hour}}$. In general, the word *per* translates to division. Now let's include the units on our product of 45 and 3.

$$\frac{45 \text{ miles}}{\text{hour}} * 3 \text{ hours}$$

$$= \frac{45 \text{ miles}}{\cancel{\text{hour}}} * \frac{3 \cancel{\text{ hours}}}{1}$$

$$= 135 \text{ miles}$$

The hours in the denominator of the first factor cancel with the hours in the numerator of the second factor, which produces miles as the resulting unit.

Consider the unit fractions $\frac{12 \text{ in.}}{1 \text{ ft}}$ and $\frac{1 \text{ yd}}{3 \text{ ft}}$. When we multiply these two fractions we obtain

$$\frac{12 \text{ in.}}{1 \text{ ft}} * \frac{1 \text{ yd}}{3 \text{ ft}} = \frac{12 \text{ in. yd}}{3 \text{ ft ft}}$$

which is not terribly useful! However, if we multiply $\frac{12 \text{ in.}}{1 \text{ ft}}$ and $\frac{3 \text{ ft}}{1 \text{ yd}}$, we obtain

$$\frac{12 \text{ in.}}{1 \text{ ft}} * \frac{3 \text{ ft}}{1 \text{ yd}} = \frac{36 \text{ in.}}{1 \text{ yd}}$$

Notice that the ft in the denominator of the first fraction cancel the ft in the numerator of the second fraction. The result is $\frac{36 \text{ in.}}{1 \text{ yd}}$, which we know to be a unit fraction because there are 36 inches in 1 yard.

The process of multiplying **unit fractions** is the same as multiplying numerical fractions. Because unit fractions are equivalent to 1, multiplying by a unit fraction does not change the value of the original expression. Using these ideas, together with the following general strategy, we can convert between different units of measurement.

Strategy

Converting Between Units

To convert a given quantity to an equivalent measurement with different units do the following.

1. Identify the units in the goal.
2. Start with the given quantity. If more than one piece of information is given, then we need to choose a quantity whose units have the same meaning as our goal. For example, if our goal is miles per hour, then we need to start with a measurement whose units are distance per unit of time such as $\frac{\text{ft}}{\text{sec}}$ or $\frac{\text{laps}}{\text{min}}$.
3. Multiply by unit fractions in a way that the units you are trying to eliminate cancel and the units you are aiming for remain. (If we are converting from square units to square units we usually need two factors of each unit fraction. Similarly, if we are converting from cubic units to cubic units we usually need three factors of each unit fraction.)
4. When the units cancel appropriately, leaving you with the units you desire, perform the multiplication and record your results.

Example 1   How many ounces are in 5 pounds?

Solution   Because both ounces and pounds are American units, we look at the American–American conversions on our conversion table following the appendices. From the table, we know that 16 ounces are in 1 pound. If we multiply 5 pounds by the unit fraction $\frac{16 \text{ oz}}{1 \text{ lb}}$, the pounds associated with the 5 cancel the pounds in the denominator of the unit fraction, leaving the correct units of ounces. Multiplying a quantity by 1 does not change its value. Because the unit fraction $\frac{16 \text{ oz}}{1 \text{ lb}}$ is equal to 1, we can multiply 5 pounds by $\frac{16 \text{ oz}}{1 \text{ lb}}$ without changing the weight.

$$5 \text{ lb}$$
$$= \frac{5 \text{ lb}}{1} * \frac{16 \text{ oz}}{1 \text{ lb}}$$
$$= 80 \text{ oz}$$

There are 80 ounces in 5 pounds.

Example 2

Solution

How many tablespoons are in a cup?

Again both units of measurement are American, so from the American–American conversions we obtain the following conversion values.

$$8 \text{ fluid ounces} = 1 \text{ cup}$$
$$1 \text{ tablespoon} = 0.5 \text{ fluid ounces}$$

We need to start with the given value of 1 cup. Next we multiply by a unit fraction that cancels the cups. If we multiply 1 cup by the unit fraction $\frac{8 \text{ fl oz}}{1 \text{ c}}$, the cups cancel, leaving fluid ounces. Then we can multiply this by the unit fraction $\frac{1 \text{ tbsp}}{0.5 \text{ fl oz}}$, and the fluid ounces cancel, leaving tablespoons, which is the desired unit. This process can be seen here.

$$1 \text{ c}$$
$$= \frac{1 \text{ c}}{1} * \frac{8 \text{ fl oz}}{1 \text{ c}} * \frac{1 \text{ tbsp}}{0.5 \text{ fl oz}}$$
$$= \frac{1 * 8 * 1}{1 * 1 * 0.5} \text{ tbsp}$$
$$= 16 \text{ tbsp}$$

There are 16 tablespoons in 1 cup.

Example 3

Solution

How many square feet are in 2.43 acres?

We use the conversions $1 \text{ mile}^2 = 640$ acres and $5280 \text{ ft} = 1 \text{ mile}$. Recall that the notation 1 mile^2 means $1 \text{ mile} * \text{mile}$.

$$2.43 \text{ acres}$$
$$= 2.43 \text{ acres} * \frac{1 \text{ mile} * \text{mile}}{640 \text{ acres}} * \left(\frac{5280 \text{ ft}}{1 \text{ mile}}\right) * \left(\frac{5280 \text{ ft}}{1 \text{ mile}}\right)$$
$$= \frac{2.43 * 1 * 5280 * 5280}{640 * 1 * 1} \text{ ft}^2$$
$$\approx 106{,}000 \text{ ft}^2$$

There are about 106,000 square feet in 2.43 acres.

There are several things to notice in Example 3. First, when we multiplied the given 2.43 acres by $\frac{1 \text{ mile}^2}{640 \text{ acres}}$, we obtained square miles. To cancel the square miles, we must multiply by two factors of $\frac{5280 \text{ ft}}{1 \text{ mile}}$ because $\text{mile}^2 = \text{mile} * \text{mile}$. Second, we rounded the results. The given information is an approximate measurement and contains three significant digits. Each conversion used was an exact conversion. We therefore rounded our answer to three significant digits.

Example 4   We know that there are 3 feet in 1 yard. Explain why there are 27 cubic feet in 1 cubic yard.

Solution   A cubic yard can be visualized as a cube with dimensions 1 yard by 1 yard by 1 yard. This is the same as a cube with dimensions 3 feet by 3 feet by 3 feet.

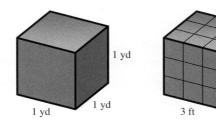

From the figure, we can see that

$$(3 \text{ feet}) * (3 \text{ feet}) * (3 \text{ feet}) = 27 \text{ ft}^3$$

are in 1 cubic yard.

We can also convert 1 cubic yard to cubic feet.

$$1 \text{ yd}^3$$

$$= 1 \text{ yd} * \text{yd} * \text{yd} * \left(\frac{3 \text{ ft}}{1 \text{ yd}}\right) * \left(\frac{3 \text{ ft}}{1 \text{ yd}}\right) * \left(\frac{3 \text{ ft}}{1 \text{ yd}}\right)$$

$$= 1 * 3 * 3 * 3 \text{ ft}^3$$

$$= 27 \text{ ft}^3$$

Example 5   How many milliliters are in 3 tablespoons?

Solution   Because milliliters is a metric unit and tablespoons is an American unit we need to find a volume conversion between the American units and metric units. From the conversion table, the only American–metric volume conversion is 1.06 quarts ≈ 1 liter. In addition to this conversion, we may need the following conversion values.

1.06 quarts ≈ 1 liter	8 fluid ounces = 1 cup
1000 milliliters = 1 liter	2 cups = 1 pint
1 tablespoon = 0.5 fluid ounces	2 pints = 1 quart

First, we start with the given 3 tablespoons. Then, we multiply by unit fractions until we obtain milliliters.

$$3 \text{ tbsp}$$

$$\approx 3 \text{ tbsp} * \frac{0.5 \text{ fl oz}}{1 \text{ tbsp}} * \frac{1 \text{ c}}{8 \text{ fl oz}} * \frac{1 \text{ pt}}{2 \text{ c}} * \frac{1 \text{ qt}}{2 \text{ pt}} * \frac{1 \text{ L}}{1.06 \text{ qt}} * \frac{1000 \text{ mL}}{1 \text{ L}}$$

$$= \frac{3 * 0.5 * 1000}{8 * 2 * 2 * 1.06} \text{ mL}$$

$$\approx 44 \text{ mL}$$

Because 3 tablespoons is most likely measured more accurately than the nearest tablespoon, we round our result to two significant digits. About 44 milliliters are in 3 tablespoons.

Example 6

A recreational swimmer can swim 10 laps in 15 minutes. If one lap is 50 meters long, determine the rate that the person is swimming in miles per hour.

Solution   In this problem we are given two different pieces of information. We know that

$$10 \text{ laps} = 15 \text{ minutes} \qquad \text{and}$$
$$1 \text{ lap} = 50 \text{ meters}$$

Each piece of given information in this problem can be written as a ratio. For example the ratio of 10 laps to 15 minutes can be written as the fraction

$$\frac{10 \text{ laps}}{15 \text{ min}} \qquad \text{or} \qquad \frac{15 \text{ min}}{10 \text{ laps}}$$

Similarly, the ratio of 1 lap to 50 meters can be written as the fraction

$$\frac{1 \text{ lap}}{50 \text{ m}} \qquad \text{or} \qquad \frac{50 \text{ m}}{1 \text{ lap}}$$

So, which information do we start with? One strategy is to start with the information whose units have the same meaning as our goal. Our goal is to determine the speed in miles per hour $\left(\frac{\text{miles}}{\text{hour}}\right)$, which is a ratio of distance to time. Therefore, we need to start with a ratio of $\frac{\text{distance}}{\text{time}}$, so we start with the ratio $\frac{10 \text{ laps}}{15 \text{ min}}$.

By beginning with laps per minute we have two units of measurement to convert. We need to convert the laps to miles and the minutes to hours. It does not matter which we start with. Both conversions can be done in a single expression.

In the following unit conversion, we first multiply by the unit fractions $\frac{50 \text{ m}}{1 \text{ lap}}, \frac{1 \text{ km}}{1000 \text{ m}},$ and $\frac{1 \text{ mile}}{1.609 \text{ km}},$ which converts the laps to miles, giving us units of miles per minute. Then, we complete the conversion by multiplying by $\frac{60 \text{ min}}{1 \text{ h}}$.

$$\frac{10 \text{ laps}}{15 \text{ min}}$$

$$= \frac{10 \text{ laps}}{15 \text{ min}} * \frac{50 \text{ m}}{1 \text{ lap}} * \frac{1 \text{ km}}{1000 \text{ m}} * \frac{1 \text{ mile}}{1.609 \text{ km}} * \frac{60 \text{ min}}{1 \text{ h}}$$

$$= \frac{10 * 50 * 60}{15 * 1000 * 1.609} \frac{\text{miles}}{\text{h}}$$

$$\approx 1.24 \frac{\text{miles}}{\text{h}}$$

The person is swimming at a rate of about $1\frac{1}{4}$ miles per hour.

Notice that even though we converted two different types of units, the conversion was still done using a single expression.

So far, we used unit analysis and unit fractions to convert between different units of measurement. Unit analysis can also be used as a problem-solving tool. Often in problem solving, we are not sure which mathematical operation to use. Should we multiply or divide two quantities, and if we divide, in which order do we perform the division? Analyzing the units in a problem can be helpful in deciding which operation to use.

Example 7

The 3000-Meter Competition. Marise just completed a 3000-meter preliminary race in 12 minutes 35 seconds. The race was held on an indoor track, and runners had to run 10 laps to complete the 3000-meter distance. She now wants to know how fast she has to run each lap in the finals competition so that her time is as good as the preliminary race. Marise is uncertain about the order to perform the division. Which one is correct?

$$\frac{10 \text{ laps}}{12 \text{ min } 35 \text{ sec}} \quad \text{or} \quad \frac{12 \text{ min } 35 \text{ sec}}{10 \text{ laps}}$$

Solution   The units in the first expression are in laps per length of time. This is not what Marise wants to know. In the second expression the resulting units are length of time per lap. Therefore, performing the division, $\frac{12 \text{ min } 35 \text{ sec}}{10 \text{ laps}}$ gives the length of time it takes to run one lap. To completely answer the question of how fast Marise should run, we first need to convert 12 minutes and 35 seconds to seconds.

$$12 \text{ min} * \frac{60 \text{ sec}}{1 \text{ min}} + 35 \text{ sec} = 755 \text{ sec}$$

Then,

$$\frac{12 \text{ min } 35 \text{ sec}}{10 \text{ laps}} = \frac{755 \text{ sec}}{10 \text{ laps}} = \frac{75.5 \text{ sec}}{\text{lap}} = 75.5 \text{ seconds per lap}$$

Marise must run each lap in about 75.5 seconds (or 1 minute and 15.5 seconds) to reach her goal.

In this section, we used **unit analysis** to convert between different units of measurement. In general, we start with the given quantity and multiply by **unit fractions** until we reach our desired units. If we are converting from square units to square units, we usually use two factors of each unit fraction. Similarly, if we are converting from cubic units to cubic units, we usually use three factors of each unit fraction.

Unit analysis can also be used as a problem-solving tool. If you are not sure whether to multiply or divide by a number, analyze the units. When the units work out correctly, you have most likely performed the correct operation.

Problem Set G

For Problems 1–15, write a single expression using **unit fractions** to determine the answer to each question. Perform all computations in one step on your calculator. Round all answers reasonably.

1. How many feet are in 3.2 miles?

2. How many millimeters are in 2.6 kilometers?

3. How many feet are in 645 centimeters?

4. How many ounces are in 1.98 American tons?

5. How many grams are in 0.375 ounces?

6. How many kilograms are in 708 milligrams?

7. How many seconds are in 1 year?

8. How many milliliters are in 0.5 cups?

9. How many square feet are in 0.125 square miles?

10. Convert 3.5 pints per minute to cubic feet per hour.

11. Convert 49.6 cents per gram to dollars per American ton.

12. Convert 2032 cubic centimeters to cubic inches.

13. How many cubic inches are in a 3.0-liter engine?

14. Mercury is a liquid with a density of 13.6 grams per cubic centimeter (at room temperature). What is the density of mercury in pounds per cubic inch?

15. If water weighs 62.4 pounds per cubic foot, what is the weight in ounces of one cubic inch of water?

16. Determine the volume of the cylinder shown in the figure. Then convert your results to cubic meters using unit fractions. (*Note:* The volume of a cylinder is equal to the product of the area of the base and the height.)

3 in.

3 ft 2 in.

17. If one acre of land is in the shape of a square, what are the dimensions of this piece of land in feet?

Geometry Reference

Definitions

angle   A figure formed by two line segments extending from a common point called the vertex

     central   An angle whose vertex is at the center of a regular polygon or a circle

     interior   An angle formed by two adjacent sides of a polygon and lying inside the polygon

circumference   The distance around the outside of a circle

congruent figures   Geometric figures having the same size and shape; orientation may be different

diagonal   A straight line joining two nonadjacent vertices of a polygon

diameter   A line joining the two sides of a circle or sphere and passing through the center

n-gon   A polygon with n sides, where n is a natural number

parallel lines   Two or more lines in a plane that never meet or intersect no matter how far they are extended

perimeter   The distance around the outside edge of a figure

polygon   A closed geometric figure made up of straight-line segments

     convex   A polygon in which none of the interior angles is greater than or equal to 180° (see illustration)

     concave   A polygon in which at least one of the interior angles is greater than or equal to 180° (see illustration)

     regular   A polygon in which all sides are the same length and all interior angles are the same size

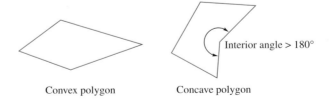

Convex polygon   Concave polygon

radius   A line from the center of a circle or sphere to the edge of the circle or sphere

similar figures   Geometric figures having the same shape but not necessarily the same size; corresponding angles are equal and corresponding sides are proportional

Formulas for Two-Dimensional Objects

A **rectangle** is a four-sided figure with four right angles.

area $= LW$

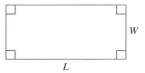

A **triangle** is a three-sided figure.

area $= \dfrac{1}{2}bh$

NOTE: The height of any figure must be perpendicular to the base.

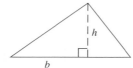

The Pythagorean Theorem

In a *right triangle,* the sum of the squares of the legs is equal to the square of the hypotenuse.

$\text{leg}_1{}^2 + \text{leg}_2{}^2 = \text{hypotenuse}^2$

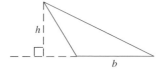

A **parallelogram** is a four-sided figure with opposite sides parallel.

area $= bh$

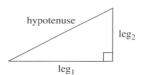

A **trapezoid** is a four-sided figure with two parallel sides, called bases.

area $= \dfrac{h(b_1 + b_2)}{2}$

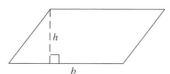

A **circle** is the set of all points in a plane that are equidistant from the center.

area $= \pi r^2$

circumference $= 2\pi r$

or

circumference $= \pi d$, where $d =$ diameter

Formulas for Three-Dimensional Objects

A **rectangular solid** is a three-dimensional figure in which all sides are rectangles.

volume $= LWH$

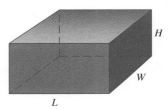

A **prism** is a polyhedron with two parallel congruent faces, called the bases of the prism. If the lateral sides are perpendicular to the bases, the prism is called a **right prism.**

volume $=$ *area of the base * height*

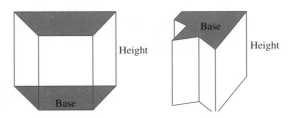

A **pyramid** is a polyhedron that has a base consisting of any polygon and triangular sides.

volume $= \dfrac{1}{3} *$ *area of base * height*

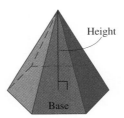

A **circular cylinder** is a three-dimensional object with two parallel faces that are circles. These are the bases of the cylinder. If the lateral surface is perpendicular to the bases, it is called a **right circular cylinder.** If not otherwise specified, a cylinder is assumed to be a right circular cylinder.

volume $= \pi r^2 h$

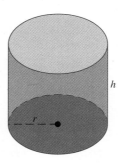

A **circular cone** has a circular base and the lateral surface forms a single point.

volume $= \dfrac{1}{3}\pi r^2 h$

surface area $= \pi r^2 + \pi r s$

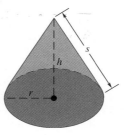

A **sphere** is the set of all points in space that are equidistant from the center.

volume $= \dfrac{4}{3}\pi r^3$

surface area $= 4\pi r^2$

Conversion Tables

AMERICAN–AMERICAN CONVERSIONS

Distance	Volume
12 inches (in.) = 1 foot (ft)	3 teaspoons (tsp) = 1 tablespoon (tbsp)
3 feet = 1 yard (yd)	1 tablespoon = 0.5 fluid ounces
5280 feet = 1 mile	8 fluid ounces (fl oz) = 1 cup (c)
8 furlongs = 1 mile	2 cups = 1 pint (pt)
6 feet = 1 fathom	2 pints = 1 quart (qt)
100 links = 1 chain	4 quarts = 1 gallon (gal)
66 feet = 1 chain	1 ft^3 ≈ 7.481 gallons

Area	Weight
640 acres = 1 mile^2	480 grains = 1 ounce (oz)
	16 ounces = 1 pound (lb)
	2000 pounds = 1 ton

METRIC–METRIC CONVERSIONS

Distance	Volume
1 kilometer (km) = 1000 meters (m)	1000 milliliter (mL) = 1 liter
100 centimeters (cm) = 1 meter	1 cm^3 = 1 milliliter
1000 millimeters (mm) = 1 meter	
10 decimeters (dm) = 1 meter	

Area	Weight
1 hectare = 10,000 m^2	1 kilogram (kg) = 1000 grams
	1000 milligrams (mg) = 1 gram (g)
	1000 kilograms = 1 metric ton

AMERICAN–METRIC CONVERSIONS

Distance	Volume	Weight
2.54 cm = 1 inch	1.06 qt ≈ 1 liter	454 g ≈ 1 lb
39.37 in ≈ 1 meter		2.2 lb ≈ 1 kg
1.609 km ≈ 1 mile		

TIME CONVERSIONS

60 seconds (sec) = 1 minute (min)
60 minutes = 1 hour (h)
24 hours = 1 day
7 days = 1 week
365 days = 1 calendar year (nonleap year)
365.242199 days = 1 tropical year

METRIC PREFIXES

Prefix	Factor		Prefix	Factor	
tera	10^{12}	T	deci	10^{-1}	d
giga	10^{9}	G	centi	10^{-2}	c
mega	10^{6}	M	milli	10^{-3}	m
kilo	10^{3}	k	micro	10^{-6}	μ
hecto	10^{2}	h	nano	10^{-9}	n
deca	10^{1}	da	pico	10^{-12}	p

Selected Answers

Chapter 1

1. a. *A* is independent.
H is dependent.

b. *H* is independent.
P is dependent.

c. *S* is independent.
T is dependent.

d. *A* is independent.
C is dependent.

e. *M* is independent.
T is dependent.

6. a.

Number of checks	Monthly charges ($)
0	3
5	$3 + 5 * 0.12 = 3.60$
10	$3 + 10 * 0.12 = 4.20$
15	$3 + 15 * 0.12 = 4.80$

b.

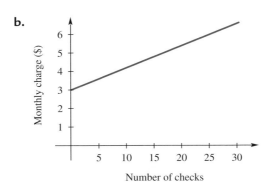

Number of checks

c. $M = 3 + 0.12n$, M = monthly charge and n = number of checks

d. $4.20, $6.24, $3.00

e. 16 checks, no

9. c. $L = \dfrac{50 - 2w}{2}$; $A = w\left(\dfrac{50 - 2w}{2}\right)$; $L = $ length, $w = $ width, $A = $ area

13. b.

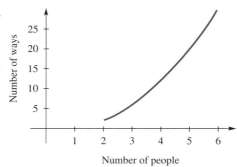

c. $w = p(p - 1)$; $w = $ number of ways, $p = $ number of people

16. a. 10,100 tiles

  b. 5050 tiles

  c. 5050

  d. $n = $ a positive whole number

  $T = $ sum of the first n numbers

  $T = \dfrac{n(n + 1)}{2}$

17. a. i. distance $= 5$

SECTION 1.2

2. a. $y = {}^{-}x$ $\qquad$ $y = {}^{-}16, {}^{-}9.4, 0, 10$

  b. $y = 5(x + 6)$ $\qquad$ $y = 110, 77, 30, {}^{-}20$

  c. $y = 5 + 6x$ $\qquad$ $y = 101, 61.4, 5, {}^{-}55$

  d. $y = 100$ $\qquad$ $y = 100, 100, 100, 100$

4. a. Input x $\qquad\qquad\qquad$ **b.** Input x
  Multiply by 3 $\qquad\qquad\qquad\qquad$ Find sum of x, 35, 34, and 40
  Add 6 $\qquad\qquad\qquad\qquad\qquad$ Divide by 4

  c. Input x $\qquad\qquad\qquad\qquad$ **d.** Input x
  Add 3 $\qquad\qquad\qquad\qquad\qquad$ Output 45
  Square the result

5. Tables b and c represent functions.

Chapter 2

SECTION 2.1

2. a. $y = \dfrac{32 - x}{4}$ $\qquad\qquad\qquad$ **c.** $R = \dfrac{2}{13 - 3K}$

  b. $m = \dfrac{104 - 24p}{-3}$ $\qquad\qquad$ **d.** $T = \dfrac{2W}{3}$

3. a. 1 $\quad$ **b.** $\dfrac{-5}{4}$ $\quad$ **c.** $\dfrac{-8}{15}$ $\quad$ **d.** $\dfrac{1}{3}$

5. a.

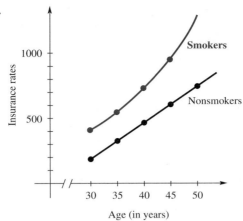

b. Nonsmokers; Insurance rates increase $27.60 for every year older.

6. a.

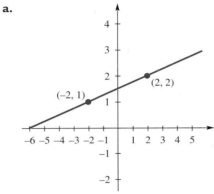

b.

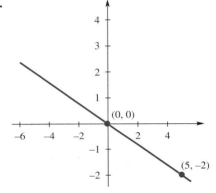

c.

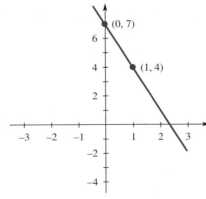

d.

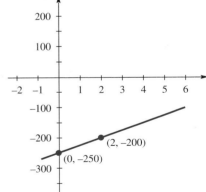

e.

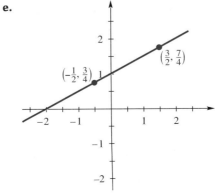

SECTION 2.2

3. b, c

5. a. $y = -\frac{4}{3}x + \frac{16}{3}$, slope $-\frac{4}{3}$, vertical intercept $\left(0, \frac{16}{3}\right)$

   b. $y = 0.4x - 4.8$, slope 0.4, vertical intercept $(0, -4.8)$

7. a. $y = \frac{2}{3}x + 5$, $(-7.5, 0)$

   b. $y = -140x + 600$, $\left(4\frac{2}{7}, 0\right)$

8. a. $m = 3$; $(0, -2)$

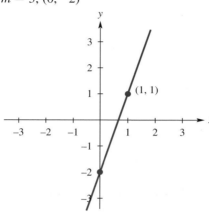

  d. $m = \frac{1}{2}$; $(0, -10)$

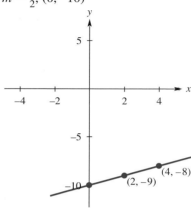

  b. $m = -5$; $(0, 10)$

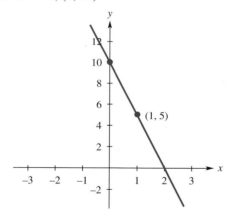

  e. $m = \frac{-4}{5}$; $(0, 3)$

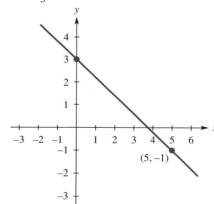

  c. $m = \frac{-3}{4}$; $(0, 24)$

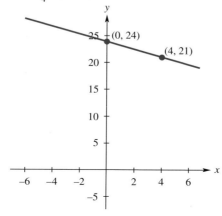

  f. $m = 1$; $(0, 0)$

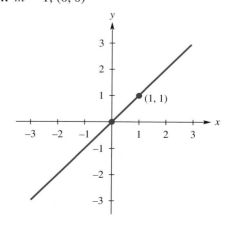

13. a. Is a line with slope of $^-7$

   b. Is a line with slope of $\sqrt{2}$

   c. Not a line

   d. Is a line with slope of $\frac{5}{2}$

   e. Is a line with an undefined slope

   f. Not a line

   g. Is a line with slope of $\frac{1}{2}$

   h. Is a line with slope of $\frac{5}{4}$

SECTION 2.3

2. a. $y = \frac{17}{40}x + 0.75$

   b. $y = {}^-62.5x + 275$

5. a. $y = \frac{-3}{2}x + 2.5$

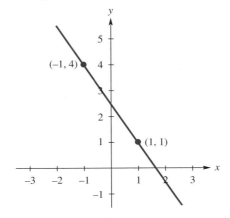

   b. $y = 4x - 24$

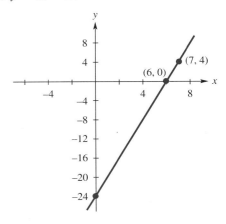

   c. $y = {}^-x + 1$

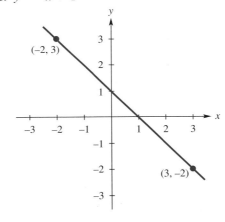

   d. $y = 1.5x + 12.3$

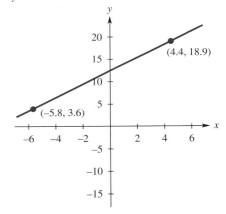

6. a. $y = 4x - 3$

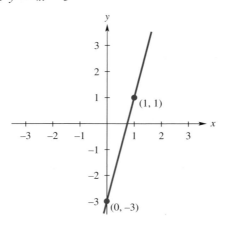

b. $y = \frac{2}{3}x + 5$

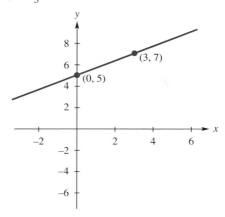

c. $y = -\frac{1}{5}x - 4$

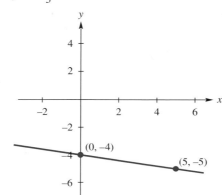

d. $y = {}^-4x + 13$

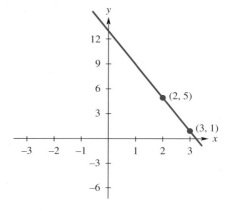

9. Many other answers are possible for each.

a. $y = {}^-2x$, $y = {}^-2x + 5$, $y = {}^-2x - 16$

b. $y = \frac{5}{4}x$, $y = 5$, $x = 4$

c. $y = 5x$, $y = 5x + 3$, $y = 5x - 1$

15. a. $P = 12.5t + 18$, where P = pollution in ppm, t = hours after 8:00 A.M.

b.

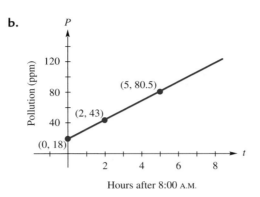

Hours after 8:00 A.M.

    c. $\frac{\text{ppm}}{\text{hour}}$; pollution is increasing at a rate of 12.5 ppm per hour.

    d. (0, 18); at 8:00 A.M. the pollution index is 18 ppm.

    e. 61.75 ppm; 211.75 ppm; answer at 11:30 A.M. makes sense because it falls between 8 A.M. and 4 P.M. Answer at 11:30 P.M. does not make sense.

    f. Be inside after 2:30 P.M.

24. a. Graph B   **b.** Graph A   **c.** Graph B   **d.** Graph C   **e.** Graph D

SECTION 2.4

1. a. Is a solution   **b.** Not a solution

3. a. $x = 5, y = 3$ **b.** $x = 2.5, y = 2$

4. a. $x = 4, y = 3$ **b.** $x = {}^{-}1.5, y = {}^{-}3$

11. a. Let h = number of hours and C = cost of rental.

    RRE: $C = 10 + 0.75h$

    TTES: $C = 2.95 + 1.22h$

    b. The cost for 15 hours is $21.25 for both companies.

13. They had 37 nickels and 53 quarters.

18. She should mix 1.3 liters of 10% acid with 0.7 liters of 30% acid.

Chapter 3

SECTION 3.1

1. a. $(^{-}6)^5$ **e.** $(^{-}5)^{-2}$

    b. $(^{-}4)^2$ **f.** $^{-}5^{-3}$

    c. $^{-}4^2$ **g.** $2n^2$

    d. $\left(\dfrac{1}{x}\right)^2$ **h.** $(2n)^{-2}$

4. a. $^{-}8$ **b.** $\dfrac{^{-}1}{8}$ **c.** $^{-}4$ **d.** $-\dfrac{1}{4}$ **e.** $^{-}1$

8. a. $\dfrac{16}{81}$ **b.** $\dfrac{81}{16}$ **c.** 1

10. a.

x	x^2	x^3	x^4	x^5
$^-3$	9	$^-27$	81	$^-243$
$^-2$	4	$^-8$	16	$^-32$
$^-1$	1	$^-1$	1	$^-1$
0	0	0	0	0
1	1	1	1	1
2	4	8	16	32
3	9	27	81	243

12. a. 9^{-2}

    b. $(^-3)^{-4}$

    c. $(^-2)^{-3}$

    d. $(^--10)^4$ or 10^4

    e. $^-4^2$

    f. $^-(^-1)^{-2}$

    g. $\left(\frac{^-1}{10}\right)^{-6}$

17. $\sqrt{(^-2)^2 - 4 * 1 * {}^-15} = 8$

19. a. $^-4x^8$

    b. $4p^8$

    c. $80R^3 - 60R^2T^3$

    d. A^2

    e. $\frac{1}{z^{10}}$

    f. m^{14}

    g. $\frac{16}{w^8}$

26. a. $\frac{1}{m^{10}}$

    b. 1

    c. $\frac{5}{x^3}$

    d. $\frac{x^3}{5}$

    e. 3

    f. 1

    g. $\frac{B^2}{A^2}$

    h. $\frac{7}{Q^{14}}$

    i. $\frac{1}{Q^{20}}$

29. a. $^-(^-2)^2(^-3)^3 = 108$

    c. $(^-0.25)^{-2} + (^-0.5)^{-3} = 8$

    b. $\dfrac{7(2.4^3 * {}^-5.1)^2}{14 * {}^-5.1} \approx {}^-487.31$

    d. $\dfrac{1}{(^-0.25)^2 + (^-0.5)^3} = {}^-16$

31. a. Area $= xh + 10h$

    b. Area $= AB + \dfrac{B^2}{2}$

SECTION 3.2

1. a. 64   **b.** 729   **c.** 4096   **d.** 530,712

4. 64,000; 125,000

8. a. 1000   **b.** $\frac{1}{17,576}$ **c.** Larger

Chapter 4

SECTION 4.1

1. a. $9A^2 - 30AB + 25B^2$ **c.** $3x - \dfrac{12}{y}$

b. $3m^3 - 2m + 1$ **d.** $1.25A + 6.75B^3$

6. $x \geq 3$

7. $x > {}^-9.5$

15. a. 2.25 min; ≈ 2.3 min; yes

b. $d =$ your distance from the shore

$$\frac{d}{20} < \frac{150}{65}; \ d < 46 \text{ meters}$$

19. Input ≥ 20

SECTION 4.2

2. a.

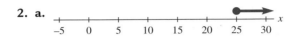

b.

c.

d.

3. a

b.

c.

d.

e.

5. a. $x > {}^-1$ **b.** $x \leq 3$ **c.** $x \leq {}^-5$ or $x \geq 10$ **d.** ${}^-2 < x < 3$

8. The distance from A to C is greater than 5 and less than 25.

12. a. $x \approx {}^-2$ or $x \approx 6$ **b.** $x < {}^-2$ or $x > 6$ **c.** ${}^-1.5 < x < 5.5$

SECTION 4.3

1. a. $|x| = 10$ **b.** $|w + 4| = 5$ **c.** $|x| \geq 5$ **d.** $|x - 7| < 4$

5. a. $4 < x < 10$ **c.** $^{-}7.5 \leq x \leq 7.5$

     b. $x < ^{-}18$ or $x > 12$ **d.** $x = 23.5$ or $x = ^{-}6.5$

11. a. $x = D$ or $x = F$

SECTION 4.4

2. a. $x = 15$ or $x = 5$ **b.** $1 < T < 11$ **c.** $^{-}37 \leq y \leq 3$ **d.** $^{-}7.5 < m < 0.5$

5. $17.7 \leq P \leq 42.3$; $103 \leq$ number planning to enlist ≤ 245

8. a. $|x| = 5$ **b.** $|x - 4| \leq 3$

SECTION 4.5

4. a. Mean ≈ 104.6, standard deviation ≈ 3.04

b.

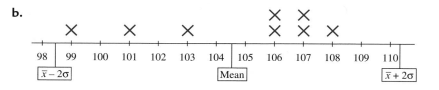

d. $\bar{x} + 2\sigma \approx 110.7$; $\bar{x} - 2\sigma \approx 98.5$

e. 100%

6. a. Game A range $= 5$, game B range $= 4$

     b. Game A standard deviation ≈ 1.7

        Game B standard deviation ≈ 1.2

Chapter 5

SECTION 5.1

1. a. $\sqrt[5]{y}$ **e.** $^{-}\sqrt[3]{125}$

     b. $(^{-}27)^{1/3}$ **f.** $^{-}x^{1/3}$

     c. $5M^{1/2}$ **g.** $y^{-1/5}$

     d. $(5M)^{1/2}$

3. a. 7 **b.** 3 **c.** 4 **d.** 2 **e.** 25 **f.** $\frac{1}{5}$ **g.** $^{-}6$ **h.** $^{-}\frac{1}{3}$

5. a. $32\sqrt[5]{x}$ **d.** $2\pi R^{1/3}$ **g.** $\sqrt[5]{^{-}x}$

     b. $(M + N)^{1/2}$ **e.** $\left(\dfrac{L}{W}\right)^{1/2}$ **h.** $\dfrac{1}{\sqrt[3]{p + q}}$

     c. $\sqrt[5]{32x}$ **f.** $^{-}\sqrt[5]{x}$ **i.** $\dfrac{\sqrt{w - 2xy}}{5y}$

8. a. $9^{1/2}$

b. $(-27)^{1/3}$

c. $(--10{,}000)^{-1/4}$ or $10{,}000^{-1/4}$

d. $-4^{1/2}$ or $-(4)^{1/2}$

e. $-(-1)^{1/2}$

f. $(-10)^{-10}$

g. $(--125)^{1/3}$ or $125^{1/3}$

10. a.

x	$x^{1/2}$	$x^{1/3}$	$x^{1/4}$	$x^{1/5}$
-4	undefined	≈ -1.59	undefined	≈ -1.32
-3	undefined	≈ -1.44	undefined	≈ -1.25
-2	undefined	≈ -1.26	undefined	≈ -1.15
-1	undefined	-1	undefined	-1
0	0	0	0	0
1	1	1	1	1
2	≈ 1.41	≈ 1.26	≈ 1.19	≈ 1.15
3	≈ 1.73	≈ 1.44	≈ 1.32	≈ 1.25
4	2	≈ 1.59	≈ 1.41	≈ 1.32

11. a. $54b^{1/4}$ **b.** $H^{1/6}$ **c.** 1 **d.** $3R^{1/12}$ **e.** $x^{1/2} + 1$

SECTION 5.2

1. a. $m = 2.5$

b. $k \approx \pm 7.746$

c. $x \doteq 5.44$

d. $y = -123$

e. $R = 655{,}360{,}000$

4. a. $p \approx -0.6229$

b. $x = \pm 3$

c. $x = 23$

d. $x = \frac{23}{9} \approx 2.556$

e. $Q = 5$ or $Q = \frac{5}{3}$

f. $x \approx 0.05946$ or $x \approx -2.059$

g. $x = \frac{23}{35} \approx 0.6571$

h. No real solution

6. a. $M = \dfrac{K^3}{G}$

b. $r = \left(\dfrac{A}{P}\right)^{1/5} - 1$

c. $A = \dfrac{d}{C_p v^2}$

d. $r = \left(\dfrac{3V}{4\pi}\right)^{1/3}$

e. $L = g\left(\dfrac{T}{2\pi}\right)^2 = \dfrac{gT^2}{4\pi^2}$

SECTION 5.3

1. a. $(2x)^{4/5}$

b. $(\sqrt{y})^5$

c. $-\sqrt[3]{x^4}$

d. $-x^{4/3}$

e. $(x - 3)^{3/4}$

f. $\pi\sqrt[3]{r^2}$

g. $\pi r^{3/2}$

2. a. $(\sqrt[4]{x})^5$ **f.** $(\sqrt{8y})^5$

b. $m^{3/2}$ **g.** $7(cd)^{5/3}$

c. $25T^{4/3}$ **h.** $x^{2/3}$

d. $(25T)^{4/3}$ **i.** $\left(\sqrt[3]{\dfrac{3v}{4}}\right)^2$

e. $8(\sqrt{y})^5$ **j.** $\dfrac{-3\sqrt{x}}{y}$

3. a. $-x^{5/3}$ **d.** $(\sqrt{16x})^3$ **g.** $-(\sqrt{x})^5$

b. $(^-y)^{5/3}$ **e.** $16(\sqrt{x})^3$ **h.** $2\pi V^{3/2}$

c. $\dfrac{n^{3/2}}{x}$ **f.** $(\sqrt{^-x})^5$ **i.** $\left(\dfrac{c}{d}\right)^{3/2}$

4. a. 125   **b.** 9   **c.** Nonreal numbers   **d.** $^-27$ **e.** $\dfrac{1}{100}$ **f.** 4   **g.** $^-8$

9. a. Positive   **b.** 1296

17. a. $w = 4$ **b.** $y \approx 0.504$ **c.** no real solution   **d.** $x \approx \pm 7.684$

Chapter 6

SECTION 6.1

1. a. Linear function   **e.** Other

b. Power function (quadratic)   **f.** Exponential function

c. Exponential function   **g.** Linear function

d. Power function   **h.** Other

2. a. Exponential function   **d.** Neither

b. Linear function   **e.** Exponential function

c. Exponential function   **f.** Linear function

7. a. $x \approx \pm 2.432$ **e.** No real solution

c. $x = 3127$ **g.** $x = ^-6$

9. a. 27 ft

b.

Number of Times the Ball Hits the Ground	Height the Ball Reaches After Hitting the Ground
0	36
1	27
2	20.25
3	15.1875

    c. $H = 36 * 0.75^t$

       t = number of times the ball hits the ground

       H = height the ball reaches after hitting the ground

    d. 13 bounces

SECTION 6.2

1. a. Exponential function

    b. Power function

    c. Linear function

    d. Other

    e. Exponential function

    f. Linear function

3. a. An increasing exponential function with vertical intercept (0, 5)

    b. A decreasing exponential function with vertical intercept (0, 10)

    c. A decreasing exponential function with vertical intercept (0, 28)

    d. An increasing exponential function with vertical intercept (0, 0.56)

    e. A line with vertical intercept (0, 6) and slope $^-3$

    f. A decreasing exponential function with vertical intercept (0, 4)

10. a.

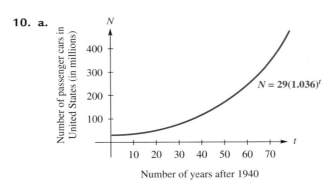

    b. ≈ 242 million

    c. 1960; 1980

Chapter 7

SECTION 7.1   **1. a.**

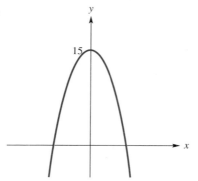

c.

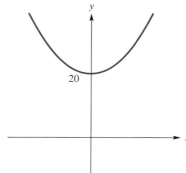

b.

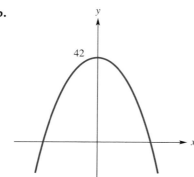

d.

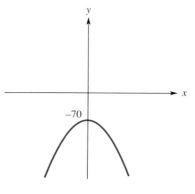

10. a.

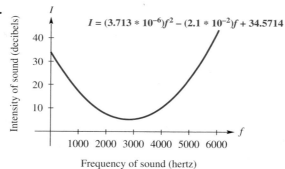

$$I = (3.713 * 10^{-6})f^2 - (2.1 * 10^{-2})f + 34.5714$$

b. The minimum frequency is about 2828 Hz.

SECTION 7.2   **1. a.** $x = \pm 4$, square root   **d.** $x = 2$ or $x = \frac{-4}{3}$, quadratic formula

b. $x = {}^{-}20$ or $x = {}^{-}25$, quadratic formula   **e.** $x \approx \pm 3.873$, square root

c. $x = 0$ or $x = 6$, factoring   **f.** $x = {}^{-}1$, linear

7. a. Other   **e.** Other

b. Quadratic   **f.** Quadratic

c. Linear   **g.** Linear

d. Other   **h.** Other

14. The diameter needs to be a little bit larger than 1.92 m.

SECTION 7.3

1. a. $m = 4$

2. a. $x \approx 2.174$ or $x \approx {}^-1.840$

6. Other answers are possible.

  a. $(x + 3)(x - 7) = 0$ **c.** $\left(x - \frac{1}{2}\right)(x + 6) = 0$

  b. $(x - 5)(x - 20) = 0$ **d.** $(x - 8)(x - 8) = 0$

SECTION 7.4

1. a. Linear   **c.** Quadratic   **e.** Quadratic

  b. Quadratic   **d.** Linear   **f.** Linear

3. a. $m = \pm\sqrt{\dfrac{7}{p}}$

  c. $R = \dfrac{S^2 T}{12}$

  e. $M = \dfrac{3A \pm \sqrt{({}^-3A)^2 - 4 * AR * 4}}{2AR}$ which simplifies to $M = \dfrac{3A \pm \sqrt{9A^2 - 16AR}}{2AR}$

  g. $A = \dfrac{{}^-4}{RM^2 - 3M}$

4. a. $r = \pm\sqrt{\dfrac{H}{8m}}$

  b. $x = \dfrac{1 \pm \sqrt{1 - 4 * R * {}^-Rp}}{2R}$ which simplifies to $x = \dfrac{1 \pm \sqrt{1 + 4R^2 P}}{2R}$

  c. $y = {}^-p\,, y = q$

  d. $m = \dfrac{{}^-(2k + 5) \pm \sqrt{(2k + 5)^2 - 4 * 2 * (5k - 9)}}{4}$ which simplifies to

       $m = \dfrac{{}^-2k - 5 \pm \sqrt{4k^2 - 20k + 97}}{4}$

5. a. $R = \dfrac{V}{I}$

  b. $I = 0, I = \dfrac{V}{R}$

  c. $m = \dfrac{{}^-RC \pm \sqrt{(RC)^2 - 4 * LC * 1}}{2LC}$ which simplifies to $m = \dfrac{{}^-RC \pm \sqrt{R^2 C^2 - 4LC}}{2LC}$

  d. $S = C - Rt$

  e. $b = \dfrac{Ma}{a - M}$

  f. $g = \dfrac{{}^-(2 - D) \pm \sqrt{(2 - D)^2 - 4}}{2}$ which simplifies to $g = \dfrac{D - 2 \pm \sqrt{D^2 - 4D}}{2}$

SECTION 7.5   **2. a.** $-3.2 < x < 2.2$ **b.** $x \le -6.5$ or $x \ge 9.8$

5. a. (M, N) and (R, T)

Chapter 8

SECTION 8.1   **1. a.** $TU \approx 5.1$ cm
                           $UV \approx 8.6$ cm

5. a. $\triangle ABC \sim \triangle HGI$ **b.** $GI \approx 1.95$ in.

9. $\triangle ABC \sim \triangle ADB \sim \triangle BDC$

11. a. $NK \approx 11.47$ in.
           $MK \approx 3.44$ in.

SECTION 8.3   **1. a.** $\frac{9}{15}$ **b.** $\frac{12}{15}$ **c.** $\frac{9}{12}$ **d.** $\frac{12}{15}$

3. a. $\frac{5}{13}$ **b.** $\frac{12}{13}$ **c.** $\frac{5}{12}$ **d.** $\frac{5}{13}$

6. ≈ 73 ft

9. $\approx 7.1\%$

13. ≈ 15 ft

15. Rafter ≈ 17.3 ft; rise ≈ 6.6 ft

SECTION 8.4   **1. a.** $\approx 40.7°$ **b.** $\approx 29.0°$ **c.** $\approx 61.9°$ **d.** $\approx 59.6°$

4. Slope distance ≈ 999 ft
      Slope angle $\approx 7.6°$

7. slope angle $\approx 4.1°$

9. percent slope $\approx 3.5\%$

14. a. 0.3420; 0.8829; 0.7265   **b.** 32°; 10°; 22°   **c.** $\approx 25°$

Chapter 9

SECTION 9.1

2. a. 4   **c.** 5   **e.** $^-47$

4. a. In 1990, the population of Little Rock, AR, is 176,000.
   b. The population of Little Rock, AR, in 1997

8. a. $x = 3.75$

SECTION 9.2

1. a. Has an inverse function
    input of inverse = amount earned in royalties
    output of inverse = number of music CDs sold
   b. No inverse function exists
   c. Has an inverse function
    input of inverse = number of minutes you spend on an exercise bike
    output of inverse = number of calories burned

4. a. Population of Little Rock, AR, in 1980
   b. Population in 2000
   c. Years after 1980 when population of Little Rock is 250,000

5. a. Cost to build an 800-square-foot house
   b. Square footage of a house that costs $100,000
   c. Square footage of a house that costs $200,000

9. a. $y = x - 2$ **b.** $y = \dfrac{x + 4}{3}$ **c.** $y = 2(x - 5)$ **d.** $y = x - \dfrac{5}{2}$

Chapter 10

SECTION 10.1

1. a. 2   **b.** 2   **c.** $^-2$ **d.** $^-3$ **e.** 0   **f.** 3

5. a. $2^5 = x$ **c.** $x^{1/2} = 5$ **e.** $e^4 = {}^-x$
   b. $\log_4(16) = m$ **d.** $4^{-1/2} = x^2$ **f.** $\log(12{,}000) = {}^-m$

6. a. $x = 81$ **b.** $t = 4$ **c.** $b = 9$

10. a. False   **b.** True   **c.** False   **d.** False

SECTION 10.2

1. a. Exponential   **c.** Linear   **e.** Logarithmic
   b. Quadratic   **d.** Other   **f.** Power

3. b. $x = 0.6$ **d.** $m = 17$ **f.** No real solution   **h.** $x \approx 0.60$

6. a. $I = 10^R$

    b. $t = \dfrac{1}{r} \ln\left(\dfrac{P}{P_0}\right)$

    c. $x = \dfrac{e^M}{e^M - 1}$

    d. $I = I_0 * 10^{L/10}$

    e. $k = \dfrac{-1}{t} \ln\left(\dfrac{A}{A_0}\right)$

    f. $t = \dfrac{\log 2}{n \log\left(1 + \dfrac{r}{n}\right)}$

    g. $k = \dfrac{\ln 2}{4} \approx 0.1733$; assuming $t \neq 0$

Appendix A

1. a. 15

    b. 0, 15

    c. $0, 15, {}^-10, {}^-\sqrt{9}$

    d. $2\frac{3}{5}, 0, 2.5, -\frac{1}{4}, 15, 0.667, {}^-10, 0.151515\ldots, {}^-\sqrt{9}$

    e. $\sqrt{7}, 0.010120123\ldots$

    f. All of them

2. a. rational, real

    b. rational, real

    c. irrational, real

    d. natural, whole, integer, rational, real

    e. rational, real

    f. natural, whole, integer, rational, real

    g. integer, rational, real

    h. irrational, real

    i. rational, real

    j. rational, real

3. a. False

    b. True

    c. True

    d. False

    e. False

    f. True

    g. False

    h. True

    i. True

4.

5. b

6. d

7. b

Appendix B

1. a. rational

    b. expression

    c. equation, terms, factors

    d. power, cubed

    e. square, product

    f. irrational

    g. irrational, approximated, rational

    h. root

    i. sum, square, square

    j. square, sum

    k. grouping symbols

    l. addition

2. a. $5x$

   b. $6 + w$

   c. 5^6

   d. $4(n + 15)$

   e. $^-(p + t)$

   f. $(p + t)^2$

   g. $p^2 + t^2$

   h. $6^3 * k^2$

   i. $\dfrac{1}{5}$

   j. $\dfrac{1}{5 + x}$

   k. $\sqrt{m}$

   l. $\sqrt{m + 25}$

   m. $3 + 2p = p^2$

   n. $tq = t + q$

   o. $5 + x^3 > 5.6$

   p. $\dfrac{7.2}{y + 3.56} \geq y - 91.3$

3. One possible reading for each part is listed. Other correct readings are possible.

   a. two minus the square of x

   b. the square of the difference of 2 and x

   c. b divided by twice a is the same as b plus 2.

   d. seven times the sum of x and 5

   e. the square root of the difference of t and 6.5

   f. the reciprocal of the sum of x and 6.5

   g. the sum of the reciprocal of x and 6.5

   h. Y equals 5 times the sixth power of the difference of x and 3.

4. a. 4 terms

   b. 7, R, P, and P or 7, R, and P^2

   c. 4, w, and $(R + P)$

   d. $7P^2$

   e. 7

   f. $^-1$

   g. $\dfrac{1}{5}$

6. a. The terms are 7 and $5(4 - 7)$.

   b. The terms are 24 and $\dfrac{8}{4 * 2}$.

   c. The one term is $5 * 2^2$.

   d. The one term is $(5 * 2)^2$.

   e. The one term is $\sqrt{5 * 9 - 3^2}$.

   f. The terms are $(^-4)^2$, 7, and 12.

   g. The terms are $^-4^2$, 7, and 12.

   h. The one term is $\dfrac{^-12}{3}(3 + {}^-5)$.

   i. The one term is $\dfrac{^-12}{3(3 + {}^-5)}$.

   j. The one term is $\left(\dfrac{4 - 12}{4}\right)^3$.

Appendix C

1. $A \approx (5, 2)$

   $B \approx (1, 5)$

   $C \approx (^-7, 4)$

   $D \approx (^-2, 0)$

   $E \approx (0, {}^-3)$

   $F \approx (2, {}^-6)$

   $G \approx (^-8, {}^-5)$

   $H \approx (^-4, {}^-50)$

   $J \approx (7, {}^-37)$

   $K \approx (5, 25)$

   $L \approx (9, 110)$

   $M \approx (0, 130)$

Appendix D

1. a. $x = 20$ **b.** $p = 4$ **c.** $x = 6$

2. a. $(1, {}^-2), (2, {}^-1), (3, 0)$ are three possibilities

b. $(1, 19), (2, 16), (3, 11)$

c. $(1, 2), (4, 4), (9, 6)$

3. a.

x	y = 12 − x
⁻3	15
⁻2	14
⁻1	13
0	12
1	11
2	10
3	9

b.

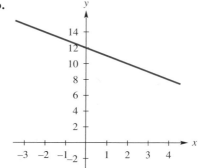

4. a.

x	$y = \dfrac{x + 1}{2}$
⁻5	⁻2
⁻3	⁻1
⁻1	0
0	0.5
1	1
3	2
5	3

b.

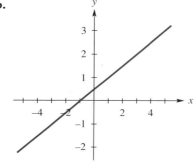

5. a.

x	y = (x − 4)(x + 4)
⁻4	0
⁻3	⁻7
⁻2	⁻12
⁻1	⁻15
0	⁻16
1	⁻15
2	⁻12
3	⁻7
4	0

b.

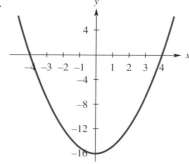

6.

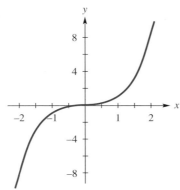

Appendix E

1. a. $z = 5.5$

b. $A = {}^-6$

c. $G = 19.5$

d. $y = 2.5$

e. $m = 174$

f. $w = 3\frac{2}{3}$

g. $x = 21$

3. $4 - 5 * x - 3$ should be $4 - 5(x - 3)$; same mistake in check.

4. b, d, f, and g

5. a. $x = 0.6$ **b.** $R = {}^-1.125$ **c.** $x = {}^-1.6$ **d.** $x = 8.5$

Appendix F

1. a. $x = \dfrac{y - b}{m}$

b. $\dfrac{P - \pi r - L}{2} = W$

c. $a = \dfrac{11k}{4b}$

d. $T = \dfrac{A}{I + r_1 - r_2}$

e. $M = 2Y - 2N$

2. $2 * \dfrac{M}{2} + N$ should be $2\left(\dfrac{M}{2} + N\right)$ which simplifies to $2 * \dfrac{M}{2} + 2 * N$; same mistake in check.

3. a. $x = \dfrac{3 + 3a}{a + 3}$

b. $x = 4P - 261$

c. $T_2 = \dfrac{T_d + 3T_3}{3}$

d. $h = \dfrac{3V}{\pi r^2}$

e. $a = \dfrac{Mb}{b - M}$

f. $t = \dfrac{C - S}{R}$

g. $y = 4\left[1 - \dfrac{(x - 2)^2}{9}\right]$

h. $C = \dfrac{{}^-1}{LM^2 + RM}$

4. a. $T = \dfrac{A - D}{1.00 - P} + D$ **b.** $3200

5. a. $A = ab + \dfrac{ac}{2} + \dfrac{\pi b^2}{8}; P = a + b + c + d + \dfrac{\pi b}{2}$

b. $a = P - b - c - d - \dfrac{\pi b}{2}$

c. $a = \dfrac{A - \dfrac{\pi b^2}{8}}{b + \dfrac{c}{2}} = \dfrac{8A - \pi b^2}{8b + 4c}$

6. 2:30:00 = 2.5 hours; 6:30:45 = 6.5125 hours

Appendix G

1. $\approx$ 17,000 ft

2. = 2,600,000 mm

3. $\approx$ 21.2 ft

4. $\approx$ 63,400 oz

5. $\approx$ 10.6 g

6. = 0.000708 kg

7. = 31,536,000 sec

8. $\approx$ 100 mL

9. $\approx$ 3,480,000 ft^2

10. $\approx 3.5 \dfrac{\text{ft}^3}{\text{h}}$

Index

INDEX OF STRATEGIES